T0099971

From Action Systems to Distributed Systems

The Refinement Approach

Chapman & Hall/CRC
Computational Science Series

SERIES EDITOR

Horst Simon
Deputy Director
Lawrence Berkeley National Laboratory
Berkeley, California, U.S.A.

PUBLISHED TITLES

COMBINATORIAL SCIENTIFIC COMPUTING
Edited by Uwe Naumann and Olaf Schenk

CONTEMPORARY HIGH PERFORMANCE COMPUTING: FROM PETASCALE
TOWARD EXASCALE
Edited by Jeffrey S. Vetter

CONTEMPORARY HIGH PERFORMANCE COMPUTING: FROM PETASCALE
TOWARD EXASCALE, VOLUME TWO
Edited by Jeffrey S. Vetter

DATA-INTENSIVE SCIENCE
Edited by Terence Critchlow and Kerstin Kleese van Dam

THE END OF ERROR: UNUM COMPUTING
John L. Gustafson

FROM ACTION SYSTEMS TO DISTRIBUTED SYSTEMS: THE REFINEMENT APPROACH
Edited by Luigia Petre and Emil Sekerinski

FUNDAMENTALS OF MULTICORE SOFTWARE DEVELOPMENT
Edited by Victor Pankratius, Ali-Reza Adl-Tabatabai, and Walter Tichy

FUNDAMENTALS OF PARALLEL MULTICORE ARCHITECTURE
Yan Solihin

THE GREEN COMPUTING BOOK: TACKLING ENERGY EFFICIENCY AT LARGE SCALE
Edited by Wu-chun Feng

GRID COMPUTING: TECHNIQUES AND APPLICATIONS
Barry Wilkinson

HIGH PERFORMANCE COMPUTING: PROGRAMMING AND APPLICATIONS
John Levesque with Gene Wagenbreth

HIGH PERFORMANCE PARALLEL I/O
Prabhat and Quincey Koziol

PUBLISHED TITLES CONTINUED

From Action Systems to Distributed Systems

The Refinement Approach

Edited by

Luigia Petre

Åbo Akademi University
Turku, Finland

Emil Sekerinski

McMaster University
Hamilton, Canada

CRC Press
Taylor & Francis Group
Boca Raton   London   New York

CRC Press is an imprint of the
Taylor & Francis Group, an **informa** business

A CHAPMAN & HALL BOOK

CRC Press
Taylor & Francis Group
6000 Broken Sound Parkway NW, Suite 300
Boca Raton, FL 33487-2742

© 2016 by Taylor & Francis Group, LLC
CRC Press is an imprint of Taylor & Francis Group, an Informa business

No claim to original U.S. Government works

Printed on acid-free paper
Version Date: 20160323

International Standard Book Number-13: 978-1-4987-0158-7 (Hardback)

Visit the Taylor & Francis Web site at
http://www.taylorandfrancis.com

and the CRC Press Web site at
http://www.crcpress.com

To Kaisa

Contents

V   Applications   187

13 Action Systems for Pharmacokinetic Modeling   189

M.M. Bonsangue, M. Helvensteijn, J.N. Kok, and N. Kokash

14 Quantitative Model Refinement in Four Different Frameworks   201

Diana-Elena Gratie, Bogdan Iancu, Sepinoud Azimi, and Ion Petre

15 Developing and Verifying User Interface Requirements for Infusion Pumps: A Refinement Approach   215

Rimvydas Rukšėnas, Paolo Masci, and Paul Curzon

Preface

FROM ACTION SYSTEMS TO DISTRIBUTED SYSTEMS – THE REFINEMENT APPROACH is a book dedicated to the memory of Kaisa Sere (1954–2012).

Kaisa lived a scientifically-intense life. She was a Professor of Computer Science and Engineering at Åbo Akademi University, Turku, Finland since 1997. Between 1993–1997 she was an Associate Professor of Computer Science at the University of Kuopio, Finland. She got her PhD in 1990 on the formal design of parallel algorithms and held a postdoctoral position at the Utrecht University in The Netherlands during 1991–1992.

She has established and led for more than 10 years the Distributed Systems laboratory at TUCS – Turku Centre for Computer Science, the graduate research institute in Informatics, Turku, Finland. She has supervised 20 PhD students and more than 60 MSc students; the majority of her former students were – some still are – members of the Distributed Systems laboratory. At the time of her passing away, she still had seven PhD students under co-supervision: to date, five of them graduated, one is graduating during the Fall 2015 and one, who has started in late 2012, is well underway.

Kaisa Sere was extremely experienced as well as very talented at attracting funding for her Distributed Systems laboratory researchers. She was granted financing for an impressive array of research projects. Some highlights include Desiré, Asynkron and FOSSE projects, funded by the Academy of Finland; EFFIMA/DiHy and EFFIMA/Digihybrid projects, funded by Tekes – the Finnish Funding Agency for Technology and Innovation; Matisse, RODIN and Deploy projects, funded by the European Union during 2000–2012. She was the vice-leader of CREST, the Finnish national Centre of Excellence for Formal Methods in Programming Research, during 2002–2007. She led the Nordic education network NODES on dependability, financed by NordForsk – the organization funding Nordic research cooperation and infrastructures.

Kaisa Sere served in numerous research evaluation committees both nationally and internationally as well as in programme committees. She has organised several summer schools, conferences and workshops; notably, she was a program chair of the Integrated Formal Methods conference in 2002 (IFM'2002) and the general chair of the international symposium on Formal Methods in 2008 (FM'2008) – the flagship conference in the field of formal methods.

The most extensive part of Kaisa Sere's research consisted in developing *Action Systems*, a formalism for modeling, analysing, and constructing *distributed systems*. Together with her co-authors, she introduced modularisation techniques for distributed systems, correct-by-construction rules for developing distributed systems, as well as a vast array of case studies demonstrating the strength of the proposed approach. All these contributions were expressed through the flexible Action Systems language, in a period when tool support was still in its infancy. Quite importantly, Action Systems are at the foundation of a current mainstream formalism called *Event-B*, which embeds Kaisa Sere's contributions to modeling, analysing, and constructing distributed systems in a tool-centred environment, the *Rodin platform*. This platform is reliable and consequently, industrial acceptance of these methodologies increases. In fact, the European project Deploy strongly promoted Event-B and the Rodin platform to industry.

Within the design of distributed systems, Kaisa Sere's main research focus was on *refinement*-based approaches to the construction of systems ranging from pure software to hardware and digital circuits. The refinement technique ensures, by mathematical proof, that the final system respects its initial requirements. Central to this is the concept of *abstraction*, by which an initial model of the system-to-develop only encompasses its fundamental requirements; such an initial model is consequently very simple. This abstraction mechanism makes it feasible to prove that the initial model satisfies the fundamental requirements as well as basic properties. Upon proving these for the system, one gradually adds the missing details in a manner that keeps the proven requirements and properties true and allows the satisfiability proof of other requirements and new properties. This correct-by-construction method for software development complements model checking approaches as well as testing and simulation approaches for ensuring the quality of systems. When used properly, it can shorten the development lifecycle of systems, in addition to certifying their quality.

Kaisa Sere was a remarkably positive, energetic, and inspirational researcher, who attracted an impressive number of collaborators. When the generous opportunity of CRC Press arose, to send a last salute to the colleague, leader, and friend that she was, a large number of her former collaborators wanted to contribute. The result of their work is the present book, FROM ACTION SYSTEMS TO DISTRIBUTED SYSTEMS – THE REFINEMENT APPROACH. We have divided the book into five parts: **I Modeling**, **II Analysis**, **III Proof**, **IV Refinement**, and **V Applications**. The names of these parts are recurring themes of Kaisa Sere's research, celebrated now by her collaborators. Some of the chapters employ the Action Systems formalism; some employ and further develop its Event-B successor; some chapters focus on analysis, some on proof, some on refinement, and some on comparing various semantical models for these; some chapters discuss model checking approaches to analysis and refinement; some chapters include applications of actions systems or of the refinement approach in pharmacokinetics, biology, and medicine; and many chapters emphasize future research directions including refinement as a central theme. Therefore, we hope that the book will address a broad audience, from graduate students to researchers and practitioners interested in applying formal methods to develop distributed systems of quality.

September 2015   Luigia Petre
                                                                   Emil Sekerinski

Acknowledgements

FROM ACTION SYSTEMS TO DISTRIBUTED SYSTEMS – THE REFINEMENT APPROACH is an edited book collecting scientific contributions mainly on the themes of distributed systems and refinement. Each chapter was thoroughly reviewed by experts in the field, resulting in 16 accepted articles out of the initial 19 submissions.

We would like to warmly thank the anonymous reviewers for their hard work on promoting scientific excellence. Our gratitude extends to the kind and professional personnel at CRC Press who helped and supported this long-term project; in particular, warm thanks to Ms Randi Cohen, Senior Acquisitions Editor and Ms Jennifer Ahringer, Senior Project Coordinator. We would also like to acknowledge the Easychair support, for handling our submission and review process; in particular, thanks to Andrei Voronkov, for making this framework available.

In the end, we would like to acknowledge the collegiality and scholarship of the chapter authors: it is your work that makes up this book! Thank you!

Luigia Petre
Emil Sekerinski

List of Figures

List of Tables

About the Editors

Dr. Luigia Petre is Associate Professor of Computer Science in the Faculty of Science and Engineering at Åbo Akademi University (Turku, Finland). She got her Ph.D. in Computer Science in 2005 on modelling techniques in formal methods. Her research interests include formal methods and their integration, wireless sensor networks, network architectures, meta-modelling, non-functional properties and time-space dependent computing. She has supervised 11 Master's students, three Ph.D. students and currently has one Ph.D. student under supervision. She was granted funding from the Academy of Finland to lead a consortium research project named FResCo during 2013-2015 and has been coordinating NODES—the Nordic Network on Dependable Systems (financed by Nordforsk), concerned with deploying a dependability curriculum for the Nordic countries, during 2007-2012. She has organized a winter school and several conferences; most notably, she has actively participated in the International Conference on Integrated Formal Methods, organising it twice in Turku (2002 and 2013), as a program committee member of this conference in 2002, 2004, 2005, 2007, 2012-2014, as a program committee chair in 2013 and as a member of the steering committee for this conference since 2014. Dr. Petre has edited two books and three special issues of international journals; she has published about 45 peer-reviewed articles.

Dr. Emil Sekerinski is Associate Professor in the Department of Computing and Software at McMaster University (Hamilton, Ontario, Canada). He got his Dr. rer. nat in 1994 from the University of Karlsruhe (Germany) on the formal development of object-oriented programs by stepwise refinement. His interests include programming languages and tools, program correctness, concurrency, components, embedded systems. He has supervised at McMaster 18 Master's student, 4 Ph.D. students, and one postdoctoral researchers. Currently he receives funding through Canada's National Science and Engineering Research Council and through the Ontario Research Fund and has received support from IBM through the Southern Ontario Smart Computing Innovation Platform. He spent sabbaticals at ETH Zürich, TU München, and TU Dresden. In 2006, he organized the FM Symposium in Hamilton and was a member of the program committee of a number of conferences, e.g. Integrated Formal Methods, Refinement Workshop, International Conference on B and Z, International Colloquium on Theoretical Aspects of Computing, Mathematics of Program Construction. Dr. Sekerinski has edited three books and published over 50 peer-reviewed articles.

List of Contributors

Jean-Raymond Abrial
Independent consultant
Marseille, France

Sepinoud Azimi
Turku Centre for Computer Science and
    Åbo Akademi University
Turku, Finland

Richard Banach
University of Manchester
Manchester, UK

Eerke Boiten
School of Computing, University of Kent
Canterbury, Kent, UK

Marcello M. Bonsangue
LIACS, Leiden University
Leiden, the Netherlands

Pontus Boström
Åbo Akademi University
Turku, Finland

Michael Butler
University of Southampton
Southampton, UK

Paul Curzon
Queen Mary University of London
London, UK

John Derrick
Department of Computer Science,
    University of Sheffield
Sheffield, UK

Alessandro Fantechi
Dipartimento di Ingegneria
    dell'Informazione, Università di Firenze
    and ISTI–CNR
Florence, Italy

Stefania Gnesi
ISTI–CNR
Pisa, Italy

Diana-Elena Gratie
Turku Centre for Computer Science and
    Åbo Akademi University
Turku, Finland

Thibaut Le Guilly
Department of Computer Science, Aalborg
    University
Aalborg, Denmark

Michiel Helvensteijn
LIACS, Leiden University
Leiden, the Netherlands

Bogdan Iancu
Turku Centre for Computer Science and
    Åbo Akademi University
Turku, Finland

Einar Broch Johnsen
University of Oslo
Oslo, Norway

Joost N. Kok
LIACS, Leiden University
Leiden, the Netherlands

Natalia Kokash
LIACS, Leiden University
Leiden, the Netherlands

Linas Laibinis
Åbo Akademi University
Turku, Finland

Paolo Masci
Queen Mary University of London
London, UK

Mats Neovius
Åbo Akademi University
Turku, Finland

Petur Olsen
Department of Computer Science, Aalborg
   University
Aalborg, Denmark

Olaf Owe
University of Oslo, Department of
   Informatics
Norway, and University of California,
Department of Computer Science
Santa Cruz, USA

Ion Petre
Turku Centre for Computer Science and
   Åbo Akademi University
Turku, Finland

Ka I Pun
University of Oslo
Oslo, Norway

Anders P. Ravn
Department of Computer Science, Aalborg
   University
Aalborg, Denmark

Atle Refsdal
SINTEF ICT
Oslo, Norway

Mauno Rönkkö
Department of Environmental Science,
   University of Eastern Finland
Kuopio, Finland

Rimvydas Rukšėnas
Queen Mary University of London
London, UK

Ragnhild Kobro Runde
University of Oslo
Oslo, Norway

Steve Schneider
University of Surrey
Surrey, UK

Arne Skou
Department of Computer Science, Aalborg
   University
Aalborg, Denmark

Gheorghe Stefanescu
University of Bucharest
Bucharest, Romania

Martin Steffen
University of Oslo
Oslo, Norway

Ketil Stølen
SINTEF ICT and University of Oslo
Oslo, Norway

S. Lizeth Tapia Tarifa
University of Oslo
Oslo, Norway

Leonidas Tsiopoulos
Åbo Akademi University
Turku, Finland

Helen Treharne
University of Surrey
Surrey, UK

Elena Troubitsyna
Åbo Akademi University
Turku, Finland

Jüri Vain
Tallinn University of Technology
Tallinn, Estonia

Marina Waldén
Åbo Akademi University
Turku, Finland

David M. Williams
VU University Amsterdam
Amsterdam, the Netherlands

Ingrid Chieh Yu
University of Oslo
Oslo, Norway

Part I

Modeling

Chapter 1

Modeling Sources for Uncertainty in Environmental Monitoring

Mauno Rönkkö

Department of Environmental Science, University of Eastern Finland

Abstract. In this chapter we discuss formal modeling of environmental monitoring. Environmental monitoring is needed to study complex environmental processes and to understand the effects of our actions in those processes. Environmental monitoring, however, is about measuring, whereby it is prone to measurement errors. When measuring, there are many sources for uncertainty. This makes it hard to estimate the total uncertainty. As the main contribution, we investigate the use of action systems in modeling an environmental monitoring system together with sources for uncertainty. We use hybrid action systems, that is, action systems with differential actions, to model both discrete-time and continuous-time dynamics. We illustrate the approach by modeling the central function of a home monitoring system, monitoring of room temperature. We also discuss, how properties of interest are validated from the model.

1.1   Introduction

Global trends, such as sustainability and cutting greenhouse gas emissions, have contributed to the increasing need for implementing novel environmental measurement and monitoring systems [278]. As pointed out by Messer et al. [242] "High-resolution, continuous, accurate monitoring of the environment is of great importance for many applications from weather forecasting to pollution regulation." In short: what you cannot measure, you cannot control.

However, as it was pointed out by Ward et al. [344] already in 1980's, implementing ever new measurement systems creates a problem. While new systems provide more data with better accuracy, data is just data unless computational methods are used to obtain information. In fact, Ward et al. call this the "data-rich but information-poor syndrome".

The state of environmental monitoring is worse still than described. For instance, according to Williams et al. [346], private weather stations produce data without any indication of measurement errors. As argued by Williams et al. such data is useless, unless advanced statistical methods are applied to it.

Assigning uncertainty estimates to measurement data is one way to improve the quality. Perhaps a more fundamental approach would be to model the sources for uncertainty. However, for low-cost sensor technologies, fault and failure models are hard to come by. Still, without such models, it is hard to estimate the potential sources for uncertainty for the measurement devices and, thus, for the measurement data. Therefore, without fault and failure models, assigning uncertainty estimates is still guesswork.

Having individual fault and failure models for the sensor, however, is not enough, as an environmental monitoring system forms a complex ensemble. Without a model taking this into account, assessing the total uncertainty is hard again. Therefore, as the main contribution, we investigate here the use of hybrid action systems to model the sources for uncertainty in an environmental monitoring system. With hybrid action systems, we can model both continuous-time and discrete-time dynamics. Therefore, we can address a whole range of sources for uncertainty, including non-deterministic dynamics of environmental variables, variations associated with measurement activities, sporadic device failures, and inaccuracies introduced by computational methods.

We introduce hybrid action systems in Section 1.2. In Section 1.3, we discuss environmental monitoring and associated sources for uncertainty. In Section 1.4, we present a case study involving monitoring of built environment. In the case study, we use hybrid action systems to model room temperature monitoring together with the most central sources for uncertainty. Room temperature monitoring is the main function in a home monitoring system which is becoming ever more popular due to an increasing need for saving energy as well as costs for heating and cooling. In Section 1.4, we also discuss provability of properties of interest from the presented hybrid action system model.

1.2 Hybrid Action Systems

Action systems were originally proposed by Back and Kurki-Suonio [27]. They are iterated systems of actions based on Dijkstra's guarded command language [113]. Action systems were later used to model control systems [73], and extended with a differential action [280] to characterize continuous-time dynamics. The resulting hybrid action systems [281] were used to develop models of hybrid systems [147] by stepwise refinement.

We shall now introduce hybrid action systems by defining conventional actions and the differential action. We then define action systems with differential actions, constituting hybrid action systems, and a parallel composition for hybrid action systems. Lastly, we define stepwise refinement use of action systems which also applies to hybrid action systems.

1.2.1 Conventional Actions

Conventional actions are used for capturing discrete-time dynamics. The meaning of an action is defined with a weakest precondition predicate transformer [113]. It returns a predicate, the weakest precondition, for a given action and a postcondition predicate. The predicate describes the largest set of states from which the execution of the action terminates in a state satisfying a given postcondition. For a postcondition q, and an action A the weakest precondition is denoted by $\mathrm{wp}(A, q)$.

An action is executed only if it is enabled. Formally, the set of states in which an action is enabled is given by $\mathrm{g}(A) \mathrel{\widehat{=}} \neg \mathrm{wp}(A, \mathit{false})$. An action is said to be disabled in states $\neg \mathrm{g}(A)$. An executing action may either terminate or continue indefinitely. The set of states from which an action terminates is given by $\mathrm{t}(A) \mathrel{\widehat{=}} \mathrm{wp}(A, \mathit{true})$. An action is said to abort in all those states from where it does not terminate, i.e., $\neg \mathrm{t}(A)$. The execution of an action is atomic. Therefore, an action is executed to its completion before any other actions are considered. Consequently, if an action aborts, no other action will be executed.

The weakest precondition semantics of the conventional actions [26, 38, 113] that we consider here are given in Table 1.1. As for the actions, "Skip" keeps the state unchanged. "Assignment" sets the values of the variables to the values of given expressions. "Non-deterministic assignment" changes non-deterministically the values of the variables so that a given condition holds. "Sequential composition" executes one action after another. "Non-deterministic choice" selects arbitrarily one enabled action and executes it. "Guarded command" executes a guarded action, if the guard holds in the current state. "Iteration" executes repeatedly an action till it becomes disabled.

1.2.2 Differential Action

The differential action was introduced by Rönkkö et. al [281]. It captures continuous-time dynamics that is described as a differential relation with respect to an evolution guard. To unfold the dynamics of the differential action, the differential relation, often expressed as a system of differential equations, needs to be solved. The solution functions describe then the continuous-time dynamics. In particular, if the solution functions satisfy the evolution guard forever, the differential action never terminates. Correspondingly, the differential action terminates, if the solution functions eventually reach a state that does not satisfy the evolution guard.

Table 1.1: Semantics of conventional actions

action	notation	wp(action, q)
Skip	**skip**	q
Assignment	$X := E$	$q[E/X]$
Non-deterministic assignment	$X := \chi.r$	$\forall \chi{:}r.\ q[\chi/X]$
Sequential composition	$A;B$	$\mathrm{wp}(A, \mathrm{wp}(B, q))$
Guarded command	$p \to A$	$p \Rightarrow \mathrm{wp}(A, q)$
Non-deterministic choice	$A \,[\!]\, B$	$\mathrm{wp}(A, q) \wedge \mathrm{wp}(B, q)$
Iteration	**do** A **od**	$\mathrm{wp}(\mu S.\ \mathrm{g}(A) \to A;S \,[\!]\, \neg\mathrm{g}(A) \to \mathbf{skip}, q)$

Above, X are the model variables, and χ are bound variables disjoint from X. Also, there are predicates p, q, and r. Lastly, A and B denote some actions and, in the iteration, S is an action variable, and μS denotes the least fixed point [38].

Let e, d, and q be predicates denoting an evolution guard, a differential relation, and a post condition over variables X, respectively. Also, let ϕ denote a continuous function over time. Then, the duration of ϕ with respect to the evolution guard e is defined as:

$$\Delta(\phi, e) \mathrel{\widehat{=}} \inf\{\tau{:}I\!R \cap [0, \infty) \mid \neg e[\phi(\tau)/X]\}$$

Such a ϕ is a solution function to the differential relation, only if its initial value equals to the current values of the variables X and, from the initial state, it satisfies the differential relation while it satisfies the evolution guard. This is captured by the predicate:

$$SF(\phi, e, d) \mathrel{\widehat{=}} \phi(0) = X \wedge \forall\, \tau{:}I\!R \cap [0, \infty).\ (e \Rightarrow d)[\phi(\tau)/X, \dot\phi(\tau)/\dot X]$$

With these two definitions, we can define the differential action, which is of form $e : \to d$, to have the weakest precondition semantics:

$$\mathrm{wp}(e : \to d, q) \quad \mathrel{\widehat{=}} \quad \forall\, \phi{:}C^1.\ SF(\phi, e, d) \wedge \Delta(\phi, e) > 0$$
$$\Rightarrow \Delta(\phi, e) < \infty \wedge q[\phi(\Delta(\phi, e))/X]$$

As noted by Rönkkö et. al [281], it is important to realize that the differential action speaks only about observations of evolutions, not of time. For instance, if x denotes a location, $0 \le x \le 10 : \to \dot x = 10$ speaks only about the location, not of the velocity or of time. To include an observation of the passage of time we need a clock variable, t. Then, $0 \le x \le 10 : \to \dot x = 10 \wedge \dot t = 1$ speaks simultaneous about the location and time. From this differential action we may also infer velocity.

1.2.3   Action Systems and Parallel Composition

An action system is an initialized block of the form [25]:

$$\mathcal{A} \mathrel{\widehat{=}} \,|\,[\,\mathbf{var}\ X{:}T;\ X := E;\ \mathbf{do}\ A_1 \,[\!]\, .. \,[\!]\, A_n\ \mathbf{od}\,]\,|\,:V$$

The expression **var** $X{:}T$ declares a set of variables X with types T. The variables consist of local and shared variables. The shared variables are indicated with an asterisk, for instance y^*. The variables imported from other action systems are listed as variables V. Thus, the union of variables X and V forms the state space of $\mathcal{A}$.

The action $X := E$ initializes all the variables X by constants E. Hence, the shared variables are initialized in the action systems where they are declared. All action systems initialize synchronously before any other action of $A_1..A_n$ is considered.

As our action systems are hybrid action systems, an action A_i of $A_1..A_n$ is either a guarded command or a differential action. After initialization, an enabled action A_i is selected non-deterministically for execution. The execution of A_i is always atomic, even for composite or differential actions. Thus, other actions can be considered only after the executed action terminates.

The entire action system $\mathcal{A}$ terminates, when all actions are disabled, i.e., $\neg \mathrm{g}(A_1 \parallel .. \parallel A_n)$. Similarly, $\mathcal{A}$ aborts, if any executed action aborts.

The parallel composition of action systems was originally given by Back [25], and it is formalized as follows. Consider the two action systems, where X and Y are disjoint variables that can be either local or shared:

$$\mathcal{A} \ \widehat{=}\ |[\mathbf{var}\ X{:}T;\ X := E;\ \mathbf{do}\ A_1 \parallel .. \parallel A_m\ \mathbf{od}]| : V$$
$$\mathcal{B} \ \widehat{=}\ |[\mathbf{var}\ Y{:}U;\ Y := I;\ \mathbf{do}\ B_1 \parallel .. \parallel B_n\ \mathbf{od}]| : W$$

Then, the parallel composition is given as:

$$\mathcal{A} \parallel \mathcal{B} = |[\mathbf{var}\ X, Y{:}T, U;$$
$$X, Y := E, I;$$
$$\mathbf{do}\ A_1 \parallel .. \parallel A_m \parallel B_1 \parallel .. \parallel B_n\ \mathbf{od}$$
$$]| : (V \cup W) - (X \cup Y)$$

Thus, the parallel composition combines the state spaces of the two action systems, merging the shared variables, and keeping the local variables distinct. Imported variables in $V \cup W$, that are defined in either of the action systems, i.e., $(X \cup Y)$, are no longer imported. The actions from the action systems are merged with a non-deterministic choice, modeling parallelism by interleaving. Note that parallel composition is both associative and commutative.

1.2.4 Refinement of Hybrid Action Systems

As hybrid action systems are action systems with differential actions, and as the differential action has weakest precondition semantics, we can apply refinement of action systems also to hybrid action systems. For the refinement of action systems, we consider a general development step where an abstract action system $\mathcal{A}$ is shown to be weakly simulated [25] by a concrete action system $\mathcal{C}$. Informally this means that $\mathcal{C}$ has the same state changes over the shared variables as $\mathcal{A}$; however, $\mathcal{C}$ may include new intermediate state changes over the local variables. Furthermore, as aborting is an undesirable behavior in a reactive setting, if $\mathcal{A}$ aborts in a state, $\mathcal{C}$ may do anything in that state. Lastly, if $\mathcal{A}$ terminates in some state, so does $\mathcal{C}$.

The refinement of action systems is formalized as data refinement [25, 38] supporting the use of a refinement relation [25]. A refinement relation is a predicate over all the variables, and it is used for relating all the states of the concrete model to some states in the abstract model. A refinement relation r holds between an abstract action A and a concrete action C, if C can reach the same states, related by r, from at least the same initial states, related by r, as A. Then, A is refined by C, and it is denoted by $A \sqsubseteq_r C$. Data refinement $A \sqsubseteq_r C$ over an action A with local variables X and shared variables Z and an action C with local variables Y and shared variables Z is defined [25] as:

$$\forall\ q{:}PRED(X, Z)\ .\ r \wedge \mathrm{wp}(A, q) \Rightarrow \mathrm{wp}(C,\ \exists\ X.\ r \wedge q)$$

As for the refinement of action systems, consider a refinement relation r, and the two action systems:

$$\mathcal{A} \; \hat{=} \; |[\mathbf{var}\ Z^*, X\!:\!V, T \; \bullet \; Z, X := O, E; \; \mathbf{do}\ A\ \mathbf{od}]|$$

$$\mathcal{C} \; \hat{=} \; |[\mathbf{var}\ Z^*, Y\!:\!V, U \; \bullet \; Z, Y := O, I; \; \mathbf{do}\ C \mathbin{[\!]} B\ \mathbf{od}]|$$

Then, $\mathcal{A}$ is weakly simulated by $\mathcal{C}$, denoted by $\mathcal{A} \preceq_r \mathcal{C}$, if the following proof obligations are shown to hold [25]:

1. the initializations do not contradict the refinement relation, i.e.,
   $r[O/Z,\ E/X,\ I/Y]$

2. the abstract actions A are refined by the concrete actions C, i.e., $A \sqsubseteq_r C$

3. the introduced intermediate actions B refine stuttering in $\mathcal{A}$, i.e.,
   $\mathbf{skip} \sqsubseteq_r B$

4. the introduced actions are self-disabling, i.e., $r \Rightarrow \mathrm{t}(\mathbf{do}\ B\ \mathbf{od})$

5. $\mathcal{C}$ does not terminate, unless $\mathcal{A}$ terminates as well, i.e.,
   $r \wedge \mathrm{g}(A) \Rightarrow \mathrm{g}(C \mathbin{[\!]} B)$

1.3   Environmental Monitoring

As mentioned in the introduction, we are already "flooded" by environmental data due to an increasing need for monitoring our environment. The data, however, comes mostly from sensors and measurement systems for which the uncertainties are poorly understood, thus, making the use of data questionable. Assigning uncertainty estimates, as done by Williams et al. [346], is one way to improve the quality of environmental data.

The major challenge in environmental monitoring, however, is that the environmental processes are poorly understood. The biggest reason for this is that the processes are complex and open, and therefore influenced by external factors. In fact, the reason for measuring the environment is to learn more about the processes.

The qualities of environmental processes are also often non-measurable. Therefore, instead of the actual qualities of interest, indicators and other variables are measured. The measured variables are then assumed to have a causal or statistical relationship to the qualities of interest. Such relations are also often approximations.

In addition, there is often more than one way to measure the environmental variables of interest. The choice of the measurement method depends on external factors and restrictions imposed by end-user application or need. Furthermore, even for the same method, there are typically several measurement devices and device manufacturers. This makes the merging of data from several data sources challenging, as discussed by Williams et. al [346].

All this is further complicated by the need to measure the environmental variables from a large spatial area and over a long period of time. Wide and continuous spatial and temporal coverage, however, is not realistic, whereby sampling is used. There are several sampling methods to choose from, and incomplete understanding of the environmental processes makes it hard to choose the "optimal" sampling method.

Lastly, all the measurements are prone to measurement errors. However, because of all the previous challenges, it is hard to ensure that all measurement errors are accounted for. In particular, it is very likely that for two different measurement settings, two different mechanisms are used to compensate for the measurement errors. Again, this makes the merging of data from several sources evermore challenging.

With hybrid action systems, we can address these sources for uncertainty by modeling the monitoring system as a whole. The uncertainties associated with the continuous-time dynamics can be embedded into differential actions. The uncertainties, faults, and failures associated with discrete-time dynamics can be modeled with the conventional actions. In this way, we can share our understanding of the dependencies formally, to support more accurate and meaningful treatment of the measurement data. We shall illustrate this approach next.

1.4 Case Study: Monitoring Room Temperature

As the case study, we consider monitoring of room temperature. It is a basic function of HVAC control system [21]. It is also the main function in a home monitoring system, such as AsTEKa [308]. Home monitoring, or monitoring of built environment in general, has recently become popular due to the availability of low-cost sensors and the ever increasing need to save energy while improving indoor air quality.

We demonstrate here how hybrid action systems are used to model some of the most prominent sources for uncertainty in a room temperature monitoring system. We start by giving a system overview. We then present a hybrid action system model for each of the subsystems. Lastly, we discuss which properties of the model are of interest and subject to model-checking and rigorous proofs.

1.4.1 System Overview

The model for the room temperature monitoring system is composed of three subsystems: the environment, the temperature sensor, and the monitoring logic. Each of the subsystems will be modeled as an action system, and the overall model is then given as a parallel composition.

The model of the environment, *Environment*, captures the absolute time, the measured time, and the absolute room temperature. In particular, the measured time is modeled as an inaccurate clock.

The model of the temperature sensor, *Sensor*, captures the measurement of the absolute room temperature each 10 seconds and the delivering of the measured value to the monitoring logic. In particular, the model embeds indirect and inaccurate temperature measurement. Furthermore, the model includes a non-periodic sensor failure typical to low-cost sensor technology.

The model of the monitoring logic, *Monitor*, exposes the measured room temperature, for instance to HVAC control, as a preprocessed value. The value is exposed as an average of the measured values over a sliding window of 60 seconds. In particular, the model compensates for the measurement inaccuracies by discarding measurement values that deviate "too much" from the current average. A counter is used to record the number of discarded measurement values for potential diagnostic purposes.

The overall model of the room temperature monitoring system is given by the parallel composition:

$$System \,\hat{=}\, Environment \,\parallel\, Sensor \,\parallel\, Monitor$$

We shall now continue by describing each of the subsystems in more detail.

1.4.2   Environment

The model of the environment captures three continuous-time dynamics: the absolute time, the measured time, and the absolute room temperature. The absolute time is considered ideal and accurate. The measured time is considered inaccurate, and it may drift. The absolute room temperature is also considered as the ideal evolving value. The room temperature is assumed to be subject to control, where the objective is to keep it within [20, 23] Celsius. Since we do not explicitly model the control here, we embed its effect to the model of the environment. We assume a control logic that avoids hysteresis. Thus, we assume that the control switches on heating, when the room temperature drops below 20 Celsius. Correspondingly, the control switches on cooling, when the room temperature exceeds 23 Celsius. Furthermore, we assume that the heating and cooling are mutually exclusive, and that the room temperature can always be maintained within the range of [19, 24] Celsius.

In the model, there are three real valued variables. The absolute time is denoted by a variable T, and its dynamics is given by $\dot{T}=1$. The inaccurate clock measuring the passage of time is denoted by a variable t. Since t may drift, its dynamics are given by $\dot{t} \in [0.999, 1.001]$. The absolute room temperature is denoted by a variable c. We assume that a room can be heated by at most 1 Celsius in 100 seconds and cooled by at most 1 Celsius in 1000 seconds. Then, the dynamics of the room temperature is given by $\dot{c} \in [-0.001, 0.01]$.

The model, $Environment$, is shown below. Initially, the absolute time and the measured time is set to 0, and the room temperature is set to 20. The dynamics of the environment is captured by three differential actions. All the evolutions are limited by the condition $t<10$. The reason for this is that the temperature sensor is to measure the room temperature each 10 seconds. The clock t is used to measure this passage of time. The first differential action captures the dynamics, where the room temperature is free to change within the allowed range of [19, 24] Celsius. The second differential action captures the case, where the room temperature drops below 20 Celsius. Then, the heating is switched on and the dynamics is given by $\dot{c} \in (0, 0.01]$. The third differential action captures the case, where the room temperature exceeds 23 Celsius. Then, the cooling is switched on, and the dynamics are given by $\dot{c} \in [-0.001, 0)$. The (non-deterministically) combined dynamics of these three differential actions captures behavior, where room temperature is free evolve within the range of [19, 24] Celsius, but the measured time t advances always to the value 10. After that all the differential actions become disabled allowing the actions of the sensor model to take place. Note, however, that the measured 10 seconds may not be the "absolute" 10 seconds, as t may drift.

$\mathcal{E}nvironment \;\widehat{=}$
$\quad |\,[\,\textbf{var}\;\; T^*, t^*, c^*\!:\!I\!R, I\!R, I\!R$
$\qquad T, t, c := 0, 0, 20;$
$\qquad \textbf{do}\;\; t<10 \,\wedge\, 19<c<24 \;:\to \dot{T}{=}1 \,\wedge\, \dot{t} \in [0.999, 1.001] \,\wedge\, \dot{c} \in [-0.001, 0.01]$
$\qquad \;\|\;\;\; t<10 \,\wedge\, c<20 \qquad\;\; :\to \dot{T}{=}1 \,\wedge\, \dot{t} \in [0.999, 1.001] \,\wedge\, \dot{c} \in (0, 0.01]$
$\qquad \;\|\;\;\; t<10 \,\wedge\, 23<c \qquad\;\; :\to \dot{T}{=}1 \,\wedge\, \dot{t} \in [0.999, 1.001] \,\wedge\, \dot{c} \in [-0.001, 0)$
$\qquad \textbf{od}$
$\quad]\,|$

1.4.3 Temperature Sensor

The model of the temperature sensor captures the indirect and inaccurate measurement of the absolute room temperature. Thus, the measurement is subject to inaccuracies caused by the indirect measurement method, device dependent static noise and drifting, as well as environmental disturbances. We assume that the measurement error caused by the indirect measurement method, the device dependent static noise, and the environmental disturbances falls within $[-1, 1]$ Celsius. We also assume that the measurement value drifts at most 0.8 Celsius in one year. Moreover, we assume that the drifting affects also the measurement accuracy so that after one year, the total accuracy of a measurement falls within $[-1.8, 1.8]$ Celsius. Such a deviation is not uncommon to low-cost sensor devices. Lastly, we assume that the sensor has a non-periodic failure rate of one failure within $[2, 24]$ hours. Thus, the sensor is certain to provide an unreliable measurement within 24 hours, but after such failure, it is certain to provide reliable measurements for at least 2 hours.

In the model, there are three variables. The number of delivered reliable measurements is denoted by a variable n. It is used to model the sporadic device failure. A flag variable s is used to indicate that room temperature has been measured, but not read, yet. The actual measurement value is denoted by a variable m. The model also refers to all environmental variables, T, t, and c. Lastly, the model refers to a monitor variable in indicating that a measurement value is available for the monitoring logic for further processing.

The model, $\mathcal{S}ensor$, is shown below. Initially, the number of delivered reliable measurements is set to 0, the reading flag s is set to *false*, and the measured value is also set to 0. The value of the monitor variable in is assumed to be *false* initially. The measurement logic is then modeled by using three actions. The first action captures the actual measurement. It can occur only 10 seconds after the most recent measurement, as indicated by t. The action models indirect and inaccurate measurement of the room temperature c with the variable m. The drifting, ι, of the measurement value over time is captured with a $\log^2(T)$ term. As discussed, the drifting also affects the measurement accuracy, ς, whereby the resulting inaccurate measurement is captured by $m := c+\iota+\varsigma.(\iota=\log^2(T) \wedge \varsigma \in [-1-\iota, 1+\iota])$.
The first action also sets the value of the flag s to *true* indicating that a measurement is taken. The second action captures the occurrence of a sporadic failure. As this may occur 2 hours after the most recent failure, and measuring occurs each 10 seconds, the number of reliable measurements, n, must be at least 720 for the failure to occur. Then, the measured value is set to any value within the range $[-50, 50]$ Celsius, and the number of reliable measurements is reset. The third action captures the case, when the measurement is delivered to the monitoring logic. Since the sensor is certain to fail once in 24 hours, there can be at most 8640 successive, reliably delivered measurements as indicated by n. The actual delivery of a measurement is indicated by setting the value of the monitor variable in to *true* and the flag variable s to *false*. Note that the choice between the actions is non-deterministic, so

the second and the third actions together capture the failure dynamics of the sensor. Note also that once the third action is executed, all the actions of the environment model remain disabled and all the actions of the sensor model become disabled allowing the actions of the monitoring logic to take place.

$$\mathcal{S}ensor \; \widehat{=}$$
$$| \, [\, \mathbf{var} \; n, s, m^* : I\!N, I\!B, I\!R$$
$$n, s, m := 0, false, 0;$$
$$\mathbf{do} \; t \geq 10 \, \wedge \, \neg \, s \, \wedge \, \neg \, in \rightarrow m := c + \iota + \varsigma.(\iota = \log^2(T) \wedge \varsigma \in [-1 - \iota, 1 + \iota]);$$
$$s := true$$
$$[\!] \quad s \, \wedge \, \neg \, in \, \wedge \, 720 {<} n \rightarrow m := \varsigma.(\varsigma \in [-50, 50]); n := 0;$$
$$[\!] \quad s \, \wedge \, \neg \, in \, \wedge \, n {<} 8640 \rightarrow n := n + 1; s, in := false, true$$
$$\mathbf{od}$$
$$] \, | : T, t, c, in$$

1.4.4   Monitoring Logic

The model of the monitoring logic captures the exposure of the measured room temperature as an averaged value over a sliding window of 60 seconds. It also captures the compensation for measurement errors by discarding values that deviate more than 2 Celsius degrees from the computed average. Such values are likely to be measurement errors. The number of discarded values is also recorded, and exposed, for potential diagnostic purposes. Note that the computation of the average value requires the storing of at most 6 measurement values, because the temperature sensor measures a value each 10 seconds.

In the model, there are five variables. The variable *in* indicates that a measurement is taken by the sensor and that the measurement value is available to the monitoring logic. The exposed averaged room temperature value is denoted by a variable *avg*. The number of discarded measurement values is denoted by a variable *dev*. The most recent measurement values for computing the average are kept in a sequence of real valued number denoted by a variable *d*. Lastly, the number of measurement values used for computing the exposed average value is denoted by a variable *n*. Thus, if the value of *n* differs from the length of the sequence *d*, denoted by $\#d$, there is a new recorded measurement value and the average value needs to be recomputed. In addition, the model refers to the sensor model variable *m*, which is the measured room temperature. The model also refers to the clock variable *t* of the environment model.

We use here two sequence operators. Consider a value *m* and a sequence *d* consisting of *n* values $\langle v_1, v_2, ...v_n \rangle$, where a value m_i in the sequence is referred to as $d(i)$. Then, the concatenation of a value *m* to the end of the sequence *d*, denoted by $d \oplus m$, is defined as $\langle v_1, v_2, ...v_n, m \rangle$. The tail operator, denoted by $tail(d)$ returns a sequence, where the first value in the sequence is omitted. Thus, for the sequence *d*, the tail operator $tail(d)$ is defined as the sequence $\langle v_2, ...v_n \rangle$.

The model, $\mathcal{M}onitor$, is shown below. Initially, the flag variable *in* is set to *false*, the (computed) average room temperature is set to 20, the number of deviating measurements is set to 0, the sequence of most recent measurement values is set to contain one value, 20, and the number of measurement values used for computing the average value is set to 1. The first action captures the case, where a reliable measurement value is taken by the sensor and made available to the monitor. Then, the measured value is appended to the end of the sequence of measurement values, *d*. This also disables the action, as the equality $n = \#d$ no longer holds. The second actions captures the case, where the sensor provides a measurement value that is deviating too much from the average. Then, the measured value is discarded, current average is appended to the end of the sequence of measurement values, and the number of rejected measurements is incremented. The third action ensures that the sequence of measured value contains only at most 6 values. This is done by removing the least recent measurement value from the sequence. Lastly, the fourth action updates the average, and resets both the clock t and the variable in. Then, all the actions of the monitoring logic become disabled, and at least one of the actions of the environment model becomes enabled, allowing the environment model to advance again.

$$
\begin{aligned}
&\mathcal{M}onitor \; \widehat{=} \\
&\quad | \, [\, \textbf{var} \; in^*, avg^*, dev^*, d, n{:}\mathbb{B}, \mathbb{R}, \mathbb{N}, \langle \mathbb{R} \rangle, \mathbb{N} \\
&\qquad in, avg, dev, d, n := false, 20, 0, \langle 20 \rangle, 1; \\
&\qquad\quad \textbf{do} \; in \, \wedge \, n{=}\#d \wedge \, | \, avg{-}m \, | \leq 2 \to d := d \oplus m \\
&\qquad\quad \| \;\; in \, \wedge \, n{=}\#d \wedge \, | \, avg{-}m \, | > 2 \to d := d \oplus avg; dev := dev{+}1 \\
&\qquad\quad \| \;\; in \, \wedge \, n{\neq}\#d \, \wedge \, \#d{>}6 \qquad \to d := tail(d); n := n{-}1 \\
&\qquad\quad \| \;\; in \, \wedge \, n{\neq}\#d \, \wedge \, \#d{\leq}6 \qquad \to n := \#d; avg := \tfrac{1}{n}\sum_{i=1}^{n} d(i); \\
&\qquad\qquad\qquad\qquad\qquad\qquad\qquad\quad\; t, in := 0, false \\
&\qquad\quad \textbf{od} \\
&\quad] \, | \, {:} m, t
\end{aligned}
$$

1.4.5 About Validation of Properties of Interest

The most important property in the presented model is the accuracy of the measurement value. The presented model accounts both external and internal sources for uncertainty. Thus, we can use it to study the effect of the accuracy to an application objective, such as HVAC control objective, by extending the application logic. In an ideal case, such an analysis could be performed by model-checking [279], for instance, by using statistical model-checking, such as UPPAAL-SMC [69]. In a generic case, numerical simulation may be required.

Similar analysis methods could also be used to analyze drifting of the measured time. The use of an inaccurate clock for measuring the passage of time may not be an issue for a control application. However, when considering applications that require measurement data over a long period of time, inaccuracies in the measurement time do cause challenges. For instance, training of an anomaly detection algorithm may strongly be affected by temporal deviations in the training data, causing malfunctioning detection or, in the worst case, failing of the training altogether. For such a case, the presented model could be used to analyze alternative methods to account for the inaccuracies in the measurement time. For instance, the presented model could be adjusted by averaging the measurement values over a longer period of time, to see if such a change compensates for the drifting of the measured time.

Perhaps the biggest advantage of the presented model is that it can be subject to refinement [32, 282, 281]. Then, for instance, prior to system upgrading, the sensor model could be refined, to account for more advanced sensor technologies. Similarly, the monitoring model could be refined to account for more detailed and versatile monitoring logic. In case the refinement proofs succeed, we know that the overall room temperature monitoring system is still provably correct despite the modeled changes.

Here, however, we explore the opposite of refinement, coarsement [300], in simplifying the validation task. In case of the room temperature monitoring, the use of the differential actions prohibits use of, for instance, theorem provers [340] in the validation. By using coarsement, we can provide a more abstract, provable correct model of the environment, where the differential actions are substituted with non-deterministic assignments. Such an action system can then be analyzed with a theorem prover. The abstract model of the environment, $\mathcal{A}n\,Environment$, is shown below. For clarity, the relation between absolute time, T, its update γ, measured time t, and its update τ is given as a relation $\Gamma(T, \gamma, t, \tau)$ which is defined as $0.999(\tau - t) \leq \gamma - T \leq 1.001(\tau - t)$. It can be proven by using data refinement that $\mathcal{A}n\,Environment \preceq_{true} \mathcal{E}nvironment$ indeed holds. A similar proof was presented and discussed in an earlier article [281]. In short, the proof relies on the fact that a differential action terminates at the boundary of its evolution guard. Hence, the updated values for the variables must be some of the boundary values.

$$\mathcal{A}n\,Environment \,\widehat{=}$$
$$|[\,\mathbf{var}\ T^*, t^*, c^*{:}\,I\!\!R, I\!\!R, I\!\!R$$
$$\quad T, t, c := 0, 0, 20;$$
$$\quad \mathbf{do}\ t{<}10 \wedge 19{<}c{<}24 \quad \rightarrow \quad T, t, c := \gamma, \tau, \varsigma.$$
$$\qquad \Gamma(T, \gamma, t, \tau) \wedge t{<}\tau{\leq}10 \wedge 19{\leq}\varsigma{\leq}24 \wedge (\tau{=}10 \vee \varsigma{=}19 \vee \varsigma{=}24)$$
$$\quad [\!]\ \ t{<}10 \wedge c{<}20 \quad \rightarrow \quad T, t, c := \gamma, \tau, \varsigma.$$
$$\qquad \Gamma(T, \gamma, t, \tau) \wedge t{<}\tau{\leq}10 \wedge \varsigma{\leq}20 \wedge (\tau{=}10 \vee \varsigma{=}20)$$
$$\quad [\!]\ \ t{<}10 \wedge 23{<}c \quad \rightarrow \quad T, t, c := \gamma, \tau, \varsigma.$$
$$\qquad \Gamma(T, \gamma, t, \tau) \wedge t{<}\tau{\leq}10 \wedge 23{\leq}\varsigma \wedge (\tau{=}10 \vee \varsigma{=}23)$$
$$\quad \mathbf{od}$$
$$]|$$

1.5 Conclusion

We proposed the use of hybrid action systems to model the sources for uncertainties in environmental monitoring. We illustrated the approach by modeling a room temperature monitoring system. In the model, we formally modeled the most prominent sources for uncertainty, including non-determinism associated with the dynamics of the environmental variables, external variations associated with measuring temperature, drifting of measurement clocks, sporadic device failures, and inaccuracies introduced by computational methods. The advantage of modeling environmental monitoring systems by using hybrid action systems, is the support for rigorous proofs about the properties of the model. Refinement can be applied to maintain the model with respect to system upgrades and changing end-user requirements. Alternatively, coarsement can be used to abstract away details to enable use of validation tools, such as theorem provers.

Acknowledgments. This research is funded by the Academy of Finland project "FResCo: High-quality Measurement Infrastructure for Future Resilient Control Systems" (Grant number 264060).

Chapter 2

Mandatory and Potential Choice: Comparing Event-B and STAIRS

Atle Refsdal

SINTEF ICT, Norway

Ragnhild Kobro Runde

University of Oslo, Norway

Ketil Stølen

SINTEF ICT, Norway and University of Oslo, Norway

Abstract. In order to decide whether a software system fulfills a specification, or whether a detailed specification preserves the properties of a more abstract specification, we need an understanding of what it means for one specification to fulfill another specification. This is particularly important when the specification contains one or more operators for expressing choice. Operators for choice have been studied for more than three decades within the field of formal methods in general, and within methods for action-refinement in particular. In this paper we focus on Event-B, a more recent method for action refinement. The STAIRS method belongs to another tradition. It originates from the UML community and is designed to provide an understanding of refinement and fulfillment for UML. STAIRS distinguishes between potential and mandatory choice, where only the latter is required to be preserved by refinement. This paper investigates the relationship between the operators for choice in Event-B and STAIRS.

2.1   Introduction

In order to decide whether a software system fulfills a specification, we need a clear understanding of the concept of fulfillment. Similarly, when a specification is developed further into a new more detailed (for example, platform-specific) specification, the essential properties captured by the original specification must still be present in the new specification. This requires an understanding of what it means for one specification to fulfill another specification.

STAIRS [156, 288] was designed to provide the UML community with this kind of understanding at a level of abstraction that is easily comprehensible for UML practitioners. STAIRS is inspired by formal methods and refinement theory. However, STAIRS is not really a formal method in the classical sense, as explained in the following. When formal methods are combined with more applied methods for software engineering, the resulting approaches may typically be classified according to whether:

- Artifacts of the applied method, typically specifications and models, are translated into the formal method and used for formal analysis.

- Artifacts of the applied method, again normally specifications and models, are annotated with formal expressions and used for formal analysis building on some unified underlying semantics.

STAIRS does not fit within this classification scheme since the emphasis of STAIRS is to provide a foundation for fulfillment within the conceptual universe of UML rather than supporting formal analysis of UML specifications and their relationships.

STAIRS addresses primarily sequence diagrams. Implicitly, STAIRS also defines the notion of fulfillment for the other UML notations for modeling dynamic behavior, where the behavior may be captured by sets of sequence diagrams. In many respects, sequence diagrams are more general than other UML notations for dynamic behavior, e.g., state machines, because sequence diagrams may be used to describe examples of required behavior, rather than the complete allowed behavior. STAIRS provides this expressiveness by offering operators for potential as well as mandatory choice.

Operators for choice have been studied for more than three decades within the field of formal methods in general, and within methods for action-refinement in particular [32]. A prominent example of a method for action-refinement is Event-B [6]. The objective of this paper is to investigate the relationship between the operators for choice in Event-B and STAIRS.

A large literature exists on Event-B. There are no fixed semantics for Event-B, instead the semantics are provided implicitly by proof obligations associated with a model [152]. Nevertheless, several papers have suggested failure-divergences inspired semantics as a formal underpinning [293, 71, 305]. This paper builds on this approach. Failure-divergences semantics was originally developed for CSP [171]. In Section 2.2 we therefore start our investigation by comparing choice in CSP to choice in STAIRS. Then we conduct a comparison of Event-B and STAIRS; first at the syntactic level in Section 2.3; then at the semantic level in Section 2.4. Finally, Section 2.5 provides a summary and draws conclusions.

Preserved by refinement	CSP		STAIRS
	environment	**system**	
No		Internal choice Demonic choice	Potential choice
Yes	External choice Angelic choice		Mandatory choice

Table 2.1: Choice types in CSP and STAIRS

2.2 Kinds of Choice

In this section, we relate the kinds of choice offered by CSP [171] and STAIRS [156]. We also classify the kinds of properties that may or may not be captured depending on the available choice operators.

In order to understand the choice operators in CSP we need to understand some underlying assumptions about the involved entities, as well as the communication model. As explained by [284, p. 13], in CSP a system[1] is completely described by the way it can communicate with its environment. Hence, CSP assumes a black-box view where internal communication within the system itself is hidden. Communication is synchronous (also known as handshake communication), meaning that "events only happen when both sides agree" [284, p. 9]. This can be understood as follows: At any given point the system offers a set of events to the environment. If the environment accepts one of these events then the system moves on, otherwise a deadlock occurs.

Choices made by the environment between available alternatives are called *external* choices and represented by the □ operator in CSP, while choices made by the system are called *internal*[2] and represented by the ⊓ operator. If one of the alternatives offered to the environment is removed, a deadlock will be introduced if the environment is willing to synchronize only on the removed alternative. Refinement in CSP therefore requires preservation of external choice. Internal choice, on the other hand, represents underspecification and may be reduced in a valid refinement step, as motivated by the following quote [171, p. 101–102]:

> Sometimes a process has a range of possible behaviours, but the environment of the process does not have any ability to influence or even observe the selection between the alternatives [...] The choice is made, as it were internally, by the machine itself, in an arbitrary or nondeterministic fashion [...]

> There is nothing mysterious about this kind of nondeterminism: it arises from a deliberate decision to ignore the factor which influence the selection [...] Thus nondeterminism is useful for maintaining a high level of abstraction in descriptions of the behaviour of physical systems and machines [...]

> A process specified as $(P \sqcap Q)$ can be implemented either by building P or by building Q. The choice can be made in advance by the implementor on grounds not relevant (and deliberately ignored) in the specification [...]

---

[1]The CSP literature typically uses the term "process".
[2]Hoare uses the term "nondeterministic or" or just "nondeterminism" for internal choice.

The term *angelic* choice (or angelic nondeterminism) is sometimes used to describe a choice that will always be made so that an undesirable result (a deadlock) is avoided if possible. Hoare explains this in terms of an implementation that, when choosing between P and Q, "minimises the risk of deadlock by delaying the choice until the environment makes it, and then selecting whichever of P and Q does *not* deadlock" [171, p. 105]. Similarly, Roscoe explains angelic choice in terms of an operator that "keeps on giving the environment the choice of action of P and Q as long as the environment picks an event they both offer" [285, p. 219]. Hence, angelic choice is a special kind of external choice. Conversely, although not used in the above references, the term *demonic choice* can be used to describe an internal choice that will (or at least can) be made so that a deadlock will occur, if possible. In [248], Morgan et al. use the terms demonic choice and internal choice interchangeably.

The two middle columns of Table 2.1 summarize the above discussion. The system column represents choices resolved by the specified system, while the environment column represents choices resolved by its environment. The column furthest to the left indicates whether choices are preserved by refinement.

According to [156], STAIRS is an approach to compositional development of UML interactions that assigns a precise interpretation to the various steps in incremental system development based on an approach to refinement known from the field of formal methods. There are a couple of ways in which STAIRS differs from CSP of immediate relevance for our discussion of choice here. First, STAIRS assumes an asynchronous communication model with infinite buffering, and is therefore not concerned with deadlock. Second, in STAIRS there is no implicit hiding of internal communication when composing specifications.

STAIRS offers two different choice operators: one for *potential* choice and one for *mandatory* choice. Along the same lines as internal choice in CSP, potential choice is motivated by the need for abstraction. This is explained by the following requirement to STAIRS stated in [156]:

> Should allow specification of potential behavior. Underspecification is a well-known feature of abstraction. In the context of interactions, "under-specification" means specifying several behaviors, each representing a potential alternative serving the same purpose, and that fulfilling only some of them (more than zero but not all) is acceptable for an implementation to be correct.

Mandatory choice, on the other hand, is motivated as follows:

> Should allow specification of mandatory behavior [...] Sometimes [...] it is essential to retain non-determinism in the implementation reflecting choice. For example, in a lottery, it is critical that every lottery ticket has the possibility to win the prizes [...] As a consequence, we need to distinguish explicit non-determinism capturing mandatory behavior from non-determinism expressing potential behavior.

Since potential choice facilitates underspecification by offering alternatives serving the same purpose, STAIRS allows potential choice to be reduced or removed by refinement. A mandatory choice, on the other hand, needs to be preserved in order to ensure that all intended behavior will be implemented. This applies regardless of whether the choice is made by the system or by the environment. The distinction between potential and mandatory choice in STAIRS is summarized in the right-hand column of Table 2.1.

Another kind of choice is probabilistic choice, meaning that each alternative should be selected according to a given probability. Probabilistic choice is beyond the scope of this paper and has therefore not been included in Table 2.1. However, mandatory choice (as understood in STAIRS) can be understood as probabilistic choice where all of the probabilities should be higher than 0, but where nothing more is known/specified about the probabilities.

The constructs for expressing choice offered by a specification language and its notion of refinement restrict the kinds of properties that can be captured. System properties are typically analyzed on the basis of system traces, each of which characterizes a possible run or execution. Properties can then be classified according to their means of falsification. Properties that can be falsified by a tester on the basis of a single trace are called *trace properties*, while properties that can be falsified on the basis of trace sets are called *trace set properties* [241]. The former include safety and liveness as originally investigated by Alpern and Schneider [14, 292]. The latter include information security flow properties and are what McLean referred to as possibilistic properties [241].

As an example, assume we want to specify a simulator to simulate user behavior for automatic testing of vending machines offering tea and coffee. The simulator should then be able to choose both alternatives, and the choice should be made internally by the simulator (thus reflecting a user's preference) rather than by its environment. Before using the simulator to automatically test a vending machine, we need to test the simulator itself. When testing the simulator, if we observe a single trace yielding tea, we cannot deduce that the simulator is not able to choose coffee or vice versa; such falsifications can only be made by considering all traces of the system.

Specification approaches allowing all choices made internally by the specified system (as opposed to its environment) to be reduced by refinement, have no means to ensure that trace set properties are preserved. This is referred to as the refinement paradox in [198]. In the following, we discuss the syntax and semantics of choice in Event-B and STAIRS in the light of trace properties and trace set properties.

2.3 Comparing Event-B and STAIRS at the Syntactic Level

The essence of an Event-B specification is a set of guarded events, where an event is enabled and may be chosen to occur when its guard is true. More than one event may be enabled at the same time, and the choice between enabled events is an external choice made by the environment. Internal choice made by the system itself is modeled more indirectly, using nondeterministic assignment to internal variables in order to influence the enabledness of other events.

As an example of how choice is treated in Event-B, in Figure 2.1 we look at the two vending machine specifications given by Butler in [71]. An Event-B specification consists of a specification name, a declaration and initialization of variables and a set of named events. Each event is on the form **when** *guard* **then** *body* **end**, where the guard is a boolean statement over the variables and the body is a (possibly non-deterministic) variable assignment. An event is said to be enabled if its guard evaluates to true, otherwise it is disabled.

The difference between the two specifications in Figure 2.1 is that in *VM*1, the internal variable $m1$ is set to *vend* after the Coin event has been executed, thus enabling both the Tea and the Coffee event, while in *VM*2, the internal variable $m2$ is set to either *tea* or *coffee*, thus enabling only one of Tea and Coffee. This means that in *VM*1, the choice

machine *VM1*	machine *VM2*
variables *m1* ∈ {*idle,vend*}	variables *m2* ∈ {*idle,tea,coffee*}
initialisation	initialisation
m1 := idle	*m2 := idle*
events	events
Coin ≙ **when**	Coin ≙ **when**
m1 = idle	*m2 = idle*
then	**then**
m1 := vend	*m2 :∈ {tea,coffee}*
end	**end**
Tea ≙ **when**	Tea ≙ **when**
m1 = vend	*m2 = tea*
then	**then**
m1 := idle	*m2 := idle*
end	**end**
Coffee ≙ **when**	Coffee ≙ **when**
m1 = vend	*m2 = coffee*
then	**then**
m1 := idle	*m2 := idle*
end	**end**

Figure 2.1: Two Event-B vending machines as specified by [71]

between Tea and Coffee is to be made by the environment, and is thus an example of external choice. In [71], it is argued that from a customer's point of view, this external choice should be preserved by refinement, meaning that *VM2* should not be a valid refinement of *VM1*. This could be achieved for instance by requiring that a refinement should preserve the enabledness of individual events.

In *VM2*, the choice between Tea and Coffee is made by the machine itself, and this is an example of internal choice. An internal choice may be refined by an external choice, as this ensures that all events enabled in the original machine will also be enabled in the refined one. Consequently, *VM1* should be a valid refinement of *VM2*.

As an example of potential choice in STAIRS, Figure 2.2 gives a sequence diagram specification of a vending machine with messages that correspond to the events in *VM1* and *VM2* from Figure 2.1. The main ingredients of a sequence diagram are a set of lifelines (depicted as vertical lines) and a number of messages (arrows) between the lifelines. In Figure 2.2, the choice operator alt is used to signify that this diagram specifies two example scenarios, both starting with the vending machine receiving a coin from the environment, followed by the vending machine providing tea in one scenario, coffee in the other. As alt is used to model potential choice, a sequence diagram where only one of these scenarios is positive, and the other one is specified as negative, would be a valid refinement of Vending Machine 1.

In STAIRS, there is no fundamental distinction between internal and external choice. The choice between sending Tea or Coffee in Figure 2.2 is an internal choice when seen from the sending lifeline VM, and an external choice when seen from the receiving lifeline Env. For a real vending machine, the choice between tea and coffee would be made by the

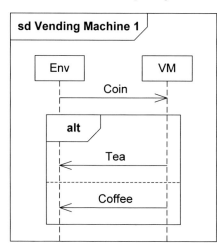

Figure 2.2: A simple vending machine with potential choice in STAIRS

user. In STAIRS, this may be modeled for instance by selection messages from `Env` to `VM` as seen in Figure 2.3. Note, however, that the choice between the two alternatives is still specified using `alt`, meaning that in a valid implementation, only one of the tea and coffee scenarios may be present.

Back to Event-B, Butler [71] argues that experience with Event-B modeling has demonstrated the need to be able to model both internal and external choice between enabled events more directly, in particular in situations where the guards are equal only as a result of abstraction. In reality, such a choice is not really external, but rather internal due to some condition not included in the abstract specification.

For instance, the choice between `Tea` and `Coffee` in *VM1* in Figure 2.1 should in some cases be seen as internal due to some condition abstracted away in *VM1*. A refinement may for instance add internal variables and guards so that coffee is always served in the morning, while tea is always served in the afternoon.

In [71], the main goal is to allow both external and internal choice to be represented directly. This is achieved by letting the specifier divide the events into groups. The intuitive interpretation is that a choice between groups of events is external, while a choice between events within a group is internal. For *VM1* in Figure 2.1, the specifier may state that the choice between `Tea` and `Coffee` is internal by grouping them together, giving the following event groups for *VM1*: $G_1 = \{\texttt{Coin}\}$, $G_2 = \{\texttt{Tea}, \texttt{Coffee}\}$.

In [71], the refinement relation is modified so that preservation of enabledness is preserved for event groups rather than for single events, thus ensuring that external choices are preserved (or increased) through refinement while at the same time allowing the amount of internal choice to be reduced. With the event groups G_1 and G_2 given above, this would mean that a valid refinement of *VM1* may choose to offer only `Coffee` (or `Tea`).

Mandatory choice is not discussed in [71], and the introduction of event groups is not sufficient to capture system choices that must be preserved by refinement, i.e., trace set properties. Assume, for illustration purposes, that the specifier wants to model a machine which arbitrarily chooses between tea and coffee at run-time (similar to the user simulator described in Section 2.2). As putting `Tea` and `Coffee` in the same event group might lead

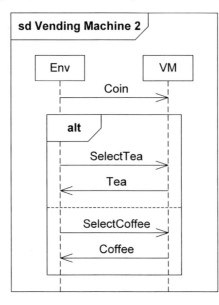

Figure 2.3: A simple vending machine with external choice in STAIRS

to an implementation offering only one of them as seen above, the only other possibility is to put them in separate event groups. However, the semantics of such a specification would allow Tea (and similarly, Coffee) to be refused only when its guard is false, meaning that neither Tea nor Coffee could be refused after Coin, so that the choice between the two remains external and not internal.

A vending machine where the internal choice is made arbitrarily as described above, may be modeled in STAIRS by using the mandatory choice operator xalt as shown in Figure 2.4. To simplify the main diagram Vending Machine 3, the diagram refers to two sub-diagrams Provide tea (also provided in Figure 2.4) and Provide coffee (not shown, but symmetrical to Provide tea). The refuse operator is used to model that a specific alternative should be considered negative, e.g., in Provide tea the vending machine should serve tea and not coffee. The main diagram Vending Machine 3 then requires the vending machine to have two mandatory behaviors, one with tea and not coffee, and one with coffee and not tea. Neither of these can be removed by refinement. Also, the mandatory choice in Vending Machine 3 in Figure 2.4 is a valid refinement of the potential choice in Vending Machine 1 in Figure 2.2, as should be expected. Further refinements may increase the mandatory behavior required by adding more xalt-operands, e.g., a third alternative providing chocolate but not tea or coffee.

2.4   Interaction-Obligations versus Failure-Divergences

In the previous section, we compared Event-B and STAIRS at the syntactic level. Semantically, an Event-B specification may be represented by a failure-divergences pair while a

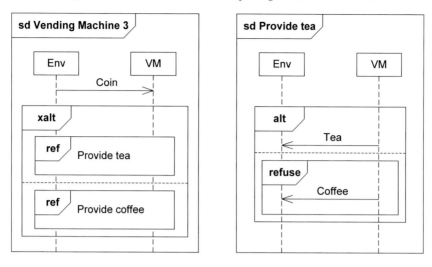

Figure 2.4: Mandatory choice in STAIRS

sequence diagram in STAIRS corresponds to a set of interaction-obligations. If the sequence diagram does not contain mandatory choice, a single interaction-obligation is sufficient. In the following, we outline the intuition behind interaction-obligations and failure-divergences and how they are related. We also explain how mandatory choice is represented semantically and discuss the relationship to external choice.

2.4.1 Interaction-Obligations

A trace in STAIRS is a finite or infinite sequence of events where an event is either the sending or the reception of a message. A trace is required to fulfill certain well-formedness conditions [287]. Informally, a trace is well-formed if, for each message, the send event is ordered before the corresponding receive event.

Let $\mathcal{H}$ denote the set of all well-formed traces. An interaction-obligation is a pair (p, n) of trace-sets which classifies the elements of $\mathcal{H}$ into three categories: the positive traces p, the negative traces n, and the inconclusive traces $\mathcal{H} \backslash (p \cup n)$. The inconclusive traces are those traces that are neither specified as positive nor as negative by the sequence diagram in question.

A pre-post specification $(pre, post)$ in Hoare-logic may be used as a first approximation of the intuition behind an interaction-obligation (p, n). Roughly speaking:

- A positive trace (in p) corresponds to an execution initiated in a state fulfilling pre that if it terminates, does so in state fulfilling $post$ (given Hoare-logic for partial correctness).

- An inconclusive trace (in $\mathcal{H} \backslash (p \cup n)$) corresponds to an execution initiated in a state fulfilling $\neg pre$.

- A negative trace (in n) corresponds to an execution initiated in a state fulfilling pre that terminates in state fulfilling $\neg post$.

It is worth noticing that while the inconclusive behavior corresponding to a pre-post specification is chaotic, meaning that anything is allowed, the inconclusive behavior of an interaction-obligation is not necessarily so. For example, if the finite trace t is inconclusive then the result $t \frown t'$ of extending t with t' is not necessarily inconclusive; $t \frown t'$ may be positive ($t \notin p \cup n \wedge t \frown t' \in p$) and another extension t'' may be negative ($t \notin p \cup n \wedge t \frown t'' \in n$). In fact, an interaction-obligation may for the same environment behavior allow inconclusive, positive as well as negative behavior. A pre-post specification on the other hand, classifies executions initiated in a state (in a pre-post setting, representing the environment behavior) as either positive or negative if it fulfills *pre* and as inconclusive otherwise.

It is also worth mentioning that for interaction-obligations as for pre-post specifications, there may be environment behaviors for which no behavior is allowed. An example of a pre-post specification of this kind is

$$(\; true \; , \; x = 0 \Rightarrow false \;)$$

It disallows any behavior for the initial state $x = 0$. In classical Hoare-logic, such a specification is not implementable because any real program has some kind of behavior whatever the environment does. In other words, no real program is partial. Hence, any implementable pre-post specification is total; it allows at least one system behavior for each possible initial state. The same is not true for interaction-obligations because sequence diagrams only specify example runs and not the full behavior of a real program. Hence, a sequence diagram only considering some input behaviors is unproblematic from a methodological point of view.

In Hoare-logic, refinement corresponds to weakening the pre-condition and strengthening the post-condition. Refinement of an interaction-obligation corresponds to reducing inconclusive behavior and redefining positive behavior as negative. Formally:

Definition 2.1. *An interaction-obligation (p', n') refines an interaction-obligation (p, n) if $p \subseteq p' \cup n'$ and $n \subseteq n'$.*

Given the mapping to pre-post specifications outlined above, reducing inconclusive behavior may be understood as weakening the pre-condition; redefining positive behavior as negative may be understood as strengthening the post-condition. Hence, refinement of interaction-obligations reflects very well refinement of pre-post specifications.

2.4.2   Failure-Divergences

In the setting of Event-B, a specification may be described by a pair (f, d) of a set of failures f and a set of divergences d. A *failure* is a pair (t, X) of a finite trace t and a set of events X that the specified system may refuse after having engaged in the external interaction corresponding to t. In other words, the specified system may deadlock after having engaged in t if offered only X or a subset of X by the environment. A *divergence* is a trace t after which the systems may diverge, meaning that it performs an infinite unbroken sequence of internal (and hence invisible) actions without any external communication happening at all, also referred to as livelock. Well-formedness constraints [171, p. 130] are imposed on failure-divergence pairs that imply that any such pair is total; it allows some behavior (possibly consisting of doing nothing) whatever the environment does.

Failures-divergences refinement corresponds to removing failures and divergences. This means set inclusion with respect to the failures and the divergences. Formally:

Definition 2.2. *A failures-divergences pair (f', d') refines a failures-divergences pair (f, d) iff $f' \subseteq f$ and $d' \subseteq d$.*

External choice cannot be reduced by refinement, as this would imply adding new failures in order to allow the specified system to refuse some of the events offered to the environment according to the more abstract specification.

2.4.3 Relating the Two Models

In the case of total correctness the relationship between an Event-B specification captured by (f, d) and a sequence diagram captured by the interaction-obligation (p, n) may be characterized as follows:

- The positive behavior p corresponds to $e \backslash d$, where $e = \{t \mid (t, \varnothing) \in f\}$.

- The inconclusive behavior $\mathcal{H} \backslash (p \cup n)$ corresponds to d.

- The negative behavior n corresponds to $\mathcal{H} \backslash (e \cup d)$.

Given the mapping above, reducing inconclusive behavior in STAIRS may be understood as reducing the set of divergences, while redefining positive behavior as negative in STAIRS may be understood as reducing the set of traces that are not divergences. This mapping is not information preserving since semantically different failure-divergences are mapped to the same interaction-obligation. In particular, the semantic difference between external and internal choice disappears.

Contrary to a failure-divergences pair, an interaction-obligation may be partial in the sense that there may exist environment behavior for which no positive system behavior is defined. It may be argued that a refinement should not impose additional constraints on the environment behavior and thereby increase partiality. Although this constraint may easily be imposed, it is not enforced by STAIRS because there are situations where this is not very practical. We may for example use the operator for potential choice to specify two different protocols for interaction with the environment and then leave it to the implementor to select which one to use. This choice will also restrict (or impose additional assumptions about) the behavior of the environment because the specified system will only work properly if the environment sticks to the selected protocol.

2.4.4 Sets of Interaction-Obligations

Sequence diagrams with mandatory choice in STAIRS is represented by a set of interaction-obligations. Informally speaking, each interaction-obligation represents an alternative that must be reflected in any correct implementation. In the most general case, refinement corresponds to:

Definition 2.3. *A set of interaction-obligations o' refines a set of interaction-obligations o if for each interaction-obligation $(p, n) \in o$ there is an interaction-obligation $(p', n') \in o'$ such that (p', n') refines (p, n).*

Hence, each interaction-obligation at the more abstract level must be refined by at least one interaction-obligation at the more concrete level. On the other hand, o' may have interaction-obligations that do not refine any of those is o. In the STAIRS literature this notion of refinement is called general refinement. A more restrictive version is limited refinement which also requires each of the more concrete interaction-obligations to be a refinement of at least one abstract at the more abstract level.

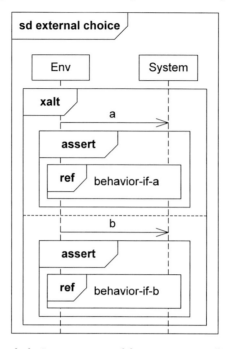

Figure 2.5: External choice represented by a sequence diagram in STAIRS

In the mapping from failure-divergences to interaction-obligations defined in the subsection above, we lost the distinction between external and internal choice. When mapping failure-divergences to sets of interaction-obligations we have the expressiveness required to keep this distinction. Roughly speaking, each external choice alternative corresponds to a separate interaction-obligation as outlined by the example in Figure 2.5. The assert, which is a standard UML 2.x operator, makes any inconclusive trace in its body negative. From the perspective of the System lifeline, the diagram captures an external choice between receiving either a or b. If neither behavior-if-a nor behavior-if-b contain xalt-operators, the semantics of the diagram is a pair of two interaction-obligations; one corresponding to receiving a and one corresponding to receiving b.

2.5   Conclusion

This paper compares Event-B (with a failure-divergences semantics) and STAIRS with particular focus on mandatory and potential choice. While the failure-divergences semantics gives a pure black-box interpretation of the specified system, STAIRS offers a white-box interpretation in terms of interaction-obligations and sets of interaction-obligations. Sets of interaction-obligations are required to capture mandatory choice while a single interaction-obligation is sufficient to model potential choice.

The main inspiration for writing this paper was Butler's proposal to capture external and internal choice directly by letting the specifier divide the events into groups. The approach seemed to resemble our proposal to capture mandatory and potential choice by sets of interaction-obligations.

Our conclusion is that it does. The expressivity offered by Butler's proposal is also provided by STAIRS. In the same sense as a single event group captures internal choice, single interaction-obligations captures potential choice. When potential choice is restricted to the specified system, internal and potential choice is the same – both represent underspecification. Moreover, in the same sense as sets of event groups capture external choice, sets of interaction-obligations capture mandatory choice. When mandatory choice is restricted to the environment, external and mandatory choice is the same – both represent nondeterminism that must be preserved by refinement.

Acknowledgments. This work has been conducted as a part of the DIAMONDS (201579) project and the AGRA (236657) project, both funded by the Research Council of Norway, and the CONCERTO (232059) project funded by the Research Council of Norway and by ARTEMIS Joint Undertaking – a public private partnership in the field of embedded systems supported under the Seventh Framework Programme of the European Commission.

Chapter 3

Modelling and Refining Hybrid Systems in Event-B and Rodin

Michael Butler

University of Southampton, UK

Jean-Raymond Abrial

Independent consultant, Marseille, France

Richard Banach

University of Manchester, UK

Abstract. We outline an approach to modelling and reasoning about hybrid systems with the Event-B method supported by the Rodin toolset. The approach uses continuous functions over real intervals to model the evolution of continuous values over time. Nondeterministic interval events are used to specify how continuous variables evolve within an operating mode. Refinement is used to constrain the choice of continuous functions and to decompose a non-deterministic interval event into a series of periodic interval events.

3.1 Introduction

Event-B is an established formalism that has been applied to a range of systems especially control systems and distributed systems [6]. A key feature of Event-B is the use of abstract modelling to represent the *purpose* of a system and the use of refinement to demonstrate conformance between the abstract models and more detailed models representing the designs that are intended to achieve the desired purpose. An Event-B machine represents a single level of abstraction and consists of state variables, invariants and events (i.e., parameterised guarded atomic actions). The application of Event-B is enabled by the Rodin toolset [2] which provides capabilities for proof obligations generation and automated and interactive proof capabilities.

Event-B has largely been used to represent and reason about discrete models. In this chapter we are interested in hybrid models, that is, models containing a mix of discrete and continuous behaviour. We focus on one kind of continuous property, namely, bounds on a continuous function. A common approach in modelling a hybrid system is to identify a number of discrete modes such that within each mode the evolution of continuous variables is specified through dynamic control laws (e.g., differential equations) [164, 266, 281]. We follow such an approach here though we abstract away from control equations, instead specifying assumptions about continuous functions; for example, we might assume that a continuous function is monotonically increasing over an interval. At the abstract level, an atomic event is used to specify the continuous behaviour within a mode; such an event nondeterministically chooses a continuous function over a time interval representing the continuous behaviour within that mode. We refer to this as an *interval* event. We use refinement for two purposes. The first is to make the choice of the continuous function more constrained (reduction of non-determinism). The second use of refinement is to introduce additional discrete steps within a mode to represent a periodic control strategy that determines, at each period, whether to remain in a mode or switch to a different mode.

We illustrate the approach we have adopted using the classic example of a controller for a water tank. The purpose of the water tank controller is to maintain the water level in the tank between a low and a high level. This is specified through bounds on a continuous function representing the evolution of the water level within a mode. In a first refinement we constrain the choice of continuous function further, specifying that it is monotonically increasing (or monotonically decreasing) within a mode. In a second refinement, we introduce periodic control events within the modes.

One objective of the work outlined here was the desire to follow an abstraction/refinement approach, starting with a model of the purpose and refining this towards an implementation strategy. An additional key aim was to understand the extent to which the existing Event-B refinement concepts and the Rodin toolset could be used to achieve the modelling and proofs of the hybrid water tank. A key enabler for mechanising the modelling and proofs in Rodin is the Theory Plug-in for Rodin [72] . The Theory Plug-in allows us to extend the mathematical language of Event-B with theories of real numbers and continuous functions over intervals. The Theory Plug-in also allows us to define new proof rules about reals and continuous functions that can be used by the Rodin proof manager.

The work presented here uses the standard refinement proof obligations of Event-B so that no changes where required in the Rodin tool. The Event-B proof rules are defined in [6]. Proving refinement in Event-B requires gluing invariants that relate abstract and concrete variables. Each event of a refining machine either refines an event of the abstract machine or refines *skip*. For an event that refines an abstract event, the guards of that refined event must entail the guards of the abstract event under the gluing invariants and the actions must maintain the gluing invariants. An event refines *skip* if it maintains the gluing invariants when no change occurs in the abstract variables.

The approach presented here was inspired by previous work on reasoning about hybrid systems using an abstraction/refinement approach. Continuous Action Systems [30] are an extension of the classical Action System approach [29] where variables are time dependent functions, that is, variables are functions over non-negative reals. Hybrid Action Systems [281] provide the ability to specify evolution of continous functions using differential equations. Inspired by Continuous Action Systems, Su et al [322] have developed an approach to modelling hybrid systems in Event-B using a combination of discrete and time varying variables. The approach followed in [322] is to start with discrete models and introduce continuous behaviour in refinement steps. When mechanising the models and proofs in Rodin, [322] approximated reals using integers (because of the lack of support for reals in Rodin at the time that the work was undertaken). Hybrid Event-B is an extension of Event-B that distinguishes *mode* variables (discrete) and *pliant* variables (continuous) [43]. Hybrid Event-B also distinguishes mode events and pliant events; mode events model instantaneous discrete changes while pliant events model continuous evolution of pliant variables over time intervals. The interval events used in this chapter are essentially intended to mimic the pliant events of Hybrid Event-B. At the time of writing there is no tool support for Hybrid Event-B.

We make a distinction between a control *goal* and a control *strategy*. We outline how both can be modelled and how refinement can be used to prove that a strategy satisfies a goal. In the case of the water tank system the control goal is as follows:

> The control goal is to maintain the water level between a high level, H, and a low level, L.

Water may flow out of the tank according to some known maximum rate. To maintain a satisfactory water level, the controller can switch on a pump which causes the water level to increase according to some known maximum rate.

> The control strategy is to sense the water level periodically and switch the pump on or off as appropriate.

We find it useful to follow the categorisation of modelling variables given in the Four-variable model of Parnas and Madey [260]. In this model, there are two main groupings of variables, *environment* variables and *controller* variables. Environment variables represent quantities in the environment of the controller. Controller variables represent quantities inside the controller machine. There are two kinds of environment variable as follows:

Monitored variables Environmental quantities whose value is not determined by the controller but that can be monitored. For example, the water level in the tank is a monitored variable.

Controlled variables Environmental quantities whose value is expected to be determined by the controller. For example the pump status (*on* or *off*) is a controlled variable.

The approach we follow is to start with a model of the control goal expressed in terms of monitored variables. We then refine this by a model of the strategy which is expressed in terms of monitored and controlled variables.

3.2   Reals and Continuous Functions

We have defined a theory of real arithmetic using the Rodin Theory feature. This introduces a new basic type, *REAL*, defines real versions of the standard arithmetic operators (addition, subtraction, multiplication and division) and defines the usual total order on reals.

With the Theory feature, operators may be defined directly (in terms of previously-defined operators), recursively (for inductive data types) and axiomatically with a collection of axioms on a group of operators. In our case we define the arithmetic operators axiomatically using standard axioms for reals.

We define an interval between two reals as follows[1] :

$$i..j \quad = \quad \{\, k \mid k \in REAL \,\wedge\, i \leq k \,\wedge\, k \leq j \,\}$$

This is an example of a direct definition in the Rodin Theory feature: the interval operator is defined using existing operators (set comprehension) for arguments i and j.

We use a standard definition of continuity. A function f is continuous at point c, written $cts(f, c)$, when it satisfies the following condition:

$$
\begin{aligned}
cts(f, c) \quad \Leftrightarrow \quad & f \in REAL \nrightarrow REAL \;\wedge \\
& \forall \epsilon \cdot 0 < \epsilon \;\Rightarrow \\
& \exists \delta \cdot 0 < \delta \;\wedge \\
& \quad \forall x \cdot x \in dom(f) \;\wedge \\
& \qquad c - \delta < x < c + \delta \\
& \quad \Rightarrow \\
& \qquad f(c) - \epsilon < f(x) < f(c) + \epsilon
\end{aligned}
$$

This states that for every neighbourhood around $f(c)$, defined by $f(c) \pm \epsilon$, there exists a δ that defines a neighbourhood around c, defined by $c \pm \delta$, that yields that neighbourhood around $f(c)$.

An interval function is continuous if it is continuous at every point in its domain so we define the set of continuous functions on the interval $i..j$, written $ctsF(i,j)$, as follows:

$$ctsF(i,j) \quad = \quad \{\, f \mid f \in i..j \to REAL \,\wedge\, \forall c \cdot c \in i..j \Rightarrow cts(f, c) \,\}$$

3.3   Modelling a Continuous Control Goal

We specify a continuous control goal in terms of monitored variables represented by continuous functions on time intervals. We use a simple form of timed automaton, where the state variables include a clock, *clk*, and monitored variables, *m*, specified as continuous functions from time zero up to *clk* (where $PosREAL = \{r \mid r \in REAL \wedge r \geq 0\}$):

$$
\begin{aligned}
clk \;&\in\; PosREAL \\
m \;&\in\; (0..clk) \to REAL
\end{aligned}
$$

This gives us two ways of specifying continuous goals in Event-B:

---

[1]For readability, we use the usual symbols for real arithmetic ($+$, $-$, $\leq$, etc.). These symbols are used for integer arithmetic in Rodin and, since operator overloading is not allowed in Rodin, we use different symbols (*plus*, *sub*, *leq*, etc.) in our REAL theory.

- *Invariants* are used to specify a property satisfied by the entire evolution of continuous variables.

- *Interval events* are used to specify a property on the continuous evolution within a behaviour mode.

Boundary constraints can be specified independently of time, that is, the value of a continuous variable at each point in its interval remains within some fixed boundaries. If we characterise the boundary as the set of values B within the boundary, then satisfaction of the boundary by all points of the continuous variable can be specified by range inclusion:

$$ran(m) \subseteq B$$

Requiring that the water level is always between the low and high marks is an example of a boundary constraint defined by the set $L..H$. Another form of pointwise property could involve the relationship between several continuous variables. For example, in a cruise control system for a car, the speed should be close to the target speed. Assuming *target* and *speed* are continuous functions on interval $i..j$, this can be specified as a pointwise predicate as follows:

$$\forall t \cdot t \in i..j \;\; \Rightarrow \;\; target(t) - \delta \;\leq\; speed(t) \;\leq\; target(t) + \delta$$

Other properties, such as monotonicity, smoothness, responsiveness and stability, span an interval and cannot be specified on individual time points. We do not consider a full range of interval properties here, but we do make use of monotonicity. For example, we define the set of *monotonically increasing* interval functions as follows:

$$mono\_inc \;\; = \;\; \{\, f, i, j \mid f \in ctsF(i,j) \,\wedge$$
$$(\forall k, l \cdot i \leq k \leq l \leq j \; \Rightarrow \; f(k) \leq f(l)) \bullet f \,\}$$

The control goal for the water tank is a boundary property (defined by constants H and L) that should always be true so we use a boundary constraint on the continuous variable to model this:

inv1 : $clk \in PosREAL$

inv2 : $wl \in ctsF(0, clk)$

inv3 : $ran(wl) \subseteq L..H$

To model the dynamics within a mode, we use a nondeterministic interval event that extends the continuous behaviour for an interval of nondeterministic length. Such an event chooses some future time t and a continuous function f over the future interval and updates the clock and the continuous variables in a way that satisfies a property P. It has the following form:

Event *UpdateMonitored* $\;\widehat{=}$

    any

        t,f

    where

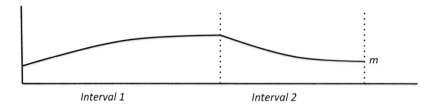

Figure 3.1: Evolution of continuous function during intervals.

> **grd1** : $t \geq clk + \epsilon$
> **grd2** : $f \in ctsF(clk, t)$
> **grd3** : $f(clk) = m(clk)$
> **grd4** : $P(f)$

then

> **act1** : $m := m \cup f$
> **act2** : $clk := t$

end

Possible continuous behaviour allowed by this event is illustrated by the graph in Figure 3.1. In the graph, the behaviour within each interval is defined by a continuous function over that interval and continuity is maintained between intervals. The function is monotonically increasing in *Interval 1* and monotonically decreasing in *Interval 2*.

Each of the guards in the *UpdateMonitored* event is essential:

- Guard *grd1* requires the next interval to have a duration of at least ϵ where ϵ is a constant. This is to prevent zeno behaviour where the time interval continually gets smaller and smaller.

- Guard *grd2* requires the next interval function f to be continuous.

- Guard *grd3* requires the endpoint of the existing interval function m and the starting point of the next interval function f to agree. This is to ensure that continuity is preserved when the interval function m is extended in action *act2*.

- Guard *grd4* requires the next interval function f to satisfy the interval property P.

We require that interval events always preserve continuity of the continuous variables (so *grd2* and *grd3* are essential). Some interval properties P are preserved under extension, that is, if the existing interval function m satisfies P and the next interval f satisfies P, that the extended interval function $m \cup f$ also satisfies P. It is easy to show that boundary properties are preserved when extending an interval function since they are time independent. Monotonicity is also reserved by interval extension, e.g., if m is monotonically increasing and f is monotonically increasing, then $m \cup f$ is monotonically increasing (provided $m \cup f$ is continuous). Clearly there are interval extensions that are not property preserving, e.g., extending a monotonically increasing function with a monotonically decreasing function does not preserve monotonicity.

In the water tank example, we use the following nondeterministic interval event to represent the required evolution of the water level during a mode:

Event *WaterLevelInterval* $\widehat{=}$

    any

        t,f

    where

        grd1 : $t \geq clk + \epsilon$
        grd2 : $f \in ctsF(clk, t)$
        grd3 : $f(clk) = wl(clk)$
        grd4 : $ran(f) \subseteq L..H$

    then

        act1 : $wl := wl \cup f$
        act2 : $clk := t$

    end

The interval property (*grd4*) for this event is a boundary property, and as just discussed, is preserved by interval extension, i.e., this event preserves the boundary invariant on the water level.

3.4 Distinguishing Modes

Our simple water tank system has two modes of operation: when the pump is on, the water level is monotonically increasing and when the pump is off, it is monotonically decreasing. In both cases the boundary property must be maintained. The interval event *WaterLevelInterval* of the previous section is an abstraction of both of these modes of operation. We construct a refined model of the water tank machine with two interval events, one for increasing the water level and the other for decreasing the water level. We require both of these to be refinements of *WaterLevelInterval* and thus they need to preserve the boundary property.

The specification of the interval event for increasing the water level is as follows:

Event *IncreaseWaterLevelInterval* $\widehat{=}$

refines *WaterLevelInterval*

    any

        t,f

    where

        grd1 : $t \geq clk + \epsilon$
        grd2 : $f \in ctsF(clk, t)$
        grd3 : $f(clk) = wl(clk)$

$$\textbf{grd4} : f \in mono\_inc$$
$$\textbf{grd5} : f(t) \in L..H$$

then

$$\textbf{act1} : wl := wl \cup f$$
$$\textbf{act2} : clk := t$$

end

Here we replace the abstract interval property $(ran(f) \subseteq L..H)$ with refined interval properties stating that f is monotonically increasing and that the endpoint of f is within the boundaries (the start-point is also bounded because of $grd3$ and the invariants specifying that wl is bounded). The refinement is valid because, if f is monotonic and its endpoints are bounded, then f is bounded at all points. This is capture by the following inference rule (here $mono = mono\_inc \cup mono\_dec$):

$$\frac{f \in cstF(i,j), \quad f \in mono, \quad f(i) \in L..H, \quad f(j) \in L..H}{ran(f) \subseteq L..H}$$

The interval event for decreasing the water level can be defined in a similar way. Note that we have not yet introduced a controller variable representing the status of the pump. At this level of refinement we remain focused on the monitored variable representing the water level.

3.5 Modelling the Control Strategy

To model the control strategy we introduce controlled variables whose value is modified at discrete steps. We make the assumption that the value of a controlled variable may influence the value of a monitored variable. For example, in the water tank we introduce a *pump* variable representing the status of the pump (*on* or *off*). If the pump is on, we assume that the water level increases monotonically while if the pump is off, then the water level decreases monotonically. For the water tank, we assume a periodic controller with a fixed period of length T that uses the following strategy at each control period:

C1 If the level is below a low threshold LT, the pump is switched on.

C2 If the level is above a high threshold HT, the pump is switched off.

C2 If the level is between LT and HT, the pump status does not change.

We assume that constants LT and HT lie in between L and T as follows:

$$\textbf{axm1} : L < LT < HT < H$$

Furthermore we assume that the rate of increase in the water level is bounded by constant RI, that is, the water level can increase by a maximum of $RI \times T$ during one fixed-length period. Similarly we assume the rate of decrease of the water level is bounded by constant

Small step intervals   Big step interval

Figure 3.2: Small step and big step intervals.

RD. The values for *LT* and *HT* are chosen such that if the water level is within *LT..HT* at a periodic control point, then the water level cannot go outside *L..H* by the next control point, i.e.,

axm2 : $L \leq LT - (RD \times T)$

axm3 : $HT + (RI \times T) \leq H$

The interval events introduced in the previous modelling level represent the evolution of a continuous variable during an entire control mode throughout which the controlled variables remain unchanged. For example, the *IncreaseWaterLevelInterval* event represents the water level increasing while the pump is on. We refer to these as *big step* events. When introducing the periodic controller in a refinement, the effect of a single interval event at the abstract level, e.g., *IncreaseWaterLevelInterval*, will be achieved by multiple sequential fixed-length periodic intervals. To model this we represent the evolution of the continuous variables *during* a mode using more fine-grained periodic interval events. The periodic events extend the continuous variables by fixed length intervals, each of size *T*. We refer to these as *small step* events. In addition to the controlled variables, we introduce additional variables in the refinement to represent the changes made by the periodic controller events: a clock variable to represent the periodic steps of time during a single mode, and continuous variables to represent the periodic evolution of the continuous variables during the mode. For the water tank example, we introduce two variables, one representing the periodic steps in time within a pump mode (clk_m), and another representing the periodic evolution of the water level during a mode (wl_m). This is illustrated in Figure 3.2: the left hand graph represents a series of periodic evolutions of the newly introduced wl_m variable within a mode; once sufficient periodic intervals have been defined by small step events, the big step event extends *wl* for a full mode interval using the periodic interval functions accumulated in wl_m.

We introduce invariants representing properties of the newly introduced variables in the refinement. For the water tank example we have the following invariants:

inv11 : $clk \leq clk_m$

inv12 : $wl_m \in ctsF(clk, clk_m)$

inv13 : $wl_m(clk) = wl(clk)$

inv14 : $wl_m(clk_m) \in L..H$

inv15 : $pump = on \Rightarrow wl_m \in mono\_inc$

inv16 : $pump = off \Rightarrow wl_m \in mono\_dec$

In interpreting these invariants, it is important to understand that the new variables clk_m and wl_m will be updated within a mode by the periodic controller events (small step events), while the original variables (which remain part of the refined model) will be updated by the refinements of the abstract nondeterministic interval events (big step events). The invariants specify that clk_m is never behind clk (*inv*11) and that wl_m is a continuous function over the interval $clk..clk_m$ (*inv*12) whose start point is fixed (*inv*13) and whose end point is bounded (*inv*14). Depending on whether the pump is *on* or *off*, determines whether wl_m is monotonically increasing or monotonically decreasing (*inv*15, *inv*16).

The periodic events for a mode will be continually executed, once per control period, while it is ok to remain within that mode. For the water tank, the following periodic event specifies the system behaviour during a single period while the pump is in the *on* mode:

Event   *PeriodicIntervalIncrease* $\widehat{=}$

>   **any**
>
>> f
>
>   **where**
>
>>   **grd1** : $pump = on$
>>   **grd2** : $wl_m(clk_m) \leq HT$
>>   **grd3** : $f \in ctsF(clk_m, clk_m + T)$
>>   **grd4** : $f(clk_m) = wl_m(clk_m)$
>>   **grd5** : $f(clk_m + T) \leq wl_m(clk_m) + (RI \times T)$
>>   **grd6** : $f \in mono\_inc$
>
>   **then**
>
>>   **act1** : $wl_m := wl_m \cup f$
>>   **act2** : $clk_m := clk_m + T$
>
>   **end**

Here the first two guards specify an enabling condition for the event, i.e., that the pump is on and that the current water level has not exceeded HT. The remaining guards specify constraints on the choice of interval function used to extend wl_m, i.e., f is an interval of length T starting from the current time (*grd*3), its starting value is the current water level *grd*4, its ending value is bounded by the rate of increase (*grd*5) and it is monotonically increasing (*grd*6). As well as extending wl_m by f, the actions of the event increase clk_m by a fixed amount T representing the fixed duration of a period.

There are two key reasons why the *PeriodicIntervalIncrease* event maintains the invariants on wl_m:

- The level at the end of the interval, $f(clk_m + T)$ is bounded above by $HT + (RI \times T)$ (from *grd*2, *grd*5) which means it is bounded above by H (*axm*3).

- The continuous composition of two monotonically increasing functions is monotonically increasing:

$$f \in cstF(i,j), \quad f \in mono\_inc,$$
$$g \in cstF(j,k), \quad g \in mono\_inc,$$
$$\frac{f(j) = g(j)}{f \cup g \ \in \ mono\_inc}$$

The periodic event for decreasing the water level is defined and verified in a similar way.

The events that represent the end of a mode are specified as refinements of the abstract nondeterministic interval events. The events are made deterministic by providing witness for the nondeterministic parameters of the abstract events and the witnesses are provided by the variables introduced to represent the evolution of time and of the continuous variables during a mode. For example, the nondeterministic interval event representing the monotonically increasing mode is refined as follows:

Event *IncreaseWaterLevelInterval* $\widehat{=}$

refines *IncreaseWaterLevelInterval*

    where

        grd1 : $pump = on$

        grd2 : $wl_m(clk_m) > HT$

    with

        t : $t = clk_m$

        f : $f = wl_m$

    then

        act1 : $wl := wl \cup wl_m$

        act2 : $clk := clk_m$

        act3 : $wl_m := \{clk_m\} \lhd wl_m$

        act4 : $pump := off$

    end

The refined event is enabled when the pump is on ($grd1$) and the water level exceeds HT ($grd2$). In the abstraction of this event, t and f are nondeterministically chosen parameters. In the refinement they are eliminated as parameters and their values are represented by deterministic witness predicates (**with** clause). The invariants of this refined model ensure that the witness values satisfy the constraints on the choice of values for the parameters in the more abstract model. The original actions of the abstract event are retained in the refinement event, and additional actions are added to reset wl_m to be a point interval on the current time and change the pump status to *off*.

3.6   Merging Big and Small Step Variables

In the refinement just outlined, we retain the variable updated by the big step events (wl) and we introduced a new variable that is updated by the small step events (wl_m). Variable wl represents the history of the water level from time zero up to the most recent big step interval while wl_m represents the recent history from the most recent big step interval up to the most recent small step interval. It is possible in a further refinement to merge the big step and small step variables into a single variable wl_s representing the full history from time zero to the most recent small step interval. This merge means that the big step clock is not required since it is no longer used in any guards and can be eliminated. The elimination and merge is characterised by the following simple invariant:

inv21 : $wl_s = wl \cup wl_m$

This merge leads to a simplification of the actions of the *IncreaseWaterLevelInterval* event: the update of *clk* is eliminated and the simultaneous update of *wl* and wl_m is realised by a *skip* action on wl_s (thus no change to wl_s is required). The only remaining action is the update to the pump status. We also take the opportunity to rename this event to *PumpOff* to indicate the control purpose that it now represents. The simplified event is specified as follows:

Event *PumpOff* $\widehat{=}$

refines *IncreaseWaterLevelInterval*

> **where**
>> **grd1** : $pump = on$
>> **grd2** : $wl_s(clk_m) > HT$
>
> **then**
>> **act1** : $pump := off$
>
> **end**

Through this merging of continuous variables, the *PumpOff* event has become an *instantaneous* event whereas previously it was an interval event. We say it is now instantaneous since it does not update a clock and it does not extend a continuous variable. We are able to simplify *PumpOff* to be instantaneous because the overall effect of the abstraction of this event is achieved by a sequence of periodic interval events. Once sufficiently many small step events have been executed to accumulate the recent history of the water level within the mode, the instantaneous mode change event is enabled and the effect specified by the abstraction of that instantaneous event will have been achieved by the accumulation of small step interval since the most recent big step events.

3.7   Derivatives

In control systems it is common to specify properties of interval functions in terms of properties of derivatives of those functions. For example, a function is monotonically increasing if its derivative is always positive, a function is linear if its derivative is a constant. Rather than defining derivatives exactly, we can characterise them axiomatically. Since not all functions have a derivative, we capture the set of differentiable functions over interval $i..j$ with a set $diff(i,j)$. We write $der(f)$ for the derivative of f. We assume (axiomatically) that all differentiable functions are also continuous and that the derivative of a differentiable function f is a continuous interval function over the same interval as f:

axm : $\forall i, j \cdot diff(i,j) \subseteq ctsF(i,j)$

axm : $\forall f, i, j \cdot f \in diff(i,j) \Rightarrow der(f) \in ctsF(i,j)$

We can capture the property that functions with positive derivatives are monotonically increasing through the following axiom:

$$\textbf{\textit{axm}} : \forall f, i, j \cdot f \in \mathit{diff}(i,j) \ \wedge \ \mathit{ran}(\mathit{der}(f)) \subseteq \mathit{PosREAL} \ \Rightarrow \ f \in \mathit{mono\_inc}$$

This axiom allows us to refine an event guard $f \in \mathit{mono\_inc}$ by a guard requiring f to be a differentiable function with positive derivative.

3.8 Concluding

We summarise our approach as follows: Continuous behaviour is specified through nondeterministic big step interval events that constrain the shape of continuous interval functions during a mode. We can refine these nondeterministic events by further constraining the shape of the continuous interval functions, e.g., we refined an interval function, that is bounded at every point, to a monotonically increasing function, that is bounded at its end points. We can also introduce periodic small step events within a mode as new events in a refinement to represent the strategy to be followed by a control system. Preservation of properties of continuous functions is key to ensuring the correctness of the refinements. For example, the most abstract interval event for the water tank preserves boundary invariants, the small step events in the water tank preserve monotonicity.

We need to be very careful in how we allow the clocks and continuous functions to be modified in order to reflect assumptions about the progression of time: time must move forwards, not backwards, and zeno behaviour must be avoided; continuous functions are extended in a forward direction only. It should be possible to enforce these idioms through a syntactic layer and this is one of the features of Hybrid Event-B [43].

Our approach fits with the abstraction/refinement approach of Event-B and the water tank development is supported by the Rodin toolset through the use of theories to define operators and proof rules for continuous functions. What is less clear is how well the approach scales to the case of high degrees of concurrency with multiple continuous functions operating to different mode intervals. This is the subject of future work.

The approach outlined here is influenced by the approach of [73] to the Steam Boiler Problem. In that paper, action systems (the basis of Event-B) are used to construct a model of the system that includes the monitored variables and the controlled variables. Although [73] uses a very simple model of discrete time whereby the environment actions model the update to monitored variables in one complete control cycle, it does encourage a *system-level* approach. By this we mean that rather than modelling the environment and controller separately, the abstract model captures the overall system and refinement is used to introduce more distinction between the environment and controller. In fact in our approach we take the abstraction one level further than in [73] by focusing on monitored variables (water level) in the abstraction and only introducing the controlled variable (the pump) through refinement.

Besides monotonicity, we have focused on expression of control goals that refer to individual time points, i.e., the monitored variables are required to satisfy some property at each time point. Treatment of properties over time intervals, as expressible in the Duration Calculus [79] or the approach of Hayes, Jackson and Jones [157], merits further investigation. We have yet to include the treatment of faults (e.g., pump or sensor failure) and fault tolerance in our approach though we believe it is sufficiently flexible to support instantaneous (e.g., sensor failure) and continuous faults (e.g., water level is decreasing when it should be increasing).

Part II

Analysis

Chapter 4

Modeling and Analysis of Component Faults and Reliability

Thibaut Le Guilly, Petur Olsen, Anders P. Ravn, and Arne Skou

Department of Computer Science, Aalborg University, Denmark

Abstract. This chapter presents a process to design and validate models of reactive systems in the form of communicating timed automata. The models are extended with faults associated with probabilities of occurrence. This enables a fault tree analysis of the system using minimal cut sets that are automatically generated. The stochastic information on the faults is used to estimate the reliability of the fault affected system. The reliability is given with respect to properties of the system state space. We illustrate the process on a concrete example using the UPPAAL model checker for validating the ideal system model and the fault modeling. Then the statistical version of the tool, UPPAAL-SMC, is used to find reliability estimates.

4.1 Introduction

Dependability of software systems in its widest meaning [22] is an area which calls for application of rigorous reasoning about programs. This is in particular the case for embedded software, where programs interact closely and continuously with a larger environment. Therefore, it has been investigated by developers of formal methods through decades. Here, Kaisa Sere with Elena Troubitsyna [299] have done seminal work which through the passed years has been continued with integration into development processes [333]. The work presented here has a similar perspective. It was inspired by model based testing of real-time systems [256] and analysis of service oriented home automation systems [92]. In these contexts, it is interesting to consider how models developed for testing or interaction analysis can be reused for safety analysis and perhaps even answer questions about the reliability of the overall system, because models are not inexpensive. It is a major development effort to build useful models of software and, even more so, of the context in which the software is embedded. The contribution in this paper is thus a systematic process, where behavior models are reused in safety analysis and reliability analysis of embedded systems.

The initial models have to describe the dynamics of a system, not only structural properties. They could come from Model Based Development (MBD) in the form of state charts or state machines for the software; for the environment there might be some form of tractable hybrid automata [164], perhaps in the form of timed automata [15] (TA). In other cases the models could come from model based testing or other analyses. Whatever the origin, we assume that they describe the behavior of ideal, correct systems which are not affected by any faults.

Since the software is embedded in physical systems, faults will inevitably occur due to wear and tear. Software component failures may also be included. Although, they may often be hidden in electronic components with built-in intelligence. Faults are essentially events that affect the behavior of the system as a whole; they may be classified in many ways as described by Avizienis in particular, see [22] for further work. However, we assume that a list of likely faults is known for the system under consideration.

In a conventional process faults are used in a safety analysis without considering behaviors, but as pointed out in e.g., [154], faults are related to dynamics of a system. A further step is to integrate the analysis with behavioral models of a system. Here we are fortunate to be able to build on the work of Shäfer [291] who uses phase automata as the system model and the duration calculus to assign semantics to fault trees, and Thums et al. [329] who use TA for modeling and Computational Tree Logic (CTL) [125] for fault tree semantics. Finally, Bozzano et al. [66] have gone a step further and automated the synthesis of fault tree from system modeled as Kripke structures, a result we extend to TA. Thus, there is a solid basis for model based safety analysis.

There is, however, a catch when augmenting an ideal model with faults and thus introducing failure modes. It should not be the case that faults provide desired functionality! This issue has been investigated by Liu and Joseph [231], who present suitable healthiness conditions. They are employed in a TA setting in [352] which is the formulation used here.

A system with failing components may be saved from failure by augmenting it with fault-tolerance mechanisms as done in [352]. Yet, this is costly, and there is still a probability of failure. The real question is: how reliable is the system? Here, a stochastic model is needed. By assigning probabilities to faults, an automata model becomes a Markov process, or if non-determinism is involved, a Markov Decision Process. Both can be handled with model checking techniques, see for instance [39, 126]. It may not be realistic to model check a larger model to get a figure for the reliability; so since the basic fault probabilities are estimates anyway, it is feasible to use the ideas of statistical hypothesis testing and get an answer with some chosen degree of likelihood. This is mechanized in Statistical Model Checking (SMC) [217] which we apply in this paper.

4.1.1   Overview

The systematic approach outlined above is presented succinctly in Section 4.2. In order to demonstrate the approach, we apply it to a concrete use case using the model checker UPPAAL [47] and its statistical version UPPAAL-SMC [69] in Section 4.3. Finally, we discuss limitations, related work, and provide a conclusion and possible future work in Section 4.4.

4.2   A Development and Analysis Process

The development process motivated in the introduction has as an objective to enable reliability assessment of reactive systems. The overall process, illustrated in Figure 4.1, is as follows:

1. Design and verify an *ideal model* of the system that correctly implements its requirements. As in any system modeling process, the models are derived from requirements and available components. Different types of requirements exist, from functional ones, that express the functionality that the system should provide to extra-functional ones, which for example constrain the time in which a function of the system should execute.

   This kind of model is well known from model based development and forms a basis for verification of correctness of a system design.

2. Augment the *ideal model* with failure modes to produce a *faulty model* and verify that they invalidate some requirements.

   This is a novel step. It aims at analyzing requirements which are orthogonal to those for the ideal model. A borderline case is safety requirements prohibiting states that can cause harm to the larger environment of the system.

3. From the augmented model a fault tree for safety analysis can be derived. It enables to detect the weak points of the system and strengthen them if necessary.

4. Associate failure modes with failure rates to obtain a *probabilistic model*. It allows an assessment of system reliability. When validation fails, the previously generated fault tree can be used to determine if the component structure should be updated or its reliability improved. This is a novel step in conventional software development, but it is known from safety analysis. Since failures are stochastic in nature, analyzing them requires an estimate of the probability of their occurrence. Exactly how they are found is a gray area, ideally they are obtained through statistical experiments; but this requires testing many similar components. This is hardly feasible for complex components, so in practice they are most likely estimates based on experience. Nevertheless,

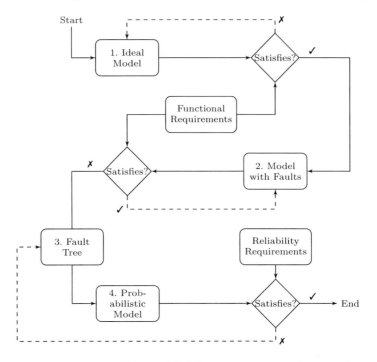

Figure 4.1: The process

it is assumed that Mean Time To Failures (MTTF) and failure rates are known in advance for both requirements and components.

This process is an iterative one, and failure in one of the validation steps implies a need to step back and modify the model or the properties specified, as illustrated by the dashed lines in Figure 4.1.

4.2.1   Ideal Model

Tractable models are usually finite state abstractions of the concrete system. Modeling the behavior of its components involves representation of their states, the transition between those, and their interactions with other components and possibly global state variables. The requirements are translated to predicates, usually in some dynamic logic in the case of reactive systems. The requirements are then checked to hold for the model. In general, two types of properties can be verified. Safety properties say that bad states are never entered, while liveness properties say that the system keeps on moving, in particular that it avoids deadlocks. Simulation is also of interest in this step, to observe the system evolution for a given period and ensure that it behaves as expected at least for the observed runs. In our setting it is an essential step in the process. Fault modeling does not make sense for an inherently faulty system, or conversely it is hard to inject specific faults with any effect in a system that does not have to satisfy any properties.

4.2.2   Modeling Faults

We recall that a failure is a transition from the system providing correct service to incorrect service. The state of the system when delivering incorrect service is called an error state. Finally, the cause of an error is called a fault. When the system delivers correct service in the absence of faults, the model is augmented with component faults, represented as transitions from normal states to error states. For convenience, error states may be duplicated. The obtained model is called a Fault-Affected Automaton [352]. We recall its definition here.

Definition 4.1. (Fault-Affected Automaton) A fault-affected (or F-affected) automaton is an automaton with identified faulty transitions to error locations.

- $L \cup ERR$ is the union of the two finite and disjoint sets of normal and error locations.

- $l_0 \in L$ is the initial location, it is a normal state, thus the system starts correctly.

Fault transitions in the set F are the only transitions moving to error locations.

In order to ensure the correctness of the fault modeling, one needs to check the following healthiness condition. Given an F-affected automaton M of a correct automaton model S:

H 4.1. $M \backslash F \approx S$, meaning that when removing faulty transitions, the obtained model is bisimilar with the correct model.

Finally, for any $f \in F$, each fault-affected model $M \backslash \{F \backslash \{f\}\}$, should invalidate at least one of the properties of S. This is to ensure that each fault is significant in the model. Insignificant faults can be ruled out using Fault Tree Analysis, described in Section 4.2.3. Note that in some cases, a combination of faults leads to an error state, while their individual occurrence does not. In that case we would consider their combination as a fault in the set F, and we would need to check that the occurrence of this composite fault leads to a violation of a system property.

Note that although these conditions are formulated for automata, they can be interpreted for transition systems in general, see [231].

4.2.3   Fault Tree Analysis

Fault Tree Analysis (FTA) is used to determine the possible causes of a system failure, and enables one to identify critical components in a system architecture. It is a top down approach in which a top level event (TLE), representing a failure, is decomposed into simpler events composed by boolean connectives. These events can be further decomposed until a level of elementary events is reached. Boolean logic is then used to analyze the possible combinations of elementary events that can lead to the failure. The basic syntax of fault trees is composed of event symbols, representing top level-, intermediate-, or basic events, and logic gates that express boolean relations between events. Research in this area has provided better semantics for the syntax of fault trees, enabling the specification of event duration and sequencing [154] for example. With advances in modeling formalisms and tools, it has been made possible to ensure the correctness and completeness of FTA with regards to models decorated with faults [291, 329]. Recent research has also shown the possibility of automatic generation of fault trees from fault-affected models [66].

Here we show how to automatically generate fault trees with minimal cut sets using CTL. In order to generate fault trees we need to compute minimal cut sets. A cut set is a

---

Algorithm 1 Minimal Cut Sets

---

    if *IsCutSet*($\varnothing$) **then**
      return $\varnothing$
    end if
    Waiting:$= \{FS \in F \mid \mid FS \mid = 1\}$
    mCS:$= \varnothing$
    while *Waiting* $\neq \varnothing$ **do**
      for all $FS \in$ *Waiting* **do**
        if *IsCutSet*(FS) **then**
          Waiting:$=$ *Waiting*$\backslash FS$
          mCS:$=$ *mCS* $\cup FS$
        end if
      end for
      Waiting:$=$ *PairwiseUnion*(*Waiting*)
    end while
    return *mCS*

---

set of failures that lead to a TLE. A minimal cut set is a cut set reduced to include only necessary and sufficient failures for the TLE to occur.

For systems modeled as state machines, we define cut sets and minimal cut sets as follows. Given a set of faults F, an initial state I, and a failure TLE, $CS \subseteq F$ is a cut set for TLE iff there exist a run of the system, starting from I, visiting all faulty states in CS before the failure state TLE. The cut set CS is minimal, iff there is no cut set CS' of TLE which is smaller, $CS' \subset CS$. Given $mCS = \{mCS_1, \cdots, mCS_n\}$ the set of minimal cut sets for a failure TLE and a set of faults F, we note that $mCS \subset \mathcal{P}(F)$.

It is possible to check if a set of faults FS is a cut set by checking if there exists a path where all faults in the cut set are active together with the failure, while faults outside of the cut set are not:

$$E\diamond \ TLE \wedge FS \wedge \neg(F \backslash FS) \tag{4.1}$$

We recall that $E\diamond \ \varphi$ means that for some paths in the model, there exists a state where φ holds.

To construct fault trees, Formula 4.1 can be used to decide the set of minimal cut sets for a given model, by iterating FS through $\mathcal{P}(F)$, starting from the sets with the smallest cardinality to ensure their minimality. This is realized by Algorithm 1.

The function *IsCutSet* checks if its argument is a cut set using Formula 4.1. The algorithm starts by checking if the empty set is a cut set. If it is, then either the fault set F is incomplete or the TLE can be reached without any error being triggered (which probably indicates an issue with the model or the specifications). The algorithm explores the power set of F, checking for each element if it is a cut set. Since the power set is explored from the bottom, a cut set is minimal, once it is found. When a cut set is found, it is removed from the *Waiting* set, as none of its supersets can be a minimal cut set. Once all sets in *Waiting* have been checked, the *PairwiseUnion* function is applied to it to move on to sets of higher cardinality.

Figure 4.2 shows an example exploration of the set of faults $\{A, B, C\}$. The algorithm finds that C is a minimal cut set in itself, thus no other sets containing C are explored. A and B are not minimal cut sets. The pairwise union function joins them into the set

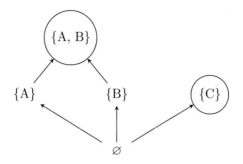

Figure 4.2: Example of power set exploration

$\{A, B\}$, which is found to be a minimal cut set. In this case the set of minimal cut sets is $mCS = \{\{A, B\}, \{C\}\}$.

In addition to identifying critical components and ruling out insignificant faults as already mentioned, FTA can be used to determine if components, or sets of components are in series or in parallel. Components in parallel will belong to one minimal cut set, while components in series will belong to separate minimal cut sets. This can be valuable as most representations of system models do not enable an easy visualization of this information.

4.2.4 Reliability Assessment

This step requires that the stochastic process is formulated with tractable distributions. The realism can always be discussed; but it is necessary to keep the model simple in order to get results. A component can fail in two ways, either temporarily (e.g., an unreliable communication channel) or permanently (e.g., a physical component breaking). Transient faults can be modeled using probabilistic branching, while permanent faults are modeled using probability distributions. Assuming constant failure rates, permanent failure transitions are modeled using an exponential distribution with parameter $\lambda = 1/MTTF$. This information is inserted into the model to determine the unreliability of the system — the probability that it fails after a given period.

Unreliability is expressed as a property over the global state space of the system. The probability of this property being verified is then estimated using statistical model checking (SMC), with two different possibilities. Firstly using hypothesis testing to validate that the probability of failure in a given time interval is less than a threshold, the time interval and threshold being part of the system requirements. Secondly using probability estimation to obtain an estimation of the system unreliability within a confidence interval.

4.3 Example

To illustrate the process, we apply it on an example system, simple enough so that models can be shown here and easily understood. We use Uppaal and its statistical version Uppaal-SMC to verify the model of the system and estimate its reliability. Note that Uppaal and Uppaal-SMC were chosen because they make it possible to apply the process on the same model, using model checking for verifying the correct modeling of faults in a first part and then using statistical model checking for evaluating reliability in a second part. We start by creating an ideal model of the example system.

4.3.1   Ideal Model

The system is a gas tank, shown in Figure 4.3. It is composed of five components:

- the tank structure,

- an input valve, controlling the incoming flow of gas in the tank,

- an output valve, externally controlled, providing gas,

- a sensor, measuring the level of gas in the tank,

- a controller, controlling the input valve based on the sensor's output.

Its function is to deliver gas when requested from its output valve. We assume this tank to have a capacity of 10L. When the gas level drops below 2L, the controller opens the input valve to refill the tank, until the level reaches 8L. If the level of the tank rises above 10L, it explodes. Obviously this is an undesirable event. Another requirement is that the tank should always be able to provide gas from its output valve, therefore it should never be empty. Now that we have specified the system and its requirements we continue to model it.

The first thing is to extract the variables from the specifications. We have:

- the level of the tank,

- the state of the sensor,

- the state of the input valve,

- the state of the output valve.

The declaration of these variables is shown in Listing 4.1. Constants are used to improve the clarity of the models. Note that the level of the tank and the sensor are initialized to a value that corresponds to a normal state of the system, where the level of the tank satisfies the requirements and the sensor reports a correct value.

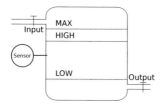

Figure 4.3: Gas tank example

Listing 4.1: Variable definitions

```
const int MAX    = 10;
const int HIGH   = 8;
const int LOW    = 2;
const int CLOSED = 0;
const int OPEN   = 1;
const int INIT_LEVEL

const int MINUTE = 1;
const int HOUR = 60*MINUTE;
const int DAY = 24*HOUR;
const int YEAR = 365*DAY;

int level  = INIT_LEVEL;
int output = CLOSED;
int input  = OPEN;
int sensor = INIT_LEVEL;
```

Listing 4.2: Channel and time definition

```
broadcast chan levelSync;
broadcast chan stop;
broadcast chan open;
broadcast chan close;
broadcast chan updateSensor;
```

Listing 4.3: Level calculation

```
void calcLevel() {
  if (input == OPEN){
    level++;
  }
  if (output == OPEN){
    level--;
  }
  if(level < 0){
    level = 0;
  }
}
```

We detail the models shown in Figure 4.4. First, we use a *ticker* (Figure 4.4a) to discretize time and enforce a time unit among the models. The tank (Figure 4.4c) updates the level variable through the function calcLevel() shown in Listing 4.3. It also triggers the update of the sensor value. The sensor reports the level of the tank, and notifies of any changes. The input valve opens or closes when told to do so. At each time unit, the output valve can be opened or closed. Since its state is externally controlled, we create two probabilistic branches with equal weight to indicate that each time the transition is taken, there is an equal probability the valve is opened or closed. Finally, the controller implements the previously introduced specifications.

Note that components can synchronize and exchange information between each other using the channels listed in Listing 4.2. A "!" indicates that the automaton is initiating the synchronization, and possibly *sending a value*, while a "?" indicates that the automaton is waiting for synchronization, and possibly *receiving a value*. Note that it is required for SMC that all channels be broadcast, meaning that more than one automaton can receive a synchronization, and that a sender can synchronize even when no receiver is waiting for synchronization.

The model is verified against the system specifications; that it should not rupture, and should never be empty. We verify them with UPPAAL using the CTL formulas shown in Table 4.1. We recall that $A\square\ \varphi$ means that for all paths and all states of the model, φ holds.

P_1 ensures that with the given design of the system, the tank cannot rupture. P_2 ensures that the tank cannot be empty. P_3 ensures that the control is satisfactory.

Theorem 4.1. The properties P_{1-3} are satisfied by the network of TA in Figure 4.4.

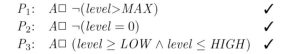

P_1: $A\Box\ \neg(level > MAX)$ ✓

P_2: $A\Box\ \neg(level = 0)$ ✓

P_3: $A\Box\ (level \geq LOW \wedge level \leq HIGH)$ ✓

Table 4.1: Specifications expressed as UPPAAL properties

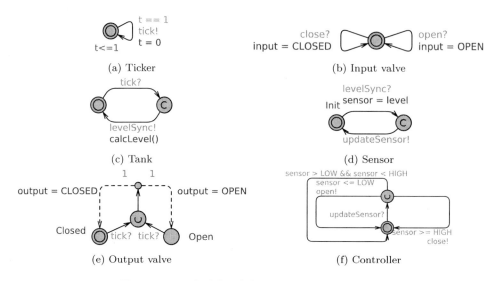

Figure 4.4: Models of the system components

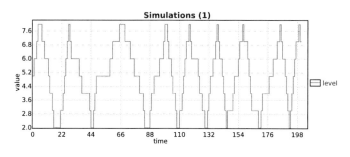

Figure 4.5: Simulation

Proof

     The properties are verified using UPPAAL.

Other features of UPPAAL can be used to analyze the model. For instance the simulator can be used to get an indication of how the model evolves. The verifier can be used to track certain values for a set of simulations and visualize how they evolve. An example of this can be seen in Figure 4.5, where the value of `level` has been tracked for one simulation. This shows that the system evolves as expected.

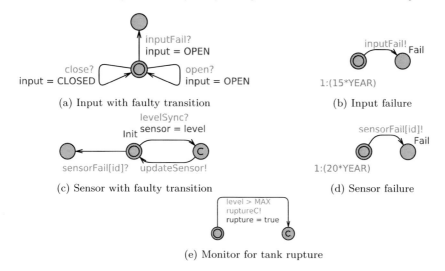

(a) Input with faulty transition

(b) Input failure

(c) Sensor with faulty transition

(d) Sensor failure

(e) Monitor for tank rupture

Figure 4.6: Fault-affected component models

4.3.2 Modeling Faults

In our example, we assume that only the input valve and the sensor can fail, and that their respective MTTF is 15 years and 20 years. We also consider that when the input fails, it stays open, and that when the sensor fails it stops reporting values, in effect stopping the controller. In order to keep the models clear, we propose to separate the modeling of the faults from the model of the components, as shown in Figure 4.6. Note that Figure 4.6e is not a fault model but an observer automaton. This is used to make queries more clear by using the *rupture* variable rather than *level>MAX*. Note also that the use of the *id* parameter in the *sensorFail[id]* synchronization is in anticipation to the instantiation of several sensors in a later step.

Theorem 4.2. None of the properties P_{1-3} are satisfied by the fault affected model of the system.

Proof

Counter examples are found using UPPAAL.

Note that we have assumed only permanent faults and simple failure models. However, transient faults can also be modeled using probabilistic branching as in Figure 4.7a which shows a model of an unreliable sensor. Figure 4.7b shows a complex observer that models a failure occurring when two out of ten measurements are erroneous. We however do not use these model in our analysis to keep the example simple.

Having a fault decorated model, we move on to FTA to obtain an overview of the combinations of faults leading to failure of the system.

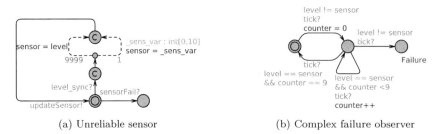

(a) Unreliable sensor   (b) Complex failure observer

Figure 4.7: Unreliable sensor and associated complex failure observer

$E \Diamond \; rupture \wedge \neg Input.Fail \wedge Sensor.Fail$ ✓
$E \Diamond \; rupture \wedge Input.Fail \wedge \neg Sensor.Fail$ ✓

Table 4.2: Output of the FTA

4.3.3 Fault Tree Analysis

We construct the fault tree using the FTA explained in Section 4.2.3. We take as TLE the rupture of the tank. The queries run by the algorithm are shown in Table 4.2. Using this analysis, we can observe that both the input valve and the sensor are single points of failure of the system. Single points of failure are usually not desirable, but can be acceptable if they are highly reliable. In order to estimate this we assess the reliability of the system.

4.3.4 Reliability Assessment

The example system can fail in two different ways. Either the tank ruptures or it becomes empty. Both are considered reliability issues, since they prevent normal operation. Only tank rupture is considered a safety issue, considering that it may endanger lives. These two cases are captured by the specifications of properties P_1 and P_2 in Table 4.1. In order to conduct the analysis, we need to set up reliability and safety requirements. We specify the following:

1. the probability of system failure within one year should be lower than 10%,

2. the probability of catastrophic event within three years should be lower than 5%.

A system failure corresponds to a transition from a state in which the system delivers correct service to one where it does not. A catastrophic event is a failure of the system that impacts the system environment, in this case the rupture of the tank. Note that a catastrophic event implies a system failure, while the opposite is not true.

We start by evaluating unreliability within one year of service. We thus ask the question,

"*What is the probability* $\mathbb{P}_M(\Diamond_{t \leq 1year}(rupture \vee level = 0))$?",

assuming a time unit in minutes. We consider that we want an uncertainty of $\varepsilon = 0.03$ and that we want a 95% confidence that the result is correct. Therefore we set the significance level to $\alpha = 0.05$.

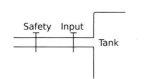

(a) Input pipe upgraded with a safety valve

(b) Safety valve model

(c) Safety valve failure model

Figure 4.8: Safety valve

With a time of 31 minutes[1], we obtain an approximation interval of $[0.0924592, 0.152456]$, which goes above the required 10%. We can therefore not guarantee that the model satisfies the requirements.

We then assess the safety of the system; that the probability of catastrophic event during a three year period is less than 5%. Given the extended time period, we use hypothesis testing, that requires a lower number of simulations than probability evaluation. We thus ask the question, given the fault affected model M of the system,

"*is* $\mathbb{P}_M(\Diamond_{t \leq 3year} rupture) \geq 0.05$?",

assuming a time unit in minutes. We set probabilistic deviations of $\delta = \pm 0.001$ and the probabilities of Type I and II errors of $\alpha = \beta = 0.05$. With a verification time of 2 hours and 24 minutes we get a positive answer, indicating that the safety requirement of the system is *not* met.

Note that we use minutes as the time unit as the number of simulations required using seconds makes the estimation process excessively long. This is obviously a drawback of SMC, since the execution time depends heavily on the length of each trace. However, compared to analytical solutions this is expected to scale better, since the generation of traces can be parallelized.

The probability of failure is too high w.r.t. the specifications. We need to strengthen the critical points of the system revealed by the FTA in order to obtain an acceptable reliability. To do so, different possibilities are available. The easiest would be to increase the reliability of the individual components. However, this is not always possible, due to cost or physical constraints. Another is to add additional safety or redundant components to the system. In this case we add a safety valve to the input and a redundant sensor.

The safety valve closes the input pipes when the tank level exceeds the level *HIGH*. This safety valve is directly connected to the sensor, and as it is simpler than the input valve, we consider its reliability to be higher (MTTF of 50 years). The safety valve, its model, and failure model are shown in Figure 4.8. The function calcLevel () of Listing 4.3 is updated to take it into account by constraining the increase of the level of the tank to when the safety valve is open.

The redundant sensor is introduced in the system by instantiating a new Sensor and SensorFail process in the model.

---

[1] All experiments are run on an i7 quad core 2.10GHz laptop with 8GB of RAM.

After introducing these additional components, the FTA is performed again. Its outputs indicate that the fault tree is composed of two minimal cut sets, one containing the failures of the input and safety valves, the other the failures of the two sensors. The absence of minimal cut set with a single component indicates the absence of single point of failure.

The statistical estimation of system reliability and safety is then performed again. We first obtain an estimation of reliability within the interval $[0.0321459, 0.0919823]$ after 24 minutes. In order to convince ourselves of the results, we use hypothesis testing that, after 39 minutes confirms the result. We are thus 95% confident that the unreliability of the system within a 1 year period is less than 10%.

The hypothesis testing for the system testing with the updated model results in a negative answer after 11 hours and 28 minutes. We thus have obtained a satisfactory model of the system with regards to its specifications.

4.4   Discussion, Conclusion, Related and Further Work

4.4.1   Discussion

The process presented in this paper relies on TA and model checking. Model checking raises the issue of the scalability of the process, due to the state space explosion problem. This is a well known problem and abstraction and optimization techniques can be used to reduce the state space and render model checking feasible. The choice of UPPAAL as the tool for model checking also reduces verification time as it seems to be the most efficient tool for model checking TA [343]. The second question regarding scalability is about the statistical model checking of the TA augmented with probabilistic transitions. The complexity of statistical model checking depends on the length of the traces to be generated, and the confidence requested for the statistical results. We have seen this in our experiments, where the combination of high time granularity and large time intervals make statistical model checking time consuming.

Regarding the choice of the tool, we note that UPPAAL-SMC is said to perform better [69] than the PRISM [212] tool for statistical model checking. However, PRISM also enables probabilistic model checking, which can be more efficient than SMC when applicable. We could thus imagine modeling the system and faults in UPPAAL and use model checking to ensure the correctness of the modeling, and perform probabilistic model checking when feasible using PRISM.

4.4.2   Conclusion

In this chapter, we presented a process to design and validate systems modeled as network of TA. The models are extended with fault transitions to error states that can be used in the application of two novel steps. We first showed how to perform an FTA based on these extended models, that helps identifying single points of failure and ensuring the correctness of the models. An algorithm to generate complete and correct fault trees with minimal cut sets was presented to facilitate FTA. The second novel step showed how to augment the fault decorated models with probabilities to perform reliability analysis of the system. We finally illustrated this process on a example, using UPPAAL and UPPAAL-SMC as tool support. Being based on model checking, the process is inherently limited in terms of the size of the system to be analyzed. However, this is a well known problem, and techniques for abstraction, reduction, and simplification are available to help reduce it.

4.4.3   Related Work

Reliability and safety are broad areas of research that have focused the attention of many researchers.

Regarding relation between formal models and FTA, we mention the work of Sere and Troubitsyna [299] who use FTA to refine formal specifications written with the action system formalism [33]. The difference with our work is that the fault tree is used to refine the system model, while we derive it from the system model and use it to validate and analyze the model.

Assessing safety through linking system reliability and components reliability using probabilistic models is also the objective of the work of McIver et al. [240]. The link is made through establishing probabilistic data refinement by simulation and is limited to sequential models. Troubitsyna takes this work further by using it in combination with the action system formalism, enabling its application to reactive systems in [331].

Regarding reliability analysis based on formal stochastic system models, we mention the work of Kwiatkowska et al. [211] using the PRISM tool [212]. Here our work emphasizes more the analysis process than the tool usability.

PRISM is also used by Tarasyuk et al. [325] to introduce reliability assessment in the Event-B refinement process. Here our work differs in that the process emphasizes fault modeling and incorporates FTA for identifying critical components.

Another use of statistical model checking in the context of reliability and safety analysis is shown by Arnold et al. [17] with the DFTCalc tool. This tool focuses on FTA to compute system reliability and MTTF. The difference in their work is that the fault tree serves as the basis of the analysis while we derive it from a formal model of the system. The advantage of DFTCalc is to have a more expressive syntax for fault tree (SPARE and Priority AND gates), but the correctness and completeness of fault trees cannot be checked. Moreover, using formal models and UPPAAL also allows for checking safety and liveness properties.

Bozzano et al. [66] present a set of algorithmic strategies that enable the generation of a fault tree with minimal cut sets. The algorithms are designed for systems modeled as Kripke structures, and are implemented in the FASP/NuSMV-SA safety analysis platform [67]. The algorithm we propose relies only on the use of reachability queries. We also mention [40, 41] who propose to use retrenchment techniques for modeling faults. A (concrete) faulty system is thus related to an (abstract) ideal system via a retrenchment relation. They then present algorithms for generating resolution trees first for timeless acyclic combinational circuits, and then for cyclic combinational circuits with clocks. Resolution trees can then be transformed into conventional fault trees, with the advantage that they provide more detailed relations between the faults, compared to the fault tree generated by our algorithm that only contain minimal cut sets.

4.4.4   Further Work

To make the process more useful and interesting for safety and reliability analysis, improving the generation of fault trees is an essential point. As already mentioned, the fault trees generated by our algorithm are *flat*, in the sense that they only provide minimal cut sets. It is however important to be able to visualize the nested relations between the faults, and generating nested trees would provide more insights. Generating such trees using a combinatorial approach is possible, but would not scale well. Applying on-the-fly algorithms such as the ones used in [66] would improve the performance and should be investigated.

Another challenge is to make the process easier to use for practitioners in the fields of reliability and safety analysis. While preparing this paper, we have recognized two direct enhancements to UPPAAL to improve its applicability in this area. The first is to enable the specification of faults in the UPPAAL GUI and the second is to implement Algorithm 1 in UPPAAL.

Currently, UPPAAL does not provide any contextual understanding of TA, their states, or transitions. Adding such contexts, by for example differentiating between system and environment models, normal and erroneous states, could facilitate modeling. Moreover, adding a notion of faults and fault affected models would enable the automation of several steps of the process presented here. The tool could for example automatically verify that requirements are met by the *ideal model*, and that each specified fault invalidates at least one of them.

With a specified set of faults and failures, Algorithm 1 could be implemented in the tool and generate a visual representation of the fault tree. The algorithm can currently be executed using the command-line interface to the UPPAAL verifier, but including it in the GUI would increase its usability. The reliability assessment could also be automated, and we are working on an extension that calculates the probability for each minimal cut set to trigger the TLE. This way the most likely path to a failure can be determined.

Adding features such as these will make UPPAAL more usable for practitioners and support the use of formal methods in industrial applications.

Chapter 5

Verifiable Programming of Object-Oriented and Distributed Systems

Olaf Owe

University of Oslo, Department of Informatics, Norway, and
University of California, Santa Cruz, Department of Computer Science, USA

Abstract Distributed and concurrent object-oriented systems are difficult to analyze due to the complexity of their concurrency, communication, and synchronization mechanisms. This paper explores a programming paradigm based on active, concurrent objects communicating by so-called *asynchronous method calls* giving rise to efficient interaction by means of non-blocking method calls, implemented by means of message passing. The paradigm facilitates invariant specifications over the locally visible communication history of each class. Compositional reasoning is supported by rules comparable to those of sequential programming, and global properties may be derived from local specifications. Reasoning about inheritance is not limited by behavioral subtyping, but allowing free reuse of code, considering also multiple inheritance. A small, illustrating example is considered.

5.1   Introduction

Today distributed systems form an essential basis of modern infrastructure. Object-oriented programming is a leading programming paradigm for such systems. It is in general required that object-oriented, distributed systems are of high quality and behave properly. However, quality assurance of object-oriented, distributed systems is a non-trivial and challenging topic. There are a number of approaches for different kinds of quality assurance. Formal methods play an important role in the systematic studies of such systems and in their semantics. Several formal approaches consider functional or single-assignment languages, which are semantically simpler than the setting of imperative, object-oriented programs. To make the result applicable to common imperative object-oriented languages, one may then consider refinement approaches. The approach considered by Kaisa Sere among others [32] is based on refinement, starting by programming in the mathematically simpler formalism of Action Systems [35], based on the paradigms of single assignment and guarded command, and making a number of refinement steps, ending up with a correct-by-construction distributed object system. This line of work focuses on refinement theory, including concepts for increasing concurrency, like superposition [35], rather than reasoning on imperative, distributed programs.

In this paper we will not consider refinements between designs at different abstraction levels, but rather consider imperative, object-oriented, distributed systems directly; and in particular formal reasoning and verification of such systems, focusing on concurrency aspects and inheritance aspects. The concurrency model has some similarities to that of Action Systems. Both are based on an execution model where each concurrent unit handles one action/process at a time and where guards may lead to several enabled actions/processes, non-deterministically chosen. Object-oriented Action Systems [62] allow non-terminating local actions, like active objects of our concurrency model. Incremental development by refinement in Action Systems has similarities to incremental program design by inheritance in our work.

The challenges in imperative, object-oriented, concurrent program reasoning are related to several factors including the programming language and constructs chosen, the specification language, and the verification system, and last but not least, tool support. The goal of this paper is to strive for simplicity in reasoning about object-oriented, distributed systems. In order to approach this goal we will make some recommendations with respect to language constructs, in particular object-oriented constructs and reasoning techniques. We will keep tool support in mind by emphasizing reasoning approaches that allow mechanized analysis.

For object-oriented systems the notions of encapsulation, inheritance, late binding, and object creation, are central, and a proper treatment of these is essential. As class hierarchies are open-ended, reasoning should be incremental with respect to addition of subclasses, rather than based on a closed-world assumption. Since distributed systems involve concurrency, it is essential to have a reasoning system that can handle concurrent units and to consider programming constructs that allow efficient interaction of concurrent units.

Simplicity of reasoning does not only concern the choice of specification language and verification system, but also the choice of programming language constructs and their semantics. In particular one should choose a programming language with compositional semantics that allows compositional reasoning. We therefore focus on programming mechanisms that have semantical advantages and allow efficient interaction. The paper builds on research results from the language development and reasoning activity around the paradigm of so-called *active concurrent objects* developed through the languages OUN [258, 257], Creol [188, 192, 191, 189, 186], and partly ABS [187]. The contribution of this paper is to reconsider relevant results from the work on active concurrent objects, while making further improvements with respect to simplicity of verifiability.

The paper is organized as follows: The next section discusses basic programming constructs for concurrent objects, introducing a core language, ending up with an example. Section 5.3 motivates and explains invariants over communication histories. Section 5.4 discusses reasoning challenges for single and multiple inheritance. Section 5.5 suggests a Hoare style verification system for classes, based on invariants over communication histories, reconsidering the example. Section 5.6 discusses extensions to future and delegation mechanisms. Finally, Section 5.7 comments on related work and makes a conclusion.

5.2 Basic Programming Constructs

For distributed systems the notion of concurrent unit together with interaction and cooperation mechanisms are crucial. The Actor model [8] is semantically appealing giving rise to compositional semantics, unifying interaction, and cooperation mechanisms through message passing. It is clearly simpler than the shared variable setting and the general thread-based model where a thread may execute code on several objects, and where non-trivial interference complicates the semantics. For object-oriented systems the natural unit of interaction is the object; in fact the term "activity" was used before the terms "class" and "object" appeared in Simula 67 [90]. Concurrent objects extend the Actor paradigm to the object-oriented setting, allowing two-way interaction by means of remote method calls rather than one-way interaction by messages. Inheritance of code is meaningful since program code is organized in methods.

Consider a remote method call, say $v := o.m(\overline{e})$, where the callee o is an external object supporting a (type-correct) method m, $\overline{e}$ is the list of actual parameters, and v is the variable receiving the return value. In the setting of concurrent objects the default way of performing such a method call is that the calling object blocks while the callee object (o) is executing the method when free to do so [91]. One may improve efficiency by adding a list of statements s that the caller can do while waiting, say by the syntax $v := o.m(\overline{e})\{s\}$, where the caller blocks after performing s and only if the return value has not arrived.

However, efficient interaction of concurrent objects requires some kind of non-blocking call mechanism allowing the caller object to do something else (like handling another incoming call) while waiting for the call to the external object to finish. With a suspension mechanism a method invocation can be suspended and placed in a queue of suspended processes while another (enabled) process may continue instead. Enabledness can be controlled by guards, either a Boolean condition b on the state of an object or the presence of a return value of a call. The guarded await statement **await** b suspends while waiting for b to become true. The await statement **await** $v := o.m(\overline{e})$ initiates the call and suspends while waiting for the call to complete [188]. This allows an object to handle a number of processes, being incoming calls or continuations of such calls, or self calls representing self activity. Local methods intended for non-terminating self activity need to have release points in order to allow periods with reactive behavior. Method invocations and method results are realized by message passing between the caller and callee objects in an asynchronous manner.

We consider objects with their own virtual processor; thus at most one process in an object can be active at a given time. This allows sequential reasoning inside a class, and suspension control is programmed within each method. Thus neither notification nor signaling from other objects or processes is needed, which simplifies reasoning. For instance in concurrent Java programs, some of the hardest bugs to locate are missing or misplaced notify statements. The await statement can be efficiently implemented since enabledness of waiting processes is only checked when no other process is active, i.e., at a suspension point or end of a method invocation. This await mechanism gives rise to passive waiting.

Concurrent objects with suspension control may be reactive, responding to method calls from the environment, or active, performing their own activity. Initial active behavior is programmed by the initialization code, for instance by calling a non-terminating local run method with suspension points allowing reactive behavior to be mixed with active behavior. By dynamic object creation, new concurrent units can be generated at any time, starting with their run method, either by themselves or by the parent object. Non-blocking method calls give independence of objects, however, sometimes synchronization may be desirable, in which case one may use the default object-oriented mechanism for blocking calls. For the case that the result of a call is not needed by the caller, we use the syntax $o.m(\overline{e})$, letting the caller continue with the next statement after sending the invocation message to the callee.

Concurrent objects interact by means of remote method calls only, and remote access to fields is not allowed. If allowed, this would create aliasing problems in reasoning. An assignment statement has the form $v := e$ where v is a simple variable (either a field variable or a local variable of the method) and e is a pure expression. Even though we have aliasing in the sense that v may refer to the same object as another variable of the same process, the classical Hoare rule for assignment is satisfied, i.e., $[Q_e^v] \, v := e \, [Q]$ where the precondition is obtained by substituting e for v. Assertions are here enclosed in square brackets (while curly brackets will occur in code). Thus reasoning is not affected by semantic alias considerations, since the (pointer) value of v is updated and not the object referred to by v.

Object-oriented languages have different approaches to class encapsulation and hiding. In addition, reasoning about object-oriented systems requires some kind of abstraction mechanism. Behavioral interfaces provide abstraction as well as encapsulation and hiding [89]. We let behavioral interfaces control which methods are visible, while hiding all fields, and let interfaces contain specification of the object behavior by means of the local communication history. Class encapsulation is then enforced by declaring object variables by interfaces, rather than by classes. Local data structures in an object can be defined by data types. Thus objects represent independent units, while data types are used for values that can be copied. A variable declared of a data type represents a value of that type, for instance a list of object identities can be represented as a data type.

The setting of concurrent objects gives rise to many logical concurrent units. However they are easily mapped to a given number of physically concurrent CPUs, since they are working asynchronously. The objects may reflect activity on parallel hardware architectures as well as distributed machines working in parallel.

A Core Language

A core language based on the concurrency model above is given in Figure 5.1. Program code is organized in classes. Classes and interfaces may include specifications in the form

$$
\begin{array}{lll}
In & ::= & \textbf{interface } F([T\ p]^*) \\
 & & [\textbf{extends } [F(\overline{e})]^+]^?\{S^*\ I\} \qquad\qquad \text{interface declaration} \\
Cl & ::= & \textbf{class } C([T\ cp]^*)\,[\textbf{implements } [F(\overline{e})]^+]^? \\
 & & [\textbf{extends } [C(\overline{e})]^+]^?\,[\textbf{inherits } [C(\overline{e})]^+]^? \\
 & & \{[T\ w\,[:=e]^?]^*\ [s]^?\ M^*\ S^*\ I\} \qquad\quad \text{class definition} \\
M & ::= & T\ m([T\ x]^*)\ B\ P^? \qquad\qquad\qquad\qquad \text{method definition} \\
S & ::= & T\ m([T\ x]^*)\ P^? \qquad\qquad\qquad\qquad\ \text{method signature} \\
B & ::= & \{[[T\ x\,[:=e]^?]^+;]^?\ [s;]^?\ \textbf{return } e\} \qquad \text{method blocks} \\
T & ::= & F\mid \mathsf{Any}\mid \mathsf{Void}\mid \mathsf{Bool}\mid \mathsf{String}\mid \mathsf{Int}\mid \mathsf{Nat}\mid\ldots \quad \text{types} \\
v & ::= & x\mid w \qquad\qquad\qquad\qquad\qquad\qquad \text{variables (local or field)} \\
e & ::= & \textbf{null}\mid \mathsf{this}\mid \mathsf{caller}\mid v\mid cp\mid f(\overline{e})\mid (e)\mid \mathbf{h} \quad \text{pure expressions} \\
s & ::= & \textbf{skip}\mid v:=e\mid v:=\textbf{new } C(\overline{e})\mid s;s \qquad \text{basic statements} \\
 & & \mid v:=e.m(\overline{e})[\{s\}]^?\mid v:=[C{:}]^?m(\overline{e}) \quad\ \text{call statements} \\
 & & \mid \textbf{await } v:=e.m(\overline{e})\mid \textbf{await }\ e \qquad\ \text{suspending statements} \\
 & & \mid \textbf{if } e\ \textbf{then } s\,[\textbf{else } s]^?\ \textbf{fi} \qquad\qquad \text{if-statements} \\
P & ::= & [A]\mid [A,\,A] \qquad\qquad\qquad\qquad\quad\ \text{pre-/postcondition} \\
I & ::= & [\textbf{inv } A]^?\,[\textbf{where } A^+]^? \qquad\qquad\quad \text{invariant specification}
\end{array}
$$

Figure 5.1: Core language syntax. We let [ ]*, [ ]+, and [ ]? denote repeated, repeated at least once, and optional parts, respectively. F denotes an interface name, C a class name, m a method name, p a formal interface parameter, cp a formal class parameter, w a field, x a method parameter or local variable, and $\overline{e}$ a (possibly empty) expression list. Expressions e are side-effect free. Specifications are written in blue. The specification [A] abbreviates [*true, A*]. Assertions A are first order formulas and may refer to the local communication history **h**.

of invariants and pre/postconditions of methods. To control code and specification reuse, we let the keyword **inherits** indicate code reuse, while the keyword **extends** indicate reuse of *both* code and specification. A class may implement a number of interfaces, specified by an **implements** clause. The code of a class (including inherited code) must satisfy its specifications (including inherited specifications), and must satisfy the **implements** clauses (including inherited ones caused by **extends** clauses). This gives rise to verification obligations. Inheritance between interfaces is controlled by **extends** clauses. We could also allow an interface $F1$ to implement another interface $F2$, meaning that $F1$ satisfies all the requirements of $F2$, but without inheriting from $F2$. Such a clause could even be stated after both interfaces are declared, for instance connecting interfaces of independently developed program parts (in a non-cyclic manner), and be exploited in later class definitions.

Interface and type names are capitalized, while class names are written in capital letters. The language is strongly typed, but here we omit typing considerations. We use a Java-like syntax; however, we use : = for assignments to not confuse equations in assertions and assignments. Assignments ($v:=e$), return statements (**return** e), if-statements, sequential composition, and object creation ($v:=$ **new** $C(\overline{e})$), are standard. Class parameters are stored in the created object (following the Simula tradition). We let class, interface, and method parameters be read-only, as well as this, referring to the current object, and caller,

```
1   interface Bank {
2     Bool sub(Nat x)
3     Bool add(Nat x) [return= true]
4     Int bal() [return= sum(h)]
5     where sum(empty) = 0, sum(h· ←add(x;true)= sum(h)+x,
6        sum(h· ←sub(x;true)=sum(h)−x, sum(h·others=sum(h) }
7   interface PerfectBank extends Bank {Bool sub(Nat x) [return= true]}
8   interface BankPlus extends Bank { inv sum(h)>=0 }
9
10  class BANK implements PerfectBank {Int bal=0;
11    Bool upd(Int x) {bal:=bal+x; return true} [inv, bal=sum(h)+x and return=true]
12    Bool add(Nat x) {Bool ok:=upd(x); return ok} [return= true]
13    Bool sub(Nat x) {Bool ok:=upd(−x); return ok} [return= true]
14    Int bal(){return bal} [return= bal]
15    inv bal=sum(h) }
16
17  class BANKPLUS implements BankPlus inherits BANK{
18    Bool upd(Int x) {Bool ok:=(bal+x>=0); if ok then ok:= BANK:upd(x) fi; return ok}
19        [inv, bal>=0 and if return then bal=sum(h)+x else bal=sum(h)]
20    inv BANK:inv and bal >=0 }
21
22  class BANKWAIT implements BankPlus extends BANK {
23    Bool upd(Int x){await bal+x>=0 ;bal:=bal+x; return true} [bal >=0]
24    inv BANKPLUS:inv }
```

Figure 5.2: A simple bank example. In assertions, **inv** refers to the current invariant, while *C*:**inv** refers to the invariant of class *C*. The auxiliary function *sum* calculates the balance from the local history **h** by means of past return events of completed *add* and *sub* calls.

referring to the caller of the current method execution. In the initialization code of a class, caller refers to the parent object.

The language supports the mechanisms for remote calls and suspension described in Section 5.2, including guarded suspension **await** b, suspending call **await** $v := o.m(\overline{e})$, interleaved call $v := o.m(\overline{e})\{s\}$, as well as *simple call* $o.m(\overline{e})$. A simple call does not wait for a result, and is useful when the result is not needed. A blocking call $v := o.m(\overline{e})$ is equivalent to an interleaved call with an empty statement list. The syntax $v := m(\overline{e})$ represents a late-bound local call and $v := C:m(\overline{e})$ a static local call, taking the method m of class C. Both are implemented in the usual stack-based manner. Dot-notation in a call, $v := o.m(\overline{e})$, may only be used when the declared interface of o supports a (type-correct) method m. It is possible to make a call to a *null* object, but no return value will ever be received by the caller in such a case. A simple call statement to *null* will terminate, but not a blocking call to *null*, while a suspended call to *null* will never be enabled. (Exceptions are not part of the core language.)

A class or interface may specify an invariant and add pre/postconditions to method declarations. Specifications may refer to the local communication history **h**, as discussed in Section 5.3. A pre- and postcondition pair is written $[P, Q]$ where P is the precondition and

```
1   interface Pin(Text pin) {
2     Bool open(Text code) [return=(user(h)=caller and code=pin)]
3     Bool close() [return=(user(h)=null)]
4     where user(empty)=null, user(h·o←open(code;_))=
5         if code=pin and user(h)=null then o else user(h),
6       user(h·o←close(;_))=if user(h)=o then null else user(h),
7       user(h·others)=user(h) }
8
9   class PIN(Text pin) implements Pin(pin) { Any u=null;
10    Bool open(Text code) {if code=pin and u=null then u:=caller fi;
11      return (u=caller and code=pin)}
12    Bool close() {if u=caller then u:=null fi; return (u=null)}
13    inv u = user(h) }
14
15  class PINBANK(Text pin) implements Bank extends PIN(pin) inherits BANK {
16    Bool sub(Nat x) {Bool ok:=false; if u=caller then ok:=upd(−x) fi; return ok}
17      [return=(u=caller)]
18    inv bal=sum(h) and u=user(h) }
```

Figure 5.3: A simple example of multiple inheritance, reusing the bank example. The interface *Any* is the most general interface, supported by any class.

Q the postcondition. The trivial precondition *true* may be omitted. An inherited method may be re-specified by simply adding the method declaration with the added pre/postcondition. The keyword **inv** identifies invariants and **where** identifies auxiliary function definitions. Auxiliary functions may be defined inductively, letting **others** in a left hand side match any remaining cases, and letting _ denote arguments in a left-hand-side pattern that are not needed in the right-hand side.

An Example

Figure 5.2 and Figure 5.3 show a simple bank example illustrating history-based specification, suspension, static, and late bound calls, as well as single and multiple inheritance. The methods *add*, *sub*, and *upd*, return a Boolean value indicating whether the transaction was successfully performed. Both *add* and *sub* are implemented by means of *upd*. Interface Bank, exporting methods *add*, *sub*, and *bal*, says that *bal* returns the *sum* calculated from the successful *add* and *sub* transactions of the local communication history **h**. Function *sum* is defined inductively over the history, according to the approach in Section 5.3. Intuitively it says that the *sum* is the sum of amounts *added* minus amounts *subtracted*, counting only successful method returns (returning *true*). The Bank interface does not require that *sub* always succeeds. There are two subinterfaces of Bank, PerfectBank, which requires that all transactions succeed, and BankPlus, which requires that the sum is never negative.

All class implementations are based on the BANK invariant expressing that the field *bal* equals the calculated sum, $bal = sum(\mathbf{h})$. Both subclasses of BANK ensure non-negative balance, BANKPLUS by letting *sub* fail (return *false*) in case of insufficient balance, and BANKWAIT by reimplementing *upd* such that it suspends until the balance is large enough.

Code reuse here is demonstrated by inhering the *bal*, *add*, and *sub* methods of class BANK in the two subclasses of BANK, while redefining the *upd* method. Specification reuse is demonstrated by the class BANKWAIT, in which *upd* inherits the specification of BANK, resulting in the specification $[\![\, bal = sum(\mathbf{h}),\ bal = sum(\mathbf{h}) + x \wedge bal\rangle = 0 \wedge return = true\,]\!]$.

As discussed below, a challenge with respect to reasoning is that the BANK specification (namely the postcondition of *upd* and support of PerfectBank) is violated by the subclass BANKPLUS. We will suggest a way to solve this challenge; and in Section 5.5 we show how to verify parts of the example.

Class PINBANK in Figure 5.3 demonstrates multiple inheritance. It inherits code from both class BANK and PIN, and inherits specifications from class PIN. It implements interface Bank and also interface Pin (implicitly through the **extends** clause).

5.3 Class Invariants

Class invariants are commonly used for specifying, and reasoning about, object-oriented systems, and are especially useful in a distributed setting with non-terminating concurrent objects [88, 243]. In a global state of such a system, we may assume that each object satisfies its invariant. This gives rise to sound compositional reasoning [312]. Interface invariants reflect parts of class invariants that are relevant to the environment. Encapsulation of inner state means that remote field access should not be allowed since several classes may implement the same interface, each with different fields. In order to capture an abstraction of the object state, one may use model variables [225] or ghost variables such as the communication history [88]. Communication histories have the advantage that they are expressive, since they include all visible events, and one need not invent specific model/ghost variables for each class. One may define functions over the history to extract information relevant in an interface or class specification.

History variables give an abstraction of the state in terms of communication events only, and distinguish different sequences of communication events. Class invariants can be formulated as predicates over the local history of the object as well as field variables, this, and class parameters. Interface invariants can be formulated as predicates over the local history of the object as well as this and interface parameters, but not field variables, since these are not visible in an interface. Interface invariants are supposed to hold at all times and should therefore be prefix closed with respect to the history, whereas class invariants need only hold at suspension and method return points.

A history is a sequence of visible events. We consider the following events:

$o \rightarrow o'.\textbf{new}\ C(\overline{e})$, o creates a new C object o' with actual class parameters $\overline{e}$.

This *creation* event is visible to o. The events for method interaction are:

$o \rightarrow o'.m(\overline{e})$, denoting a call to method m with actual parameters $\overline{e}$, with o as caller and o' as callee. This event is caused by o and is visible to o only.

$o \rightarrowtail o'.m(\overline{e})$, denoting start of processing of the call $o'.m(\overline{e})$ with o as caller. This event is caused by the callee and is visible to o' only.

$o \leftarrow o'.m(\overline{e};e)$, denoting the generation of the return value e resulting from the call $o'.m(\overline{e})$. This event is caused by the callee and visible to o' only.

$o \twoheadleftarrow o'.m(\overline{e};e)$, denoting the reception of the return value e by o of the call $o'.m(\overline{e})$. This event is performed by the caller and is visible to o only.

Thus the following events are *caused* by an object o: $o \rightarrow o'.m(\overline{e})$, $o' \twoheadrightarrow o.m(\overline{e})$, $o' \leftarrow o.m(\overline{e};e)$, $o \twoheadleftarrow o'.m(\overline{e};e)$, and $o \rightarrow o'.\mathbf{new}\, C(\overline{e})$. The events are referred to as *call, start, return, get*, and *creation* events, respectively. All four communication events are needed since method communication is asynchronous and the events above may happen at different times. However, for each call the events must happen in the order given above. This ordering is captured by a notion of global *wellformedness*, formalized in Section 5.3.

Local specification and reasoning in a class or interface are done with *local histories*, i.e., the sequence of events visible to this object. In contrast, global specification and reasoning are done in terms of *the global history*, i.e., the sequence of all events. It follows that the local histories of distinct objects are disjoint in the sense that they do not share events. This allows independent reasoning of each class, and compositional reasoning by means of a simple composition rule stating that the invariant of a global system is the conjunction of all the invariants of the objects involved, adding wellformedness [115].

In specifications, auxiliary functions may be defined inductively over the local history h, using *empty* and append right ($\cdot$) as the history constructors. In this way the last event, which reflects the current activity, is explicit. For instance the ends-with predicate (**ew**) can be defined by the two cases (where x and y range over events):

$$empty\; \mathbf{ew}\; y = false, \quad h \cdot x\; \mathbf{ew}\; y = (x = y)$$

using infix notation. The history without the last event can be defined for non-empty histories by $old(h \cdot x) = h$. Projection of the history by a set of events s, denoted h/s, is defined by

$$empty/s = empty, \quad (h \cdot x)/s = \mathbf{if}\; x \in s\; \mathbf{then}\; (h/s) \cdot x\; \mathbf{else}\; h/s$$

We let h/o denote the projection of h to events caused by o, and we let h/F denote the projection of h to the alphabet of an interface F, i.e., restricting $\twoheadrightarrow$ and $\leftarrow$ events to methods of F. Similarly, h/C denotes the projection of h to the alphabet of a class C, i.e., restricting $\twoheadrightarrow$ and $\leftarrow$ events to methods of C. And $h/(o{:}F)$, denoting the history of o as seen through the interface F, is a shorthand for $h/o/F$; and $h/(o{:}C)$, denoting the history of o as seen through the class C, is a shorthand for $h/o/C$. In practice, return and get events are essential in invariant specifications, describing output and input, respectively, of the specified object, while call and start events are often not needed (unless synchronization aspects are specified), as in the Bank example. In an interface or class specification we may write $o \leftarrow m(\overline{e};e)$ rather than $o \leftarrow \mathsf{this}.m(\overline{e};e)$, skipping the redundant this.

Invariant Refinement and Satisfaction

The invariant of a class C is written $I_C(h, \overline{w})$ where h is the local history and $\overline{w}$ its fields. The invariant of an interface F is written $I_F(h)$ where h is the local history. An interface invariant $I_F(h)$ will only restrict events in its own alphabet. In a subinterface or class with a wider alphabet, the invariant is therefore understood as $I_F(h/F)$, which ensures that the invariant does not restrict events outside its alphabet. Thus the invariant $I_F(h)$ of an interface F is inherited as $I_F(h/F)$ in a subinterface of F. For a class C implementing F, one must verify that $I_C(h, \overline{w}) \Rightarrow I_F(h/F)$, in which case C is said to *satisfy* F. Similarly, a class invariant $I_C(h, \overline{w})$ is inherited as $I_C(h/C, \overline{w})$ in a subclass *extending* C. Furthermore, for an inductively defined function $f(h)$, the equation $f(h \cdot \mathbf{others}) = f(h)$ is added to make the definition complete and to adjust for any alphabet extension when the function is used in a subinterface or subclass.

For each class C one must prove that the class invariant is established by the initializing code, that it is maintained by each method exported through a C interface, that the invariant holds at each suspension point, and that the class satisfies each interface of the class.

In an interface or class, postconditions can easily be expressed by invariants over the history. A postcondition $[Q(h, \mathbf{return}, \text{caller}, \overline{x})]$ of a method $m(\overline{x})$ abbreviates the invariant $h \, \mathbf{ew} \, (\text{caller} \leftarrow \text{this}.m(\overline{x};\mathbf{return})) \Rightarrow Q(h, \mathbf{return}, \text{caller}, \overline{x})$ where $\mathbf{return}$ is a special variable to be used in postconditions to refer to the returned value of a method. In a class postconditions may refer to fields. In an interface, (abstract) fields may be extracted by means of auxiliary functions over the history. In the examples we use postconditions as a notational convenience. Auxiliary function definitions are inherited downwards in subclasses, subinterfaces, and implementing classes, so that they are available for specification and reasoning.

Composition

For simplicity we here ignore interface and class parameters in the discussion. Given an invariant I_F over the local history of an interface F, we define the *object invariant* of an object o as *seen through* F by

$$I_{o:F}(H) \triangleq I_F(H/(o:F))_o^{\mathsf{this}}$$

where the projection $H/(o:F)$ denotes the history H reduced to the set of events generated by o and visible through F. Next the *object invariant* of an object o of class C is defined as

$$I_{o:C}(H) \triangleq \bigwedge_{F \in C.implements} I_{o:F}(H)$$

where $C.implements$ is the list of interfaces implemented by C according to the class definition. Finally the global invariant of a system with dynamically created objects initiated by an initial object *system:System* (where the *System* interface may include minimal requirements and primitives of the underlying operating system) is defined by

$$I_{global}(H) = \mathit{wf}(H) \wedge \bigwedge_{(o:C) \in ob(H)} I_{o:C}(H)$$

where $ob(H)$ is the set of object identifiers created in H by a $\mathbf{new}$ event (including the initial object):

$$
\begin{aligned}
ob(empty) &= \{system{:}System\} \\
ob(H \cdot (o' \to o.\mathbf{new} \ C(\overline{e}))) &= ob(H) \cup (o{:}C)) \\
ob(H \cdot \mathbf{others}) &= ob(H)
\end{aligned}
$$

and where $\mathit{wf}(H)$ is the welldefinedness predicate expressing that for each call the communication events obey the natural order *call, start, return, get*, that generated object identifiers are fresh, and that all identifiers used by an object o have been seen by the object:

$$
\begin{aligned}
wf(empty) &\triangleq true \\
wf(H \cdot (o \to o'.\textbf{new } C(\overline{e})) &\triangleq wf(H) \wedge o \in ob(H) \wedge id(\overline{e}) \subseteq id(H/o) \wedge o' \notin id(H) \\
wf(H \cdot (o \to o'.m(\overline{e})) &\triangleq wf(H) \wedge o \in ob(H) \wedge (\{o'\} \cup id(\overline{e})) \subseteq id(H/o) \\
wf(H \cdot (o' \twoheadrightarrow o.m(\overline{e})) &\triangleq wf(H) \wedge o \in ob(H) \\
&\qquad \wedge \#H/\{o' \twoheadrightarrow o.m(\overline{e})\} < \#H/\{o' \to o.m(\overline{e})\} \\
wf(H \cdot (o' \leftarrow o.m(\overline{e};e)) &\triangleq wf(H) \wedge o \in ob(H) \wedge id(e) \subseteq id(H/o) \\
&\qquad \wedge \#H/\{o' \leftarrow o.m(\overline{e};\_)\} < \#H/\{o' \twoheadrightarrow o.m(\overline{e})\} \\
wf(H \cdot (o \leftarrow o'.m(\overline{e};e)) &\triangleq wf(H) \wedge o \in ob(H) \\
&\qquad \wedge \#H/\{o \leftarrow o'.m(\overline{e};e)\} < \#H/\{o \leftarrow o'.m(\overline{e};e)\}
\end{aligned}
$$

where $\#$ denotes sequence length and the id function gives the set of all identifiers (including *null*) appearing in an expression or history. (In the set $\{o \leftarrow o'.m(\overline{e};\_)\}$ _ may be any value.)

5.4 Inheritance

The semantics of inheritance and late binding has been a research topic for many years, with [342] as an early investigation. Class invariants are used to specify the semantics of a class, and a (logically) strong invariant is usually called for in order to verify that the class satisfies the specifications of the given interfaces [89]. Inheritance is an essential element of object-oriented programming, giving flexible reuse of code. Reasoning is often restricted to a form of behavioral subtyping, implying that a superclass invariant must be respected and maintained by all subclasses [230]. The advantage is that behavioral subtyping allows modular reasoning about late bound calls using the specification of the class of the callee, or the enclosing class in case of a local call. However, this severely limits programming; for instance a perfect bank class, like BANK in Figure 5.2, cannot be used to derive a bank class with negative balance protection, like BANKPLUS in the example (nor versions with interest or charges), since this represents a non-conservative extension violating the BANK invariant. Hence flexibility of code reuse is lost if behavioral subtyping is accepted, unless class invariants are very weak. Thus behavioral subtyping is not an acceptable form of reasoning about object-oriented systems in our setting.

The notion of *lazy behavioral subtyping* [118, 119, 120] allows better flexibility, by talking about two kinds of properties for each class, S (for specifications) and R (for requirements), letting S_C denote properties about class C, and R_C minimal properties to be respected by all subclasses of C. And S_C is used to verify interface satisfaction as explained. In order to prove that the class satisfies S_C, one must in general reason about late-bound local calls in the method bodies of the class, in which case R_C requirements are needed, and R_C must then be enriched with minimal properties of these calls. Lazy behavioral subtyping allows reasoning support for extensible class hierarchies. Each class can be analyzed separately as long as its superclasses have been analyzed earlier. The invariant and pre/postconditions in C form the S_C specifications, while the R_C requirements (initially empty) are generated by adding minimal properties as needed for reasoning about late-bound local calls in the class. Both S_C and R_C properties must be verified in C, and by adding all verified properties to S_C, the property set R_C will be a subset of S_C, i.e., $R_C \subseteq S_C$.

By adding language support for *statically bound local calls*, which are in general useful for explicit code reuse, fewer R requirements are needed since reasoning about these can be done by S specifications. A static local call, say $v: = C{:}m(\bar{e})$, which binds to the method m of C, is meaningful in a subclass of C, say C'. We may reason about this call using S_C, and we may enrich S_C if needed. Thus in the presence of statically bound local calls, there is less need for late-bound local calls. And for programs without late-bound local calls, there is no longer a need for R requirements, we only need interface specifications and class specifications. Separation logic also offers S- and R-like requirements for the setting of sequential Java or JML style programs with non-trivial aliasing but without history specifications [82, 233, 259]. This approach is not able to handle the challenge of the Bank example.

For programs with late-bound local calls, we still have a problem with code reuse when a redefined method violates an R requirement. To solve this problem, we insist that all object variables are typed by interfaces. At run-time an object variable will refer to an object of a class implementing its declared interface, when not *null* [257, 192]. This property, called the *interface substitution principle*, is guaranteed by static type checking. Thus a callee is typed by an interface rather than a class, and reasoning relies on the interface specification. We distinguish reuse of code from reuse of specifications: Interface specifications are inherited by subinterfaces, i.e., $I_F(h)$ is inherited as $I_F(h/F)$ in a subinterface of F. Class invariants need not be inherited, and a subclass of a class implementing interface F may violate F. We let class invariants and interface support be stated independently for each class; however, if a class implements F, it implicitly implements any superinterface of F.

A subclass **class** $C1$ **extends** $C2$ inherits all code and specifications from the class $C2$ (with the usual projection of histories), thus supporting all interfaces that $C2$ implements. This corresponds to (a kind of) behavioral subtyping. A subclass **class** $C1$ **inherits** $C2$ inherits all code, but not specifications, from $C2$. This allows free code reuse, as long as one can verify the stated interface clauses of $C1$. One may inspect the proof outlines of $C2$ and see which ones are valid for $C1$, and the case of lazy behavioral subtyping occurs when all pre- and postconditions used in a proof outline for a late-bound local call in $C2$ are supported by any redefined version of the called methods. Moreover, there is no need to record R requirements, since subclasses no longer need to respect R requirements. Instead of pushing requirement downwards in the class hierarchy, one needs to look upwards in the class hierarchy when verifying a class, reconsidering inherited code. Thus we omit the downward inheritance of requirements and obtain more fine-grained verification control. This approach to reasoning can be called *behavioral interface subtyping*. The interface substitution principle is satisfied, and our approach is able to handle the challenge of the Bank example.

Lemma 5.1. Our language satisfies the interface substitution principle, and reasoning by means of behavioral interface subtyping is sound, i.e., one may reason about an object variable v declared of an interface F by means of the F invariant, assuming each class C is verified as described above, possibly involving superclasses but not any subclasses of C.

Proof outline. We prove that our language, which includes late binding and free (type-correct) method redefinition, satisfies the interface substitution principle, and that each object referred to at run-time by a variable v declared of an interface F satisfies the F specification. Each object variable v is declared of an interface F. At run-time the initial value of v is *null* or the value of an actual parameter, and its value may be changed by **new** statements, assignments, and call statements assigning the result of the call to the variable. Static type checking ensures that all these statements assign to v a value which is of type F

or a subinterface of F, or is *null*, and that each actual object parameter is of the interface (or a subinterface) of the corresponding formal parameter. Given operational semantics, this can be proved more formally by means of subject reduction, as in [192] (which also deals with the complications of call labels). An object variable will therefore at run-time be null or refer to an object of a class with an interface being F or a subinterface. It remains to prove that such a class satisfies F, and that this involves only C and superclasses.

We may assume that all classes are verified. For simplicity we assume that all pre/post-conditions are expressed through the invariant. And we assume a sound reasoning system for proving class invariants. For each class C it is required to verify that the invariant is established and maintained, and to verify the interfaces of C, using the invariant. A C object may at run-time perform methods defined in C or inherited methods, called remotely, and these may make local calls to methods in C or a superclass, including static calls, which may only bind to superclass methods, statically known. Also the binding of late-bound method calls in C or superclass methods are statically known when the executing object is a C object. Thus C satisfaction of F depends only on methods in C and its superclasses, and the binding of all local calls can be resolved at verification time. It is therefore possible to prove satisfaction of F by normal static analysis, ensuring that all methods in C as well as inherited ones maintain the C invariant, and that suspension points in locally called super-class methods respect the C invariant. If needed any local calls to superclass methods can be reverified to deal with these calls. Thus the verification depends on C and its superclasses, and not on any subclass of C. This ensures that any C object will satisfy the interfaces of C. Since an interface F inherits all requirements of any superinterface, we have that a class satisfying F also satisfies a superinterface of F. We may conclude that at run-time an object variable v will be *null* or refer to an object of a class satisfying its declared interfaces. ∎

Inheritance in the Bank Example

In the Bank example, class BANKWAIT extends BANK, while class BANKPLUS inherits BANK without claiming to respect its specifications. For class BANKPLUS it is easy to see that only the postcondition **return** $= true$ of method sub and that of upd are violated. And it is easy to see that the verification of the invariant $bal = sum(h)$ in BANK is valid in class BANKPLUS. The new specification of upd and the added invariant conjunct $bal \geq 0$ must be verified, which is straightforward. Support of interface $BankPlus$ (and thereby interface $Bank$) follows from this. Notice that with lazy behavioral subtyping, class BANKPLUS would not satisfy the R_{BANK} requirement; thus a reasoning approach requiring inheritance of R requirements would not be able to handle this kind of code reuse. We avoid verification problems in this case, as opposed to the case of previous work on lazy behavioral subtyping as well as the work on separation logic.

Multiple Inheritance

The notion of multiple inheritance has been an ideal in object-oriented programming due to its expressive power with respect to code reuse. However, programming with multiple inheritance can be confusing [238], and reasoning about multiple inheritance is challenging, especially if not restricted. In order to control programming with multiple inheritance and late binding, *binding healthiness* [120] ensures that a call textually occurring in a given class C may only bind to a class related to C, either below C or above C. To solve the diamond binding problem, i.e., to bind a method in case of several alternative definitions, we use the ordering given by the order of the inheritance list. In addition, explicit class qualification of late-bound local calls, similar to that for static binding, allows fine-grained control. Synchronization interference between multiple superclasses is avoided, due to local concurrency control. With healthiness, lazy behavioral subtyping extends to the case of multiple inheritance. When analyzing a given class C, added R specifications concern C and future subclasses of C, whereas S specifications may be added to C or some superclass of C. Classes are analyzed in some order consistent with the subclass order, typically the order in which they are defined. As before, R specifications are not needed when restricting local calls to static ones, and in this case healthiness is guaranteed.

Reasoning with behavioral interface subtyping also extends to multiple inheritance. Healthiness is not required, since reasoning is done for the case that the executing object is of the class considered. Any subclass must be considered separately, reusing reasoning results when possible, as in the case of single inheritance. Figure 5.3 gives an example where class PINBANK inherits both class PIN and BANK, not respecting the specifications of the latter, since interface PerfectBank is violated. Thus the challenge of Figure 5.2 reappears.

5.5   Local Reasoning

The disjointness of local histories of distinct objects allows sequential style local reasoning inside a class, while global reasoning is possible with the composition rule. In particular, the Hoare rules for basic statements are standard, including assignment, **if**-statements and skip, assuming expressions are side-effect free and well defined. Standard consequence, conjunction, and adaptation rules apply. The consequence rule may be adjusted to assume local wellformedness of local histories, which implies that any caller and this are non-null. One may reason about local calls as normal, using the relevant class specifications. Thus for a local static call we have

$$[\overline{x}' = \overline{e} \wedge L \wedge P^{\overline{x},\text{caller},h}_{\overline{e},\text{this},h\cdot(\text{this}\rightarrow\text{this}.m(\overline{e}))}] \; v\!:=\! C\!:\!m(\overline{e}) \; [L \wedge \exists v' . \; Q^{\overline{x},\text{caller},v,\textbf{return},h}_{\overline{x}',\text{this},v',v,h\cdot(\text{this}\leftarrow\text{this}.m(\overline{x}';v))}]$$

where L is a predicate not referring to fields of C, h, nor v, and given that method $m(\overline{x})$ of C (possibly inherited) satisfies the pre- and postcondition pair $[P, Q]$ in the current class (using the current invariant for release points). The quantifier on v' is only needed if Q refers to v. Handling of recursive calls can be done as usual. For a late-bound local call occurring in a class C reasoning can be done in the same manner, i.e., a call $v\!:= m(\overline{e})$ (as well as $v\!:= this.m(\overline{e})$) is treated as the static call $v\!:= C\!:\!m(\overline{e})$. And the same rule applies to a remote call $v\!:= o.m(\overline{e})$ given that the pre/post pair $[P, Q]$ follows from the specification of the interface of o, and that $[P, Q]$ does not refer to the (disjoint) history of that interface. Note that all methods exported through an interface must maintain the invariant. Dot-notation is not allowed on non-exported methods; and these need not satisfy the invariant. Note that this means that suspending calls (which require the invariant) are not available for non-exported methods.

Rule history in Figure 5.4 expresses that the local history is monotonic, i.e., the past can never change. Rule await guard expresses that the statement **await** b will not change local variables (but during suspension fields may be updated by other processes), and when it terminates the waiting condition is satisfied and the invariant has been reestablished provided it held before. The rule for simple call expresses that the partial correctness semantics

history	$[h_0 = h]\ s\ [h_0 \le h]$

await guard	$[I \wedge L]\ \mathbf{await}\ b\ [b \wedge I \wedge L]$

simple call	$[Q^h_{h \cdot (\mathsf{this} \to o.m(\overline{e}))}]\ o.m(\overline{e})\ [Q]$

call-while

$$\frac{[o' = o \wedge \overline{e}' = \overline{e} \wedge P]\ s\ [\forall\, v'\,.\, Q^{v,h}_{v',h \cdot (\mathsf{this} \leftarrow o'.m(\overline{e}';v'))}]}{[o \ne \mathsf{this} \wedge P^h_{h \cdot (\mathsf{this} \to o.m(\overline{e}))}]\ v\colon = o.m(\overline{e})\{s\}\ [Q]}$$

await call

$$[o' = o \wedge \overline{e}' = \overline{e} \wedge L \wedge I^h_{h \cdot (\mathsf{this} \to o.m(\overline{e}))}]\ \mathbf{await}\ v\colon = o.m(\overline{e})$$
$$[L \wedge h\ \mathbf{ew}\,(\mathsf{this} \leftarrow o'.m(\overline{e}';v)) \wedge \exists\, v\,.\, I^h_{old(h)}]$$

new

$$[\forall\, v'\,.\, fresh(v',h) \Rightarrow Q^{v,h}_{v',h \cdot (\mathsf{this} \to v'.\mathbf{new}\ C(\overline{e}))}]\ v\colon = \mathbf{new}\ C(\overline{e})\ [Q]$$

method

$$\frac{[P^{\overline{y}}_{\overline{y}'} \wedge \overline{y} = \overline{U}_{default}]\ s\ [Q^{\overline{y},\mathbf{return},h}_{\overline{y}',e,\ h \cdot (\mathsf{caller} \leftarrow \mathsf{this}.m(\overline{x};e))}]}{[P^h_{h \cdot (\mathsf{caller} \to \mathsf{this}.m(\overline{x}))}]\ T\ m(\overline{T\ x})\{\overline{U\ y};s;\mathbf{return}\ e\}\ [Q]}$$

Figure 5.4: Hoare style rules for non-standard constructs. Primed variables represent fresh logical variables, and $fresh(v',h)$ expresses that v' does not occur in h. L denotes a condition on local variables. For each class one must verify that the class invariant I holds after initialization and is maintained by all methods exported through an interface. The **await** rules ensure that it holds upon suspension.

of the call $o.m(\overline{e})$ is equivalent to that of the assignment $h\colon = h \cdot (\mathsf{this} \to o.m(\overline{e}))$. The rule call-while expresses that the partial correctness semantics of a non-local call $v\colon = o.m(\overline{e})\{s\}$ is equivalent to that of the statements

$$h\colon = h \cdot (\mathsf{this} \to o.m(\overline{e}));s;v'\colon = \mathbf{some};h\colon = h \cdot (\mathsf{this} \leftarrow o.m(\overline{e};v'));v\colon = v'$$

where v' is a fresh variable representing the locally unknown result, letting $v'\colon = \mathbf{some}$ denote a non-deterministic assignment. The call-while rule may be derived from this using the rule $[\forall\, v\,.\, Q]\ v\colon = \mathbf{some}\ [Q]$. The blocking call $v\colon = o.m(\overline{e})$ is equivalent to $v\colon = o.m(\overline{e})\{\mathbf{skip}\}$. Thus we may derive the rule

$$[\forall\, v'\,.\, o \ne \mathsf{this} \wedge Q^{v,h}_{v',h \cdot (\mathsf{this} \to o.m(\overline{e})) \cdot (\mathsf{this} \leftarrow o.m(\overline{e};v'))}]\ v\colon = o.m(\overline{e})\ [Q]$$

With respect to partial correctness, the suspending call **await** $v\colon = o.m(\overline{e})$ is equivalent to calling while suspending, i.e., $v\colon = o.m(\overline{e})\{\mathbf{await}\ true\}$, and the await call rule can be derived from this. The method rule expresses that the body of a method $T\ m(\overline{T\ x})\{\overline{T\ y};s;\mathbf{return}\ e\}$ can be seen as the statements

$$h\colon = h \cdot (\mathsf{caller} \to \mathsf{this}.m(\overline{x}));\{\overline{U\ y};s;\mathbf{return}\colon = e\};h\colon = h \cdot (\mathsf{caller} \leftarrow \mathsf{this}.m(\overline{x};\mathbf{return}))$$

(with **return** and caller as local variables). In the rule, $\overline{y}'$ is a list of logical variables used to handle possible name clashes between local variables $\overline{y}$ and variables occurring in P or Q, and uninitialized local variables are initialized by default values of the respective types

($\overline{U_{default}}$). The rule for object creation expresses the obvious extension of the local history and (local) freshness of the identity of the new object (and v' is needed if v occurs in e).

Verification of the Bank Example

Consider the inherited method *sub* of class BANKWAIT. We need to verify the invariant and the postcondition:

$$[I(h)] \ Bool \ ok\!: = upd(-x);\textbf{return} \ ok \ [I(h) \wedge \textbf{return} = true]$$

where $I(h)$ is the class invariant ($bal = sum(h) \wedge bal \geq 0$). Left-constructive reasoning according to the method rule gives

$$[I(h)] \ ok\!: = upd(-x) \ [I(h \cdot (\text{this} \leftarrow \text{this}.sub(x;ok))) \wedge ok = true]$$

which by definition of *sum* reduces to

$$[I(h)] \ ok\!: = upd(-x) \ [bal = sum(h) - x \wedge bal \geq 0 \wedge ok = true]$$

It suffices that $upd(x)$ satisfies the specification

$$[\![bal = sum(h), \ return = true \wedge bal = sum(h) + x \wedge bal \geq 0]\!]$$

which is easily verified by the rule for local calls, using the redefined *upd* (with *ok* receiving the method result). Moreover, these verification tasks could easily be automated by a tool.

A client object may call methods on a bank object b. If b is declared to be of interface PerfectBank, we obtain $[true] \ ok\!: = b.sub(a) \ [ok = true]$, and $[true] \ v\!: = b.bal() \ [v \geq 0]$ if b is of interface BankPlus. Reasoning with the compositional rule is required to obtain further information about the calls.

Verification of the PinBank Example. For class PIN it is straightforward to verify that the invariant is maintained by each method, and that it holds initially. We may also show that the postcondition given in interface Pin for *close* is established by the implementation of *close* in PIN, given the invariant as precondition. And the one for *open* is trivial to verify.

Consider the method *sub* of class PINBANK. We need to verify that the body of *sub* satisfies its postcondition and that it maintains the given invariant. In the first verification task we rely on the postcondition $\textbf{return} = true$ of *upd* from BANK. The invariance proof of $bal = sum(h)$ reduces to the one in BANK, and the conjunct $u = user(h)$ is maintained since it is not affected by *sub*.

5.6   Discussion of Future-Related Mechanisms

The notion of *future* has been suggested as a mechanism to increase concurrency and reduce waiting in method bodies [351, 229]. A future may be seen as a reference to a location where a method result will be stored when available. Futures may be *first-order* in the sense that they may be passed as parameters. Thereby information sharing is possible and one may return a future rather than waiting for the result itself. Reasoning about ABS with first-order futures is studied in [114, 116]. Futures are in particular useful when a callee depends on remote calls to other objects. However, there is a cost of using futures, both at the programming level and at the specification and reasoning level. At the programming level, an interface must decide if and where to use futures, since this affects the signatures of the interface methods. These decisions are not easy to make at an early stage, and it is difficult to change these decisions at a later stage in the programming, since they affect other classes as well. At the specification and reasoning level, one will need to talk about future identities and it is in general not trivial to make the connection between a call event and the corresponding get event with the result of the call. Thus the simple call-response paradigm is no longer syntactically reflected in the histories. Reasoning rules must deal with future identities. And as seen in [116] there is a cost with respect to compositional reasoning, related to the "get" rule for accessing a future value.

Our experience is that first-order futures are often not needed, and therefore the drawbacks mentioned are quite expensive in a reasoning perspective since futures appear in the communication events even when not used in the program. The mechanism of *delegation* is in several ways similar to futures. In the body of a method $m(\overline{x})$ one may delegate the rest of the method body to another call (possibly remote), letting for instance the statement **delegate** $o.n(\overline{e})$ mean that the current call terminates without producing a result, while delegating to the remote call $o.n(\overline{e})$ to send a result back to the caller of m. Type checking must ensure that the result type of n is the same, or better (i.e., a subinterface or subtype), than that of m. However, interface declarations are not affected by issues related to delegation. Thus delegation may be used in code when suitable without prior planning in interfaces.

Our current setting may accommodate delegation with some adjustments of the communication events, adding more information to global events, and strengthening the notion of wellformedness to make up for the difference in local and global events. In local reasoning, events may be as before, except that a delegation call must be indexed with the current call, and in this case no return event should occur. In global reasoning, all communication events could be indexed with the initial call, starting with a call event $(o \rightarrow o1.m(\overline{e}))_{(o\rightarrow o1.m(\overline{e}))}$ and ending with a get event $(o \leftarrow o1.m(\overline{e};r))_{(o\rightarrow o1.m(\overline{e}))}$. Thus $(o \rightarrow o1.m(\overline{e}))_c$ may be followed by $(o \rightarrow\!\!\!\!\rightarrow o1.m(\overline{e}))_c$, and $(o \rightarrow\!\!\!\!\rightarrow o1.m(\overline{e}))_c$ may be followed by either a delegation event $(o1 \rightarrow o2.n(\overline{t}))_c$ or a return event $(o \leftarrow o1.m(\overline{e};r))_c$. The delegation event is like a call event and may be followed by $(o1 \rightarrow\!\!\!\!\rightarrow o2.n(\overline{t}))_c$, while a return event $(o' \leftarrow o1'.m'(\overline{e'};r))_{(o\rightarrow o1.m(\overline{e}))}$ may be followed by $(o \leftarrow o1.m(\overline{e};r))_{(o\rightarrow o1.m(\overline{e}))}$, closing the cycle. A normal 4 event call cycle is now generalized to a call cycle of length $4 + 2n$ where n is the number of delegations in the cycle. If the initial call is redundant (i.e., the caller, callee, method, and actual input parameters are given by the main event), the index may be omitted. Thus if delegation is not used, events are written as in the setting without delegation. We conclude that the cost of delegation is less than that of the future mechanism.

A simpler approach is to simulate delegation by the statements **await** *dummy*: = $o.n(\overline{e})$;**return** *dummy*, where *dummy* is a fresh local variable, not used in the invariant I nor postcondition Q. This is like a delegation except that the delegating object needs to pass on the result. However, this may be efficiently implemented. We denote these statements **delegate** $o.n(\overline{e})$, whereas the statements *dummy*: = $o.n(\overline{e})$;**return** *dummy*, denoted **return** $o.n(\overline{e})$, represent a blocking version of delegation. Return may also be generalized to local calls. For flexible programming, we modify our language so that the last statement in a method body is either a return or delegation statement, or a (nested) if construct where each branch ends with a return or delegation statement. For instance, method *upd* of class BANKPLUS can be written {**if** $bal + x\rangle = 0$ **then return** BANK:$upd(x)$ **else return** *false* **fi**}. Reasoning rules can be derived from the definition above:

$$\text{delegation} \quad \frac{h\,\mathbf{ew}\,(\mathsf{this}\rightarrow o.n(\overline{e})) \wedge I \Rightarrow \forall\,\mathbf{return}.\;Q^h_{h\cdot(\mathsf{this}\leftarrow o.n(\overline{e};\mathbf{return}))\cdot(\mathsf{caller}\leftarrow\mathsf{this}.m(\overline{x};\mathbf{return}))}}{[I^h_{h\cdot(\mathsf{this}\rightarrow o.n(\overline{e}))}]\;\mathbf{delegate}\;o.n(\overline{e})\;[Q]}$$

where m is the enclosing method and $\overline{x}$ the formal parameters. For return-call we obtain

$$[o \neq \mathsf{this} \wedge Q^h_{h\cdot(\mathsf{this}\rightarrow o.n(\overline{e}))\cdot(\mathsf{this}\leftarrow o.n(\overline{e};\mathbf{return}))\cdot(\mathsf{caller}\leftarrow\mathsf{this}.m(\overline{x};\mathbf{return}))}]\;\mathbf{return}\;o.n(\overline{e})\;[Q]$$

for remote calls (local calls are similar to earlier). This allows reasoning with a delegation-like construct without changing the structure of events nor the notion of wellformedness.

5.7   Conclusion

We have motivated and presented the main elements of a concurrency model and reasoning framework based on active, concurrent objects. From a programming point of view the model allows efficient and simple programming of distributed and multi-core systems, resulting in a high degree of parallelism, and with semantically simple synchronization mechanisms, allowing local synchronization control within each method. Reasoning about such systems has been studied in [98, 264, 119, 9, 115, 117, 114]; [224, 9] survey research outside this framework. The current work makes several improvements on this setting: In the compositional reasoning our treatment is based on interface invariants rather than directly on class invariants, simplifying the treatment of prefix closure of class invariants. Secondly, the axiomatic semantics is simplified by the statement for interleaved call introduced here. Finally, we have shown how lazy behavioral subtyping can be replaced by what we call behavioral interface subtyping, avoiding inheritance of requirements to subclasses, thereby allowing free code reuse (as long as the specified interfaces are supported by each class). As seen in the example, this makes a significant difference in practice.

The approach is modular since each class can be analyzed separately, verifying the given specifications and implements-claims, and subclasses can be added incrementally. In a subclass one may need to verify new properties of inherited or statically called methods of a superclass C. By separating code inheritance from specification inheritance we obtain control of code reuse. Behavioral subtyping corresponds to inheriting while respecting superclass specifications, lazy behavioral subtyping corresponds to inheriting while respecting only pre/postconditions of redefined methods, and free code reuse corresponds to inheriting without respecting superclass specification.

We have avoided the use of futures/call labels in order to make the language more high-level. This has the advantage that the connection between postconditions and history-based invariants is direct, not depending on future identities. The interleaved call statement has allowed us to present a label-free version of the language with the expressiveness of normal (i.e., pair-wise) use of call labels. The presented rules for the different call mechanisms are simpler than in earlier work on future/label-free communication [119], due to the treatment of interleaved call and guarded suspension. Reasoning is as for sequential programs, apart from side-effects on the local history. Some topics have not been considered, like typing considerations, dynamic class updates [258, 191] and constructs for object grouping, which allow clusters of concurrent objects to be seen as a single object from the outside [190].

We have not included soundness and completeness of the Hoare style reasoning system, but the basic primitives are modeled by (partly non-executable) assignments where all updates on the history are explicit. The given rules can be derived from this model.

A number of tools have been developed for the presented concurrency model, mainly under the HATS project, including compilers and a reasoning framework in KeY [7]. In order to exploit the paradigm presented here in Java, a Java library has been developed allowing Java programming with the described communication primitives and concurrency model [239].

Acknowledgments

The author has known Kaisa Sere since the late 80ies and is indebted to her for many years of discussions on Action Systems and the concurrency model of this paper, and its predecessors, especially through the yearly NWPT workshops.

The reviewers and Charlie McDowell have contributed constructively to the presentation and discussions of the paper. This work relates to the EU projects FP7-610582 *Envisage: Engineering Virtualized Services* (http://www.envisage-project.eu) and FP7-ICT-2013-X *UpScale: From Inherent Concurrency to Massive Parallelism through Type-based Optimizations* (http://www.upscale-project.eu).

Chapter 6

A Contract-Based Approach to Ensuring Component Interoperability in Event-B

Linas Laibinis

Åbo Akademi University, Turku, Finland

Elena Troubitsyna

Åbo Akademi University, Turku, Finland

Abstract. The design by contract approach enables rigorous development of component-based software systems. In particular, it allows us to ensure component interoperability. However, defining the contracts themselves is often a challenging task, especially in the development of decentralised systems with complex component interdependencies. In this paper, we propose a refinement-based approach facilitating definition of contracts and ensuring component interoperability. The approach is based on the Event-B formalism and its modularisation extension. In the Event-B refinement process, we gradually introduce a representation of inter-component communication, distribute the global state space between the components, and decouple them. Finally, we decompose the resulting system specification into independent modules by formally defining module interfaces, obtaining at the same time component contracts. This allows us to propagate the system level properties into the component contracts and ensure component interoperability. The proposed approach is illustrated by an example – an auction system.

6.1   Introduction

Ensuring component interoperability constitutes one of the main challenges in the component-based development approach [161]. The approach relies on a composition of reusable software components (or services) to implement a desired functionality [323]. While composing them, we should ensure that the components are compatible not only at the interface but also at the semantic level. Essentially, it requires for the components to share the knowledge about their globally observable behaviour and properties. The most popular approach to representing such knowledge is by defining component contracts.

Usually a contract is expressed as an *assumption-guarantee* pair [170]. The *assumption* part postulates the properties that the component's environment should satisfy, while the *guarantee* part defines the properties that must be satisfied by the component itself.

The design by contract approach proposed by Meyer [244] facilitates structuring of software into encapsulated modules with contracts regulating component interactions. While the benefits of the contract-based approach are evident, defining the contracts themselves is often a challenging task, especially in the development of decentralised systems with complex component interdependencies. In this paper, we propose a refinement-based approach facilitating definition of contracts and ensuring component interoperability.

The formal basis of the proposed approach is within Event-B [6] and its modularisation extension [182]. Event-B is a formal approach for designing correct-by-construction distributed systems. The main development technique of Event-B – refinement – allows the designers to transform an abstract specification into a detailed model through a chain of correctness-preserving model transformations. Each refinement step is verified by proofs guaranteeing that the refined model preserves the externally observable behaviour and does not introduce new deadlocks. Modelling and verification in Event-B is automated by an industrial-strength tool – the Rodin platform [327].

Refinement allows us to formally define relations between models representing the system behaviour at different levels of abstraction. Hence it constitutes a suitable mechanism for establishing relationships between the system-level properties and the behaviour of system components.

The main idea behind our approach is to derive component contracts from a formal Event-B specification of the overall system. We start from an abstract centralised specification of the system. The initial model relies on the shared global state and abstracts away the communication between the components. The interoperability properties are fairly transparent at this level and easy to define. In the refinement process, while elaborating on the system behaviour and properties, we introduce a representation of inter-component communication, distribute the global state space between the components, and decouple them. Finally, we decompose the obtained system specification into independent modules. While deriving the module interfaces, we at the same time define their contracts. Since decomposition is a special kind of refinement step, we guarantee that the components remain interoperable under the derived contracts.

We believe that the proposed approach facilitates the rigorous development of complex component-based systems and enhances confidence in the correctness of the overall system design. It allows us to propagate the system level properties into the component contracts and ensure component interoperability.

The paper is structured as follows: in Section 6.2, we present our main modelling frame-

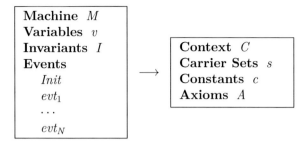

Figure 6.1: Event-B machine and context components

work – Event-B and its modularisation extension. In Section 6.3, we demonstrate how to derive contracts from Event-B models using the modularisation extension. In Section 6.4, we give a small illustrative example – an auction system. Finally, in Section 6.5, we overview the proposed approach and discuss the related work.

6.2   Background: Event-B

In this section, we overview our modelling framework – Event-B. The Event-B formalism [6] is a state-based formal approach that promotes the correct-by-construction development paradigm and formal verification by theorem proving. Event-B is a specialisation of the B Method that facilitates modelling of event-based (reactive) systems by incorporating the ideas of the Action Systems formalism [35] into the B Method.

6.2.1   Modelling and Refinement in Event-B

In Event-B, a system specification (model) is defined using the notion of an *abstract state machine* [6]. An abstract state machine encapsulates the model state, represented as a collection of model variables, and defines operations on this state. Therefore, it describes the dynamic part (behaviour) of the modelled system. Usually a machine also has the accompanying component, called *context*, which contains the static part of the model. In particular, a context can include user-defined carrier sets, constants and their properties, which are given as a list of model axioms. A general form of Event-B models is given in Figure 6.1.

The machine is uniquely identified by its name M. The state variables, v, are declared in the **Variables** clause and initialised in the *Init* event. The variables are strongly typed by the constraining predicates I given in the **Invariants** clause. The invariant clause might also contain other predicates defining properties that should be preserved during system execution.

The dynamic behaviour of the system is defined by the set of atomic events specified in the **Events** clause. Generally, an event can be defined as follows:

$$\textbf{any } lv \textbf{ where } g \textbf{ then } S \textbf{ end},$$

where lv is a list of new local variables (parameters), the guard g is a state predicate, and

Statement (S)	$BA_S(s, c, lv, x, y, x')$
skip	$x' = x \;\wedge\; y' = y$
$x := E(s, c, lv, x, y)$	$x' = E(s, c, lv, x, y) \;\wedge\; y' = y$
$x :\in Set$	$x' \in Set \;\wedge\; y' = y$
$x : \mid P(s, c, lv, x, y, x')$	$P(s, c, lv, x, y, x') \;\wedge\; y' = y$
$S_1 \parallel S_2$	$BA_{S_1} \;\wedge\; BA_{S_2}$

Figure 6.2: Before-after predicates

the action S is a statement (assignment). In the case when lv is empty, the event syntax becomes **when** g **then** S **end**. If g is always true, the syntax can be further simplified to **begin** S **end**.

The occurrence of events represents the observable behaviour of the system. The guard defines the conditions under which the action can be executed, i.e., when the event is *enabled*. If several events are enabled at the same time, any of them can be chosen for execution nondeterministically. If none of the events is enabled then the system deadlocks.

In general, the action of an event is a parallel composition of statements (assignments). The statements can be either deterministic or non-deterministic. A deterministic assignment, $x := E(x, y)$, has the standard syntax and meaning. A nondeterministic assignment is denoted either as $x :\in Set$, where Set is a set of values, or $x : \mid P(x, y, x')$, where P is a predicate relating initial values of x and y to some final value of x'. As a result of such an assignment, x can get any value belonging to Set or according to P.

Semantics of an Abstract Model. The semantics of Event-B actions is defined using before-after (BA) predicates [6]. A before-after predicate describes a relationship between the system states before and after an execution of an event action. The definitions of a BA for different action statements are shown in Figure 6.2. Here x and y are disjoint lists (partitions) of state variables, and x', y' represent their values in the state after the action execution. Moreover, lv refers to the local event variables, where s and c stand for respectively the sets and constants defined in the model context.

The notion of a BA predicate can be easily generalised to formally define model events. For an event e of the form **any** lv **where** g **then** S **end**, its BA predicate is as follows:

$$BA_e(s, c, x, y, x') \;=\; \exists lv.\; g(s, c, lv, x, y) \;\wedge\; BA_S(s, c, lv, x, y, x')$$

The semantics of a whole Event-B model is formulated as a number of *proof obligations*, expressed in the form of logical sequents. Below we present only the most important proof obligations that should be verified for the initial and refined models. The full list of proof obligations can be found in [6].

An initial Event-B model should satisfy the event feasibility and invariant preservation properties. For each event of the model – e_i – its feasibility means that, whenever the event e_i is enabled, its before-after predicate (BA) is well-defined, i.e., there exists some reachable after-state:

$$A(s, c),\; I(s, c, v),\; g_{e_i}(s, c, lv, v) \;\vdash\; \exists v' \cdot BA_{e_i}(s, c, lv, v, v') \qquad \text{(FIS)}$$

where A are the model axioms, I is the model invariant, g_{e_i} is the event guard, d are the model sets, c are the model constants, lv are the local event variables, and v, v' are the variable values before and after event execution.

Each event e_i of an initial Event-B model should also preserve the given model invariant:

$$A(s, c),\ I(s, c, v),\ g_{e_i}(s, c, lv, v),\ BA_{e_i}(s, c, lv, v, v') \vdash I(s, c, v') \tag{INV}$$

Since the initialisation event *Init* has no initial state, local variables, and guard, its proof obligation is simpler:

$$A(s, c),\ BA_{Init}(s, c, v') \vdash I(s, c, v') \tag{INIT}$$

Semantics of a Refined Model. Event-B employs a top-down refinement-based approach to system development. Development starts from an abstract system specification that models the most essential functional requirements. While capturing more detailed requirements, each refinement step typically introduces new events and variables into the abstract model. These new events correspond to stuttering steps that are not visible at the abstract level.

To verify the correctness of a refinement step, we need to prove a number of proof obligations for the refined model. Intuitively, those proof obligations allow us to demonstrate that the refined machine does not introduce new observable behaviour, more specifically, that concrete states are linked to the abstract ones via the given (gluing) invariant of the refined model. All proved properties of the abstract model are automatically "inherited" by the refined one. For brevity, we omit here discussion of these proof obligations. They can be found in [6].

The Event-B refinement process allows us to gradually introduce implementation details, while preserving functional correctness during stepwise model transformation. The model verification effort, in particular, automatic generation and proving of the required proof obligations, is significantly facilitated by the provided tool support – the Rodin platform [327].

6.2.2   Modelling Modular Systems in Event-B

Recently the Event-B language and tool support have been extended with a possibility to define modules [182, 277] – components containing groups of callable operations. Modules can have their own (external and internal) state and the invariant properties. The important characteristic of modules is that they can be developed separately and then composed with the main system.

Module Structure. A module description consists of two parts – *module interface* and *module body*. Let M be a module. The module interface is a separate Event-B component. It allows the user of the module M to invoke its operations and observe the external variables of M without having to inspect the module implementation details. The module interface consists of the module interface description MI and its context MI_Context. The context defines the required constants c and sets s. The interface description consists, respectively, of the external module variables w, the external module invariant M_Inv(s, c, w), and a collection of module operations, characterised by their pre- and postconditions, as shown in Figure 6.3. The primed variables in the operation postcondition stand for the variable values after operation execution, while the predefined variable *res* refers to the operation result to be returned.

In addition, a module interface description may contain a group of standard Event-B events under the **PROCESS** clause. These events model the autonomous module thread of control, expressed in terms of their effect on the external module variables. In other words, the module process describes how the module external variables may change between operation calls.

INTERFACE $MI(id)$
   SEES $MI\_Context$
   VARIABLES w
   INVARIANT $M\_Inv(s, c, w)$
   INITIALISATION ...
   PROCESS
      PE_1 = **any** vl **where** $g_1(s, c, vl, w)$ **then** $S_1(s, c, vl, w, w')$ **end**
      ...
   OPERATIONS
      O_1 = **any** p **pre** $Pre_1(s, c, p, w)$ **post** $Post_1(s, c, p, w, w', res)$ **end**
      ...
END

Figure 6.3: Module interface

A module development always starts with the design of an interface. After an interface is formulated, it cannot be altered in any manner. This ensures correct relationships between a module interface and its body, i.e., that the specification of an operation call is recomposable with an operation implementation. A module body is an Event-B machine. It implements each operation described in the module interface by a separate group of events. Additional proof obligations are generated to verify the correctness of a module. They guarantee that each event group faithfully satisfies the given pre- and postconditions of the corresponding interface operation.

Importing of a Module. When the module M is imported into another Event-B machine, this is specified by a special clause **USES** in the importing machine, N. As a result, the machine N can invoke the operations of M as well as read the external variables of M listed in the interface MI.

To make a module interface generic, in MI_Context we can define some abstract constants and sets (types). Moreover, the interface MI itself may be parameterised with the constant *id*, which is used as a unique identifier for a module instance within the interface. All such data structures become module parameters that can be instantiated when a module is imported. The concrete values or constraints needed for module instantiation are supplied within the **USES** clause of the importing machine. Alternatively, the module interface can be *extended* with new sets, constants, and the properties that define new data structures and/or constrain the old ones. Such an extension produces a new, more concrete module interface. Via different instantiation of generic parameters the designers can easily accommodate the required variations when developing components with similar functionality. Hence module instantiation provides us with a powerful mechanism for reuse.

We can create several instances of a given module and import them into the same machine. Different instances of a module operate on disjoint state spaces. Identifier prefixes can be supplied in the **USES** clause to distinguish the variables and operations of different module instances or those of the importing machine and the imported module. Alternatively, the pre-defined set can be supplied as an additional parameter. In the latter case, module instances are created for each element of the given set. The syntax of USES then becomes as follows:

$$\textbf{USES } \langle \textit{module interface} \rangle \textbf{ as } \langle \textit{prefix\_} \rangle$$

or

$$\textbf{USES } \langle \textit{module interface} \rangle [\langle \textit{constant set} \rangle].$$

Semantics of a Module Interface. Similarly to a machine component, the semantics of an interface component is defined by a number of proof obligations. The module initialisation must establish the module invariant M_Inv:

$$M\_Init(s, c, mv') \vdash M\_Inv(s, c, mv') \tag{MOD\_INIT}$$

Let us assume $Oper_i$, $i \in 1..N$, is one of module operations. The module invariant M_Inv should be preserved by each operation execution:

$$M\_Inv(s, c, mv), \ Pre_i(s, c, p, mv), \ Post_i(s, c, p, mv, mv', res) \vdash M\_Inv(s, c, mv') \tag{MOD\_INV1}$$

where Pre_i and $Post_i$ are respectively the precondition and postcondition of $Oper_i$.

Let us assume Ev_j, $j \in 1..K$, is one of module process events. The module invariant M_Inv should be also preserved by each such event:

$$M\_Inv(s, c, mv), \ BA_j(s, c, lv, mv, mv') \vdash M\_Inv(s, c, mv') \tag{MOD\_INV2}$$

where BA_j is the before-after predicate of Ev_j.

Finally, there is a couple of feasibility proof obligations for each $Oper_i$, $i \in 1..N$. Firstly, the operation precondition should be true for at least some of the parameter values:

$$M\_Inv(s, c, mv) \vdash \exists p. \ Pre_i(s, c, p, mv) \tag{MOD\_PARS}$$

Secondly, at least some operation post-state containing the required result must be reachable:

$$M\_Inv(mv), \ Pre_i(p, mv) \vdash \exists (mv', res). \ Post_i(p, mv, mv', res) \tag{MOD\_RES}$$

Semantics of an Operation Call. A machine importing a module instance operates on the extended state consisting of its own variables v and the module variables mv. The module state can be updated in event actions only via operations calls. The semantics of an event containing an operation call is as follows.

Let us consider the model event E_c that contains a call to the module operation Op with the given arguments $args$, i.e., it is of the form

$$\textbf{any } lv \textbf{ where } g \textbf{ then } S[Op(args)] \textbf{ end}.$$

The BA predicate of such an event can be defined as follows:

$$
\begin{aligned}
BA_{E_c}(s, c, v, mv, v', mv') \ = \ & \exists (lv, res, new\_mv). \ g(s, c, lv, v, mv) \ \wedge \\
& Post(s_{MI}, c_{MI}, args, mv, new\_mv, res) \ \wedge \\
& BA_{S*}(s, c, lv, v, mv, res, v') \ \wedge \ (mv' = new\_mv),
\end{aligned}
$$

where $S*$ is S with all the occurrences of $Op(args)$ replaced by res, while s_{MI} and c_{MI} are respectively the sets and constants defined in the module interface context. Once this is done, we can rely on the existing proof semantics to verify the invariant preservation, event simulation, and other required properties.

Moreover, we need an additional proof obligation to ensure call correctness by checking that the operation precondition holds at the place of an operation call:

$$g(s, c, lv, v, mv), \ Inv(s, c, v, mv), \ M\_Inv(s_{MI}, c_{MI}, mv) \vdash Pre(s_{MI}, c_{MI}, args, mv) \tag{CALL\_CORR}$$

The modularisation extension of Event-B facilitates formal development of complex systems by allowing the designers to decompose large specifications into separate components and verify system-level properties at the architectural level. Next we demonstrate how to define contracts based on the modularisation extension of Event-B.

COMPONENT CLASS $C(id)$

EXTERNAL VARIABLES v

INVARIANT $Inv(v)$

INITIALISATION $Init(v)$

ACTIONS

$\quad A_1 \;=\; \mathbf{params}\; w \;\mathbf{pre}\; Pre_1(v, w) \;\mathbf{post}\; Post_1(v, w, v')$

$\quad A_2 \;=\; \mathbf{params}\; w \;\mathbf{pre}\; Pre_2(v, w) \;\mathbf{post}\; Post_2(v, w, v')$

$\quad ...$

$\quad A_n \;=\; \mathbf{params}\; w \;\mathbf{pre}\; Pre_n(v, w) \;\mathbf{post}\; Post_n(v, w, v')$

END

Figure 6.4: Component contract

6.3 From Event-B Modelling to Contracts

6.3.1 Contracts

Contracts constitute a widely-used approach to ensuring component interoperability [161, 244]. Via their contracts, the components share the knowledge about their globally observable behaviour and properties. In the component-based frameworks, a contract is usually defined as an *assume-guarantee* pair [170]. The *assumption* part defines the properties that the environment must satisfy, while the *guarantee* part expresses the properties that should be satisfied by the component itself.

The development methodology based on refinement allows us to express system-wide properties, which can significantly facilitate definition of component contracts. These properties are defined as model invariants. Our goal is to find a mechanism for propagating the relevant system-level properties into the component contracts.

To ensure component interoperability, in our definition of a contract, we will describe the component interface, interactions between the component and its environment, as well as abstractions for the expected autonomous behaviour of a component. The generic form of a contract for a class of components C is presented in Figure 6.4.

Since one of the strengths of the component-based development is the support for component reuse, a single contract can represent a family (class) of components that might differ by their implementations, internal behaviour, etc. Hence, in our definition of a component contract given in Figure 6.4, we explicitly state that the contract is defined for a class of components.

The **EXTERNAL VARIABLES** clause defines the globally observable state of components (represented by a collection of variables v). The **INVARIANT** clause defines the types of the external variables as well as the properties always maintained over them. The initial state of a component is constructed according to the state predicate defined in the **INITIALISATION** clause. The **ACTIONS** part of the contract regulates the dynamic behaviour of the component. It is defined as a pre- and post-condition pair for each operation of the component that changes its externally observable state. The operations might have parameters that are defined in the **params** clause. Finally, the class itself may be parameterised with the constant *id* to be used as a unique identifier for class instances.

It is easy to note, that the definition of a contract closely resembles the definition of a module interface given in Figure 6.3. Indeed, a module interface defines the globally observable state of the component, its properties in the **INVARIANT** clause and callable operations as pre- and postcondition pairs. The only difference is in representing the internal (autonomous) component behaviour defined in the **PROCESS** clause. Next we will address this issue and establish a correspondence between the definitions of a module interface and a contract.

6.3.2   From a Module Interface to a Component Contract

In our definition of the module, the **PROCESS** clause defines the autonomous internal behaviour of the component, i.e., specifies how the external component state may change between operation calls. The behaviour is modelled by a set of events. To transform this representation into the pre-postcondition format, let us give an alternative definition of an Event-B event.

Essentially, an event e of the form **any** a **where** G_e **then** BA_e **end** is a relation describing the corresponding state transformation from σ to σ', such that

$$e(\sigma, \sigma') \;=\; \exists \rho \cdot \mathcal{I}(s, c, \sigma) \,\wedge\, G_e(s, c, \rho, \sigma) \wedge BA_e(s, c, \rho, \sigma, \sigma'),$$

where ρ represents the local event state. Here we treat the model invariant $\mathcal{I}$ as an implicit event guard. Note that, due to the possible presence of nondeterminism, the successor state σ' is not necessarily unique.

In other words, the semantics of a single model event is given as a binary relation between pre- and post-states of the event. To represent this relationship, we define two functions before and after in a way similar to [179, 326]:

$$\mathsf{before}(e) = \{(\rho, \sigma) \mid \mathcal{I}(s, c, \sigma) \wedge G_e(s, c, \rho, \sigma)\}$$

and

$$\mathsf{after}(e) = \{\sigma' \mid \exists \rho, \sigma \cdot \mathcal{I}(s, c, \sigma) \wedge G_e(s, c, \rho, \sigma) \wedge BA_e(s, c, \rho, \sigma, \sigma')\}.$$

One can see that, for a given event e and any state $\sigma \in \Sigma$, e is enabled in σ if and only if $(\rho, \sigma) \in \mathsf{before}(e)$ for some possible value of the local variables ρ. Essentially, the functions before and after define the domain and range of the underlying semantic event definition as a before-after relation between the states.[1]

The alternative representation of an event allows us to express both externally callable operations and autonomic component behaviour in the pre- and postcondition form. To be more precise, each callable interface operation

$$Op_i \;\;=\;\; \textbf{any } p_i \textbf{ pre } Pre_i(...) \textbf{ post } Post_i(...) \textbf{ end}$$

is directly translated into the action

$$A_i \;\;=\;\; \textbf{params } p_i \textbf{ pre } Pre_i \textbf{ post } Post_i,$$

while each event in the **PROCESS** clause

---

[1]We slightly modified the before definition comparing to its original one, where $\mathsf{before}(e) \;\;=\;\; \{\sigma \mid \exists \rho \cdot \mathcal{I}(s, c, \sigma) \wedge G_e(s, c, \rho, \sigma)\}$, to make it more suitable for defining component contracts.

$$Ev_j \quad = \quad \textbf{any } lv_j \textbf{ where } G_j(...) \textbf{ then } BA_j(...) \textbf{ end}$$

is mapped into the following action:

$$A_j \quad = \quad \textbf{params } lv_j \textbf{ pre } \mathsf{before}(Ev_j) \textbf{ post } \mathsf{after}(Ev_j),$$

where before and after are the functions defined above.

We have established the correspondence between the definitions of a module interface and a contract. We believe that the modularisation extension of Event-B has provided us with a suitable basis for deriving component contracts. Indeed, it allows us to easily derive the definitions of all parts of a contract from the corresponding definition of a module interface. Since the component contracts are derived from a specification of the overall system, our approach supports "interoperability by construction". It ensures that the components composed to achieve the specified system functionality are interoperable provided they comply with the derived contracts. In the next section, we illustrate the proposed approach by an example – an auction system.

6.4   Example: An Auction System

In this section, we illustrate the proposed approach by an example – a simple electronic auction. We start from a centralised abstract system model of an auction, then introduce an abstract model of the communication mechanism for sending and receiving different types of requests, and, finally, decompose the model into specifications of the interfaces (contracts) of the involved components.

6.4.1   Initial Model

The initial auction specification describes the activities of components of three different types: a seller, a buyer, and a manager. There could be any number of sellers and buyers participating in the system. However, there should be only one manager, which keeps the information about the current auction state and controls validity of the auction operations.

The auction specification describes the following allowed scenario for a seller, a buyer, and a manager. The scenario is initiated by a seller, who announces an item to be sold at the auction. Once the item announcement is received by the manager, the bidding process for this particular item starts. Any active buyer can make its bid. However, only higher bids (for a particular item) are accepted by the manager. After a predefined number of bids, the bidding process is stopped and the winner (i.e., a bidder with the highest bid) is decided by the manager. Once the payment from the winner is received, the item is declared officially sold, and the corresponding seller and buyer are notified about the transaction.

We can depict the scenario as the following chain of operations (events) involving a seller, a buyer, and the manager:

$Put\_new\_item(Seller) \longrightarrow Get\_new\_item(Manager) \longrightarrow Make\_bid(Buyer)$
$\longrightarrow Take\_bid(Manager) \cdots \longrightarrow Make\_bid(Manager) \longrightarrow Take\_bid(Manager)$
$\longrightarrow Declare\_winner(Manager) \longrightarrow Make\_payment(Buyer)$
$\longrightarrow Receive\_payment(Manager) \longrightarrow Item\_sold(Manager)$
$\longrightarrow Selling\_confirmed(Seller) \longrightarrow Buying\_confirmed(Buyer)$

We specify all these steps as the corresponding events in our Event-B model of the auction system. Each event also has an annotation indicating to which component it belongs. Moreover, we distribute the program variables among the involved components as well. Most variables are associated with the manager, except for `buyer_log` and `seller_log`, which belong to a (collective) seller and a (collective) buyer respectively.

SYSTEM *Auction*
SEES *Auction_Context*
VARIABLES
    bids, *bids_left*, *winner*, *paid_items*, *item_seller*, *buyer_log*, *seller_log*, *done*
INVARIANT ... $buyer\_log \subseteq paid\_items \land seller\_log \subseteq paid\_items \land$
    $\forall ii.\ ii \in ITEM \land done(ii) = TRUE \Rightarrow (ii \in buyer\_log \Leftrightarrow ii \in seller\_log)$
INITIALISATION ...
EVENTS

Put_new_item = .../* seller */	$Make\_payment$ = /* buyer */
Get_new_item = .../*manager */	$Receive\_payment$ = .../* manager */
$Make\_bid$ = ../* buyer */	$Item\_sold$ = $\cdots$ /* manager */
$Take\_bid$ = /* manager */	$Selling\_confirmed$ = .../* seller */
$Declare\_winner$ = /* manager */	$Buying\_confirmed$ = /* buyer */

END

For brevity, in the invariant part, we show only a couple of the most important correctness properties of the auction system: (i) all bought and sold items should be paid for, and (ii) once the auction is done for a particular item, it should be recorded in both buyer's and seller's logs.

Below, we show two auction events in detail.

Get_new_item =
any ss, ii **where**
    $ss \in SELLER$
    $ii \in ITEM$
    $done(ii) = FALSE$
then
    $bids(ii){:} = (NO\_BUYER \mapsto 0)$
    $bids\_left(ii){:} = MAX\_BIDS$
    $item\_seller(ii){:} = ss$
end

$Receive\_payment$ =
any bb, ii **where**
    $bb \in BUYER$
    $ii \in ITEM$
    $bb = winner(ii)$
    $ii \notin paid\_items$
then
    $paid\_items{:} = paid\_items \cup \{ii\}$
end

The event *Get_new_item* specifies the manager's reaction after getting a new item to sell. As a result, the bidding process is initiated. The event *Receive_payment* models recording of the received buyer payment by the manager.

In the initial model, the involved components can directly read each other's state in the event guards. In the first refinement step, we decouple these components by explicitly introducing communication between them. The components communicate by sending and receiving requests. The received requests are stored in their input buffers, while the requests to be sent are put in their output buffers. The system model becomes more decentralised.

However, the buyer and seller buffers are still collectively modelled by the corresponding arrays variables.

After an introduction of communication, the manager events *Get_new_item* and *Receive_payment* receive the following form:

$Get\_new\_item =$
any rr, ss, ii **where**
 $\quad rr \in Selling\_Req$
 $\quad rr \in manager\_input$
 $\quad ss = Seller(rr)$
 $\quad ii = Item(rr)$
 $\quad done(ii) = FALSE$
then
 $\quad ...$
 $\quad manager\_input: =$
 $\qquad manager\_input\backslash\{rr\}$
end

$Receive\_payment =$
any bb, ii **where**
 $\quad ...$
then
 $\quad ...$
 $\quad manager\_output: = manager\_output$
 $\qquad \cup \{Pay\_Confirmation(item\_seller(ii) \mapsto ii)\}$
end

The first event now models the manager reaction on the received selling request in its input buffer, while the second event creates a payment confirmation request to be sent to the item seller and places it into the output buffer.

In addition, the refined model contains the events, sole purpose of which is to transport requests from the output buffer of one component to the input buffer of the recipient component. Essentially, these events specify the behaviour of middleware that is responsible for implementing component communication.

We specify the middleware events that represent communication between the sellers and the managers as follows:

$Seller\_to\_Manager =$
any ss, rr **where**
 $\quad ss \in SELLER$
 $\quad rr \in seller\_output(ss)$
 $\quad rr \notin manager\_input$
then
 $\quad seller\_output(ss): = seller\_output(ss)\backslash\{rr\}$
 $\quad manager\_input: = manager\_input \cup \{rr\}$
end

$Manager\_to\_Seller =$
any ss, rr **where**
 $\quad ss \in SELLER$
 $\quad rr \notin seller\_input(ss)$
 $\quad rr \in manager\_output$
then
 $\quad seller\_output(ss): = seller\_output(ss) \cup \{rr\}$
 $\quad manager\_input: = manager\_input\backslash\{rr\}$
end

The respective events for the buyer-manager communication are added as well.

In the second refinement step, we decompose the system specification by explicitly introducing the manager component as well as the buyer and seller components for each element of the given sets *BUYER* and *SELLER*. We will rely on the modularisation extension of Event-B to decompose the system model. The system model is refined into that of the middleware, which calls, when needed, operations from the corresponding introduced components.

To perform such a decomposition refinement step, we need to

1. Define separate module interfaces for the manager, a buyer, and a seller;

2. Distribute the system state (not belonging to the middleware) among the introduced components;

3. In each interface, define the callable operations for accessing the component input and output buffers;

4. Distribute the events describing the autonomous component behaviour among the components, defining them as the corresponding module processes;

5. Define the gluing invariants, relating the array variables modelling the collective knowledge (e.g., buyer and seller logs) with the respective variables belonging to single module instances. Essentially, this allows us to propagate system-level properties into the definition of the derived module interfaces.

In Figure 6.5, we present the defined interface for the seller component. In a similar way, the interfaces for the manager, buyer, and seller components are defined. Having the seller interface defined, it is rather straightforward to obtain the corresponding class contract for all sellers (see Figure 6.6).

INTERFACE *Seller*(*id*)

  VARIABLES *input, output, log*

  INVARIANT $(\forall rr \in output.\ Item(rr) \notin log) \ \wedge \ (\forall rr \in input.\ Id(rr) = id) \ \wedge$
                      $(\forall rr \in input \cap Selling\_Req.\ Item(rr) \notin log) \ \wedge \ \cdots$

  INITIALISATION $input = \varnothing \ \wedge \ output = \varnothing \ \wedge \ log = \varnothing$

  PROCESS

    Put_new_item   =   **any** *ii* **where** $ii \in ITEM \ \wedge \ ii \notin log$
                                  then $output := output \cup \{Sell\_Request(id \mapsto ii)\}$ **end**

    Selling_confirmed   =   **any** *rr* **where** $rr \in input \cap Pay\_Confirmation$
                                  then $log := log \cup \{ii\} \ \| \ input := input \backslash \{rr\}$ **end**

  OPERATIONS

    Add_request   =   **any** *rr* **pre**
                            $rr \in Pay\_Confirmation \wedge rr \notin input \wedge Id(rr) = id$
                      post
                          $input' = input \cup \{rr\}$
                      end

    Take_request   =   **any** *rr* **pre**
                            $rr \in Selling\_Req \cup Selling\_Ackn \wedge rr \in output$
                      post
                          $output' = output \backslash \{rr\}$
                      end

END

Figure 6.5: Interface component

The core refined model now consists only of the state and events of the system middleware. The additional clause **USES** creates one instance of the manager component as well as a number of instances of the seller and buyer components – one per each element of the given sets *SELLER* and *BUYER*.

SYSTEM *Auction_2nd_refinement*
USES *Manager* **as** *manager, Seller*[*SELLER*], *Buyer*[*BUYER*]
VARIABLES ...
INVARIANT ...
INITIALISATION ...
EVENTS
Seller_to_Manager = .../* middleware */
Manager_to_Seller = .../*middleware */
Buyer_to_Manager = ../* middleware */
Manager_to_Buyer = /* middleware */
END

COMPONENT CLASS $Seller(id)$
　　VARIABLES $input, output, log$
　　INVARIANT $(\forall\, rr \in output.\; Item(rr) \notin log)\; \wedge\; (\forall\, rr \in input.\; Id(rr) = id)\; \wedge$
　　　　　　　　$(\forall\, rr \in input \cap Selling\_Req.\; Item(rr) \notin log)\; \wedge\; \cdots$
　　INITIALISATION $input = \varnothing\; \wedge\; output = \varnothing\; \wedge\; log = \varnothing$
　　ACTIONS
　　　　Put_new_item　　　　$=$　**params** ii **pre**
　　　　　　　　　　　　　　　　　　$ii \in ITEM\; \wedge\; ii \notin log$
　　　　　　　　　　　　　　post
　　　　　　　　　　　　　　　　$output' = output \cup \{Sell\_Request(id \mapsto ii)\}$
　　　　　　　　　　　　　　end
　　　　$Selling\_confirmed$　　$=$　**params** rr **pre**
　　　　　　　　　　　　　　　　　$rr \in input \cap Pay\_Confirmation\; \wedge$
　　　　　　　　　　　　　　　　　$Id(rr) = id\; \wedge\; Item(rr) \notin log$
　　　　　　　　　　　　　　post
　　　　　　　　　　　　　　　　$log' = log \cup \{ii\}\; \wedge\; input' = input \backslash \{rr\}$
　　　　　　　　　　　　　　end
　　　　$Add\_request$　　　　$=$　**params** rr **pre**
　　　　　　　　　　　　　　　　　$rr \in Pay\_Confirmation \wedge rr \notin input\; \wedge\; Id(rr) = id$
　　　　　　　　　　　　　　post
　　　　　　　　　　　　　　　　$input' = input \cup \{rr\}$
　　　　　　　　　　　　　　end
　　　　$Take\_request$　　　　$=$　**params** rr **pre**
　　　　　　　　　　　　　　　　　$rr \in Selling\_Req \cup Selling\_Ackn\; \wedge\; rr \in output$
　　　　　　　　　　　　　　post
　　　　　　　　　　　　　　　　$output' = output \backslash \{rr\}$
　　　　　　　　　　　　　　end
END

Figure 6.6: The seller class contract

The invariant clause must also include the gluing invariants between the abstract model variables and the variables of the decentralised system, e.g.,

$$\forall\, ss.\; ss \in SELLER\; \Rightarrow\; input(ss)\; =\; seller\_input(ss)$$
$$\forall\, ss.\; ss \in SELLER\; \Rightarrow\; log(ss)\; =\; seller\_log(ss)$$

In other words, these invariants relate the array variables of a centralised system before the refinement step, $seller\_input$ and $seller\_log$, with the corresponding variables, $input$ and log, of the seller components.

　　Below we show in detail a couple of middleware events. Note that now they are defined only in terms of the operation calls to the interface operations of the involved components.

$Seller\_to\_Manager =$	$Manager\_to\_Seller =$
any ss, rr **where**	**any** ss, rr **where**
$ss \in SELLER$	$ss \in SELLER$
$rr \in output(ss)$	$rr \notin input(ss)$
$rr \notin man.input$	$rr \in man.output$
then	**then**
$Take\_request(ss)(rr)$	$manager\_Take\_request(rr)$
$manager\_Add\_request(rr)$	$Add\_request(ss)(rr)$
end	**end**

The derived interfaces of the buyer and the seller modules represent their contracts. The auction system can be developed by composing the seller, buyer, and middleware components. To ensure interoperability of the implemented components, the designers need to verify that the components comply to the contracts defined above.

While developing the auction system, we have employed the strategy that is usually used for the development of distributed systems by refinement [184, 181, 183]. Namely, we start from an abstract centralised system model. In such a model, components can directly access each other's state. This allows us to simplify definition of the system-level invariant properties, including the ones that express interoperability conditions. In a number of refinement steps, we introduce a representation of the detailed functional requirements as well as communication mechanisms between the components. Finally, we perform model decomposition and arrive at a decentralised system model. Such a model typically consists of the components that communicate with each other via the communication support provided by the introduced middleware component. The decomposition refinement step also allows us to derive the corresponding contracts of the system components.

6.5   Conclusions

In this paper, we have presented a rigorous approach to ensuring component interoperability. The main idea of the approach was to derive contracts of the constituting components from the overall system specification in Event-B. The modularisation extension of Event-B allowed us to decompose the system specification into the independent components – modules. A module interface defines a family of components that might vary in implementation and the internal behaviour but nevertheless comply to the contract defined by the module interface. As a next step, we have demonstrated how to transform such a module interface into the corresponding contract of a component.

Our ideas were illustrated by an example – an auction system. An application of the proposed approach allowed us to formally define interoperability of the buyer, seller, and middleware components, i.e., ensure their correct interactions during the auction activities.

Our approach has been inspired by the seminal works of Kaisa Sere and her group on refinement, modularisation, and decomposition in the action systems formalism. The fundamental theoretical aspects of the refinement approach to the development of distributed systems were presented by Back and Sere in [32]. The modularisation idea for the action systems framework was proposed in [33]. The systems approach to development by refinement that enables propagation of the system-level properties into the models of components have been proposed in [73]. Since Event-B has adopted many aspects of the theoretical foundations of the action systems framework, in our approach we followed the decomposition-based development approach defined by the action systems framework. We extended the original ideas by the notions of module interface and component contracts.

The concept of contracts has been exploited in various domains and development approaches. Majority of the approaches aim at facilitating compositional component-based development. A foundational work on rigorous aspects of such development was done within the EU FP6 SPEEDS project [313]. The project studied theoretical aspects of formal component-based design. It proposed a component-based framework that defines component properties as extended transition systems and different compositional approaches associated with it.

A contract-based top-down design methodology was proposed by Quinton and Graf [272]. This work explored different forms of conformance of a component to a contract that is defined by corresponding notions of refinement. The authors also define refinement relation between contracts. There is also a vast body of research on contracts that is based on behavioural typing, where a contract is an abstraction of the component behaviour as a transition system, see, e.g., [68, 78]. As a result of this research, many contract languages were proposed, which define various ways contracts can be composed and compared, while the behavioural subtyping relation, which can be seen as refinement, is used for compliance of the component behaviour to a contract.

In our work, we pursued another goal – we aimed at deriving contracts from a refined and decomposed system specification in Event-B. The refinement process allows us to preserve the global system properties, while gradually decoupling the components and formally defining their interfaces. As a result, the derived contracts ensure that any component conforming to its contract is interoperable with the other components of the system.

Defining contracts for complex component-based systems is still a challenging problem. We believe that the proposed approach helps to alleviate it. Indeed, it allows us to derive component contracts from a specification that rigorously defines the overall interoperability conditions for components and their environments. As a future work, we are planning to experiment with the notion of probabilistic contracts, in particular, in the domain of fault tolerant systems.

Part III

Proof

Chapter 7

Meeting Deadlines, Elastically

Einar Broch Johnsen
University of Oslo

Ka I Pun
University of Oslo

Martin Steffen
University of Oslo

S. Lizeth Tapia Tarifa
University of Oslo

Ingrid Chieh Yu
University of Oslo

Abstract.
Cloud computing offers a pay-on-demand scalable infrastructure for data processing. Resource-aware services can exploit this infrastructure to elastically adapt to client traffic according to internal resource policies which balance provided QoS with the accrued costs of deployment. This paper presents initial work on worst-case response time analysis for services which distribute tasks to virtual machine instances with different processing speeds. We extend JML-like interfaces with response time annotations and develop a Hoare-style proof system to reason about response time guarantees for services expressed in a simple object-oriented language in which dynamically created objects differ in processing capacity. The simplified setting considered in this paper does not consider loops, concurrency, or reflection; we briefly discuss how these restrictions could be lifted.

7.1   Introduction

A cloud consists of virtual computers that are accessed remotely for data storage and processing. The cloud is emerging as an economically interesting model for enterprises of all sizes, due to an undeniable added value and compelling business drivers [74]. One such driver is *elasticity*: businesses pay for computing resources when needed, instead of provisioning in advance with huge upfront investments. New resources such as processing power or memory can be added to a virtual computer on the fly, or an additional virtual computer can be provided to the client application. Going beyond shared storage, the main potential in cloud computing lies in its scalable virtualized framework for data processing. If a service uses cloud-based processing, its capacity can be automatically adjusted when new users arrive or depending on the input size and required response time of different jobs. Another driver is *agility*: new services can be deployed on the market quickly and flexibly at limited cost. This allows a service to handle its users in a flexible manner without requiring initial investments in hardware before the service can be launched.

Today, software is often designed while completely ignoring deployment or based on very specific assumptions, e.g., the size of data structures, the amount of random access memory, and the number of processors. For the software developer, cloud computing brings new challenges and opportunities [151]:

- **Empowering the Designer.** The elasticity of software executed in the cloud gives designers far reaching control over the execution environment's resource parameters, e.g., the number and kind of processors, the amount of memory and storage capacity, and the bandwidth. In principle, these parameters can even be adjusted at runtime. The owner of a cloud service cannot only deploy and run software, but also control trade-offs between the incurred cost and the delivered quality-of-service.

- **Deployment Aspects at Design Time.** The impact of cloud computing on software design goes beyond scalability. Deployment decisions are traditionally made at the end of a software development process: the developers first design the functionality of a service, then the required resources are determined, and finally a service level agreement regulates the provisioning of these resources. In cloud computing, this can have severe consequences: a program which does not scale usually requires extensive design changes when scalability was not considered a priori.

To realize cloud computing's potential, software must be *designed for scalability*. This leads to a new *software engineering challenge*: how can the validation of deployment decisions be pushed up to the modeling phase of the software development chain without convoluting the design with deployment details?

The EU project Envisage addresses this challenge by extending a design by contract approach to service-level agreements for resource-aware virtualized services. The functionality is represented in a *client layer*. A *provisioning layer* makes resources available to the client layer and determines how much memory, processing power, and bandwidth can be used. A *service level agreement* (SLA) is a legal document that clarifies what resources the provisioning layer should make available to the client service, what they will cost, and the penalties for breach of agreement. A typical SLA covers two different aspects: (i) the mutual legal obligations and consequences in case of a breach of contract, which we call the *legal contract*; (ii) the technical parameters and cost figures of the offered services, which we call the *service contract*.

This paper discusses some initial ideas about applying program verification techniques to models of virtualized services. We consider response time aspects of service contracts and extend JML-like interfaces with response time annotations. This is formalized using μABS; μABS is a restricted version of ABS [187], an executable object-oriented modeling language developed in the Envisage project to specify resource-aware virtualized services [13, 195, 194]. Whereas ABS is based on concurrent objects and asynchronous method calls, the work discussed in this paper is restricted to sequential computation and synchronous method calls. In future work, we plan to alleviate these restrictions.

Paper organization. Section 7.2 introduces service interfaces with response-time annotations; Section 7.3 introduces the syntax of μABS, the modeling language considered in this paper; Section 7.4 demonstrates the approach on an example; Section 7.5 develops a Hoare-style proof system for μABS; Section 7.6 discusses related work; and Section 7.7 concludes the paper by a discussion of the limitations and possible extensions of the current work.

7.2 Service Contracts as Interfaces

Service level agreements express non-functional properties of services (service contracts), and their associated penalties (legal contracts). Examples are *high water marks* (e.g., number of users), *system availability*, and *service response time*. Our focus is on service contract aspects of client-level SLAs, and on how these can be integrated in models of virtualized services. Such an integration would enable a formal understanding of service contracts and of their relationship to the performance metrics and configuration parameters of the deployed services. Today, client-level SLAs do not allow the potential resource usage of a service to be determined or adapted when unforeseen changes to resources occur. This is because user-level SLAs are not explicitly related to actual performance metrics and configuration parameters of the services; for example, the user-level SLAs may be concerned with end-user response times but not about the number of virtual machine instances on which the service is deployed. The integration of service contracts and configuration parameters in service models enables the design of resource-aware services which embody application-specific resource management strategies [151].

The term *design by contract* was coined by Bertrand Meyer referring to the contractual obligations that arise when objects invoke methods [244]: only if a caller can ensure that certain behavioral conditions hold before the method is activated (the precondition), it is ensured that the method results in a specified state when it completes (the postcondition). Design by contract enables software to be organized as encapsulated services with interfaces specifying the contract between the service and its clients. Clients can "program to interfaces"; they can use a service without knowing its implementation. We aim at a design by contract methodology for SLA-aware virtualized services, which *incorporates SLA requirements in the interfaces at the application-level* to ensure the QoS expectations of clients.

We consider an object-oriented setting with service-level interfaces given in a style akin to JML [70] and Fresco [347]; **requires**- and **ensures**-clauses express each method's functional pre- and postconditions. In addition, a *response time guarantee* is expressed in a **within**-

```
type Photo = Rat; // size of the file

interface PhotoService {
    @requires ∀ p:Photo · p ∈ film && p < 4000;
    @ensures reply == True;
    @within 4*length(film) + 10;
    Bool request(List<Photo> film);
}
```

Figure 7.1: A photo printing shop in μABS.

clause associated with the method. The specification of methods in interfaces is illustrated in Figure 7.1.

7.3   A Kernel Language for Virtualized Computing

ABS supports modeling the deployment of objects on virtual machines with different processing capacities [187, 13, 195]. The μABS language simplifies ABS by letting each object have a dedicated processor with a given processing capacity. Thus, objects are dynamically created instances of classes such that the resource capacity of each object reflects the provisioning contract between that object and its resource provider. In contrast to ABS, communication between named objects is synchronous, which means that a method call blocks the caller until execution has finished. For simplicity in this paper, the objects share a thread of execution where at most one task is *active* and the others are waiting to be executed on the task stack. Although μABS is currently restricted to sequential execution, execution is elastic in the sense that several objects may provide instances of the same service at different speeds (and different associated costs for the service provider) and the choice of service instance for a task can be dynamically decided by the service based on, e.g., the deadline of the task and the accumulated cost of running the service.

μABS is strongly typed: for well-typed programs, invoked methods are understood by the called object. μABS includes the types Capacity, Cost, and Duration which all extend Rat with an element infinite: Capacity captures the processing capacity of virtual machines per time interval, Cost the processing cost of executions, and Duration time intervals.

Figure 7.2 presents the syntax of μABS. A *program P* consists of interface and class definitions, and a main block $\{\overline{T\ x};\ sr\}$. Interfaces IF have a name I and method signatures Sg. Classes CL have a name C, optional formal parameters $\overline{T\ x}$, and methods $\overline{M}$. A method signature Sg has a list of specifications $\overline{Spec}$, a return type T, a method name m, and formal parameters $\overline{x}$ of types $\overline{T}$. In specifications (see Section 7.2), assertions ϕ express properties of local variables in an assertion language extending the expressions e with logical variables and operators in a standard way; a reserved variable *reply* captures the method's return value. A method M has a signature Sg, a list of local variable declarations $\overline{x}$ of types $\overline{T}$, and statements sr. Statements may access local variables and the formal parameters of the class and the method.

Statements are standard, except **job**(e) which captures an execution requiring e processing cycles. A job abstracts from actual computations but may depend on state variables.

Syntactic categories	Definitions
C, I, m in Names	$P \quad ::= \quad \overline{IF} \; \overline{CL} \; \{\overline{T \; x}; sr\}$
s in Statement	$T \quad ::= \quad C \mid I \mid \mathsf{Capacity} \mid \mathsf{Cost} \mid \mathsf{Duration} \mid \mathsf{Bool} \mid \mathsf{Rat}$
x in Variables	$IF \quad ::= \quad \mathbf{interface} \; I \, \{\, \overline{Sg} \,\}$
k in Capacity	$Sg \quad ::= \quad \overline{Spec} \; T \; m \; (\overline{T \; \overline{x}})$
c in Cost	$Spec \quad ::= \quad \mathbf{@requires} \; \phi; \mid \mathbf{@ensures} \; \phi; \mid \mathbf{@within} \; \phi;$
d in Duration	$CL \quad ::= \quad \mathbf{class} \; C \, (\overline{T \; \overline{x}}) \, \{\, \overline{M} \,\}$
b in Bool	$M \quad ::= \quad Sg \, \{\, \overline{T \; x}; sr \,\}$
i in Rat	$sr \quad ::= \quad s; \mathbf{return} \; e \mid \mathbf{return} \; e$
	$s \quad ::= \quad s; s \mid x = rhs \mid \mathbf{job}(e) \mid \mathbf{if} \; e \, \{s\} \; \mathbf{else} \, \{s\}$
	$rhs \quad ::= \quad e \mid \mathbf{new} \; C(\overline{e}) \; \mathbf{with} \; e \mid e.m(\overline{x})$
	$e \quad ::= \quad \mathbf{this} \mid \mathbf{capacity} \mid \mathbf{deadline} \mid x \mid v \mid e \; op \; e$

Figure 7.2: μABS syntax for the object level. Terms $\overline{e}$ and $\overline{x}$ denote possibly empty lists over the corresponding syntactic categories.

Right-hand sides rhs include expressions e, object creation **new** $C(\overline{e})$ **with** e and synchronous method calls $e.m(\overline{x})$. Objects are created with a given *capacity*, which expresses the processing cycles available to the object per time interval when executing its methods. Thus, different instances may have different capacities (and consequently, their usage may have different costs in the metered setting of elastic computing). Method calls in μABS are *blocking*. Expressions e include operations over declared variables x and values v. Among values, b has type Bool, i has type Rat (e.g., 5/7), k has type Capacity, c has type Cost, and d has type Duration. Among binary operators op on expressions, note that division c/k has type Duration. Expressions also include the following reserved read-only variables: **this** refers to the object identifier, **capacity** refers to the processing speed (amount of resources per time interval) of the object, and **deadline** refers to the local deadline of the current method. (We assume that all programs are well-typed and include further functional expressions and data types when needed in the example.)

Time. μABS has a dense time model, captured by the type Duration. The language is not based on a (global) clock, instead each method activation has an associated local counter **deadline**, which decreases when time passes. Time passes when a statement **job**(e) is executed on top of the task stack. The effect of executing this statement on an object with capacity k, is that the local deadline of every task on the stack decreases by c/k, where c is the value resulting from evaluating e. The initial value of the **deadline** counter stems from the service contract; thus, a local counter which becomes negative represents a breach of the local service contract. For brevity, we do not present the formal semantics.

7.4 Example: A Photo Printing Shop

Let us consider a *photo shop* service which *retouches* and *prints* photos. It is cheaper for the photo shop service to retouch and print photos locally, but it can only deal with low resolution photos in time. For larger photos, the photo shop service relies on using a faster

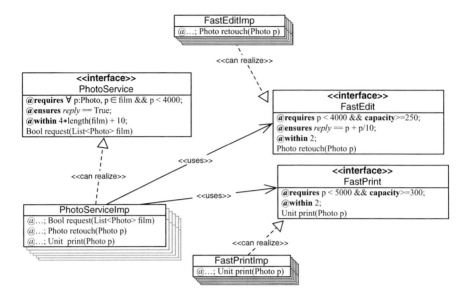

Figure 7.3: A class diagram for a photo printing shop

and more expensive laboratory in order to guarantee that all processing deadlines are met successfully.

In this example, a film is represented as a list of photos and, for simplicity, a photo by the size of the corresponding file. As shown in the class diagram of Figure 7.3, an interface **PhotoService** provides a single method **request** which handles customer requests to the photo shop service. The interface is implemented by a class **PhotoServiceImp**, which has methods **retouch** for retouching and **print** for printing a photo, in addition to the **request** method of the interface. For faster processing, two interfaces **FastEdit** and **FastPrint**, which also provide the methods **retouch** and **print**, may be used by **PhotoServiceImp**. The sequence diagram in Figure 7.4 shows how a photo is first *retouched*, then *printed*. The services of retouching and printing are done locally if possible, otherwise they are forwarded to and executed by objects with higher capacities.

The μABS model of the example (Figure 7.5) follows the design by contract approach and provides a contract for every method declaration in an interface and method definition in a class. These specifications are intended to guarantee that a **request** to a **PhotoService** object will not break the specified contract. If we consider the contract for **request** in more detail, we see that the response time of a **request(film)** call depends on the length of the film and assumes that the size of every photo contained in the **film** is smaller than **4000**. The implementation of the **request** method is as follows: Take the first photo in the **film** (by applying the function **head(film)**) and check if this photo is low resolution compared to the capacity of the **PhotoService** object, represented by a size smaller than **500** and a capacity of at least **100**, respectively. In this case, the retouch can be done locally, otherwise retouch is done by an auxiliary **FastEdit** object. A similar procedure applies to printing the retouched photos. Thus, photos of small sizes are retouched and printed locally, while photos with bigger sizes are sent to be retouched and printed externally. The ability to send tasks to the laboratory which meets the deadline at the lowest cost, expresses elasticity in the setting of

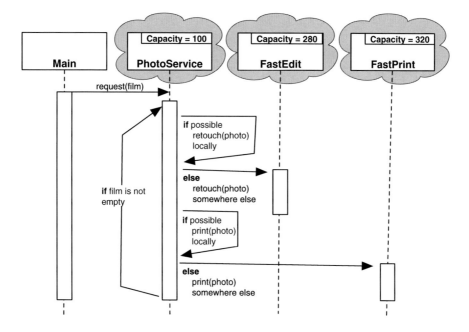

Figure 7.4: A sequence diagram for a photo printing shop

this example. The implementations of the different methods are abstractly captured using **job** statements. The expressions *e* inside the **job** statements, could be further refined or calibrated using SACO [11].

7.5 Proof System

Virtual machines are subject to failures. We interpret non-termination of a **job** statement as an underlying failure and restrict our analysis to partial correctness. The proof system for μABS is formalized as Hoare triples [170, 16] $\{\phi\}$ *s* $\{\psi\}$ with a standard partial correctness semantics: if the execution of *s* starts in a state satisfying the precondition ϕ and the execution terminates, the result will be a state satisfying the postcondition ψ. In this paper, we are particularly interested in assertions about the *deadline* variables of method activations.

The reasoning rules for μABS are presented in Figure 7.6. Reasoning about sequential composition, conditional, and assignment statements is standard, and captured by the rules COMP, COND, and ASSIGN, respectively. Time only passes when **job**(e) is executed; **job**(e) has a duration e/cap when executed on an object with capacity cap. The assertion in Rule JOB ensures that this duration is included in the response time after executing **job**(e). The subsumption rule allows to strengthen the precondition and weaken the postcondition. For method definitions, the premise of Rule METHOD assumes that the execution of *sr* starts in a

```
type Photo = Rat; // size of the file

interface FastEdit {
    @requires p < 4000 && capacity>=250; @ensures reply == p + p/10; @within 2;
    Photo retouch(Photo p);}
class FastEditImp {
    @requires p < 4000 && capacity>=200; @ensures reply == p + p/10; @within 2;
    Photo retouch(Photo p) {job(200); return (p + p/10)}}

interface FastPrint {
    @requires p < 5000 && capacity>=300; @within 2;
    Unit print(Photo p);}
class FastPrintImp {
    @requires p < 5000 && capacity>=250; @within 2
    Unit print(Photo p) {job(250);return unit}}

interface PhotoService {
    @requires ∀ p:Photo, p ∈ film && p < 4000;
    @ensures reply == True; @within 4*length(film) + 10;
    Bool request(List<Photo> film);}
class PhotoServiceImp(FastEdit edit,FastPrint print) {
    @requires ∀ p:Photo, p ∈ film && p < 4000;
    @ensures reply == True; @within 4*length(film)+1;
    Bool request(List<Photo> film) {
        Photo p = 0;
        if (film != Nil){
            p = head(film);
            if (p < 500 && capacity>=100){ p = this.retouch(p);}
            else{p = edit.retouch(p);}
            if ( p < 600 && capacity>=100){this.print(p);}
            else{print.print(p);}
            this.request(tail(film));}
        else{ job(1);}
        return (deadline >= 0) }

    @requires p < 500 && capacity>=100; @ensures reply == p + p/20; @within 1;
    Photo retouch(Photo p) {job(100); return (p + p/20)}

    @requires p < 600 && capacity()>=100; @within 1;
    Unit print(Photo p) { job(100); return unit}}
```

Figure 7.5: A photo printing shop in μABS

state where the **requires**-clause ϕ is satisfied and that the expected response time (*deadline*) is larger than expression e, where e is the specified response time guarantee from the **within**-clause. When the execution of sr terminates, the result will satisfy the **ensures**-clause ψ and the expected response time remains non-negative. For method invocations in Rule CALL, the specification of the method is updated by substituting the formal parameters $\overline{fp}$ by the input expressions $\overline{e}$. The logical variables for the return value of the method (*reply*) and of the expected response time are renamed with fresh variables α and β, respectively. To avoid name clashes between scopes, we assume renaming of other variables as necessary. Object creation (in Rule NEW) is handled similarly to assignment. The precondition ensures that the newly created object of a class C with capacity e correctly implements interface T, where T is the type of x. (Note that the class instance may or may not implement an interface, depending on its capacity.) If a method has a return value, expression e in the return statement will be assigned to the logical variable *reply* in Rule RETURN, and can be handled by the standard assignment axiom in Rule ASSIGN.

Although μABS does not currently include loops, the language supports recursion

(METHOD)
$$\frac{\{\phi \wedge deadline \geq e\} \ sr \ \{\psi \wedge deadline \geq 0\}}{\textbf{@requires} \ \phi; \ \textbf{@ensures} \ \psi; \ \textbf{@within} \ e;}$$
$$\mathsf{T''} \ m \ (\overline{\mathsf{T}} \ \overline{x}) \ \{\overline{T' \ x'}; sr\}$$

(RETURN)
$$\frac{\{\phi\} \ s; reply = e \ \{\psi\}}{\{\phi\} \ s; \textbf{return} \ e \ \{\psi\}}$$

(COMP)
$$\frac{\{\phi\} s_1 \{\psi'\}}{\{\psi'\} s_2 \{\psi\}} {\{\phi\} s_1; s_2 \{\psi\}}$$

(COND)
$$\frac{\{\phi \wedge b\} s_1 \{\psi\}}{\{\phi \wedge \neg b\} s_2 \{\psi\}} {\{\phi\} \ \textbf{if} \ b \ \{s_1\} \ \textbf{else} \ \{s_2\} \ \{\psi\}}$$

(SUBSUMPTION)
$$\frac{\{\phi'\} \ s \ \{\psi'\}}{\phi \Rightarrow \phi' \quad \psi' \Rightarrow \psi} {\{\phi\} \ s \ \{\psi\}}$$

(ASSIGN)
$$\{\phi[x \mapsto e]\} \ x = e \ \{\phi\}$$

(JOB)
$$\{\phi[deadline \mapsto deadline - (e/\mathrm{cap})]\} \ \textbf{job}(e) \ \{\phi\}$$

(NEW)
$$\frac{fresh(\alpha)}{\phi' = \phi[x \mapsto \alpha]}$$
$$T = typeOf(x)$$
$$\phi' \Rightarrow implements(C, T, e)$$
$$\overline{\{\phi'\} \ x = \textbf{new} \ C(\overline{e}) \ \textbf{with} \ e \ \{\phi\}}$$

(CALL)
$$fresh(\alpha, \beta) \quad T = typeOf(e)$$
$$\phi' = \phi[x \mapsto \alpha, deadline \mapsto deadline - \beta]$$
$$\phi' \Rightarrow requires(T, m)[\overline{fp} \mapsto \overline{e}]$$
$$\phi_1 = ensures(T, m)[\overline{fp} \mapsto \overline{e}, reply \mapsto \alpha]$$
$$\phi_2 = within(T, m)[\overline{fp} \mapsto \overline{e}, deadline \mapsto \beta]$$
$$\overline{\{\phi' \wedge \phi_1 \wedge \phi_2\} \ x = e.m(\overline{e}) \ \{\phi\}}$$

Figure 7.6: Proof system for μABS

through interfaces with associated contracts. This suggests how loops can be handled in the proof system, in terms of loop invariants which express execution time for the remaining iterations of the loop.

Example 7.1. We show in Equation 7.3 the skeleton of the proof for the method `request` in Figure 7.5 by using the proof system presented in Figure 7.6. Let sr refer to the method body of `request` and let s denote sr without the return statement. In addition, we introduce the following abbreviations for formulas:

$$\begin{aligned}
\psi &= reply == \texttt{True}, \\
\psi_1 &= \psi \wedge deadline \geq 0, \\
\phi &= \forall \, p:\text{Photo}, \ p \in film \ \wedge \ p{<}4000, \text{ and} \\
e &= 4 * \texttt{length}(film) + 10
\end{aligned} \tag{7.1}$$

and assume that

$$\psi_2 = reply == deadline \geq 0 \wedge deadline \geq 0 \tag{7.2}$$

is the postcondition of the assignment $reply = deadline \geq 0$.

By Rule METHOD, the assertions ϕ and $deadline{>}e$ serve as precondition to the whole method body sr, where ϕ and e are given in the **requires**- and **within**-clauses associated with the definition of the method `request` in Figure 7.5. The postcondition of the method body consists of ψ, which is specified in the **ensures**-clause as $reply == $ True, and the expression $deadline \geq 0$. Rule RETURN converts the **return**-statement into a statement where the expression $deadline \geq 0$ is assigned to the logical variable $reply$. Then, by the assignment axiom ASSIGN, and with the postcondition ψ_2 assumed in Equation 7.2, the precondition ψ_3 is the postcondition with the logical variable $reply$ substituted with the expression $deadline \geq 0$, and thus $\psi_3 = $ True $\wedge\ deadline \geq 0$. By using Rule SUBSUMPTION, the postcondition ψ_2 is weakened to the given postcondition ψ_1. By Rule COMP, the assertion ψ_3 is also the postcondition of the statement s.

$$
\frac{
\dfrac{\vdots}{\{\phi \wedge deadline > e\}\ s\ \{\psi_3\}} \quad
\dfrac{\{\psi_3\}\ reply = deadline \geq 0\ \{\psi_2\} \quad \psi_2 \Rightarrow \psi_1}{\{\psi_3\}\ reply = deadline \geq 0\ \{\psi_1\}}
}{
\dfrac{\{\phi \wedge deadline > e\}\ s;reply = deadline \geq 0\ \{\psi_1\}}{\{\phi \wedge deadline > e\}\ s;\mathbf{return}(deadline \geq 0)\ \{\psi_1\}}
}
\tag{7.3}
$$

@**requires** ϕ; @**ensures** ψ; @**within** e;
Bool request(List$\langle$Photo$\rangle$ film)$\{sr\}$

For brevity, the rest of the proof is omitted. The proof can be completed by repeatedly applying the corresponding rules from the presented proof system.

7.6   Related Work

The work presented in this paper is related to the ABS modeling language and its extension to virtualized computing on the cloud, developed in the **Envisage** project [12]. ABS [187] and its extensions with time [56], deployment component and resource-awareness [195] provide a formal basis for modeling virtualized computing. ABS has been used in two larger case studies addressing resource management in the cloud by combining simulation techniques and cost analysis, but not by means of deductive verification techniques; a model of the Montage case study [103] is presented in [194] and compared to results from specialized simulation tools and a large ABS model of the Fredhopper Replication Server has been calibrated using SACO [11] (a cost analysis tool for ABS) and compared to measurements on the deployed system in [99, 13]. The main difference between resource-aware models in ABS [195] and the work presented in this paper, is an elimination of non-determinism; in addition to the restriction to sequential programs discussed above, μABS unifies objects and deployment components, such that objects cannot compete for the resources on a server. As we extend μABS towards ABS in future work, this seems like a reasonable restriction to enable more precise reasoning, but it could mean that object creation could fail if the targeted deployment component lacks resources! Related techniques for modeling deployment in embedded real-time systems may be found in an extension of VDM++ [339]. In this extension, static architectures are explicitly modeled using CPUs and buses. The approach uses fixed resources targeting the embedded domain. Whereas ABS has been designed to support compositional verification based on traces [115], neither ABS nor VDM++ supports deductive verification of non-functional properties today.

Assertional proof systems addressing timed properties, and in particular upper bounds on execution times of systems, have been developed, the earliest example perhaps being [303]. Another early example of reasoning about real-time is Nielson's extension of classical Hoare-style verification to timed properties of a given program's execution [253, 252]. Soundness and (relative) completeness of the proof rules of a simple while-language are shown. Shaw [302] presents Hoare logic rules to reason about the passage of time, in particular to obtain upper and lower bounds on the execution times of sequential, but also of concurrent programs.

Hooman's work on assertional reasoning and Hoare logic [172] for concurrent programs covers different communication and synchronization patterns, including shared-variable concurrency and message passing using asynchronous channels. The logic introduces a dense time domain (i.e., the non-negative reals, including ∞) and conceptually assumes a single, global clock for the purpose of reasoning. The proof system is developed for a small calculus focussing on time and concurrency, where a **delay**-statement can be used to let time pass. This is comparable to the **job**-expression in our paper, but directly associates a duration with the job. In contrast, we associate a cost with the job, and the duration depends on the execution capacity of the deployed object where the statement is executed. Timed reasoning using Dijkstra's weakest-precondition formulation of Hoare logic can be found in [149]. Lamport's *temporal logic of actions* TLA [214, 1] has likewise been extended with the ability to reason about time [213]. Similar to the presentation here, the logical systems are generally given by a set of derivation rules in a pre-/post-condition style. Similar to our work, these approaches are compositional in that timing information for composed programs, including procedure calls, is derived from that of more basic statements. While being structural in allowing syntax-directed reasoning, these formalisms do not explore timed interfaces as part of the programming calculus as we have done here. Thus, these approaches do not support the notion of design-by-contract compositionality for non-functional properties that has been suggested in this paper.

Complementing the theoretical development of proof systems for real-time properties, corresponding reasoning support has been implemented within theorem provers and proof-assistants, for instance for PVS in [134] (using the duration calculus), and HOL [142]. An interesting approach to *compositional* reasoning about timed system is developed in [135]. As its logical foundation, the methodology uses TRIO [139], a general-purpose specification language based on first-order linear temporal logic. In addition, TRIO supports object-oriented structuring mechanisms such as classes and interfaces, inheritance, and encapsulation. To reason about open systems, i.e., to support modular or compositional reasoning, the methodology is based on a rely/guarantee formalization and corresponding proof rules are implemented within PVS. Similarly, a rely/guarantee approach for compositional verification in linear-time temporal logics is developed in [334, 196]. A further compositional approach for the verification of real-time systems is reported in [173], but without making use of a rely/guarantee framework.

Refinement-based frameworks constitue another successful design methodology for complex system, orthogonal to compositional approaches. Aiming at a correct-by-construction methodology, their formal underpinning often rests on various refinement calculi [23, 250, 249]. Refinement-based frameworks have also been developed for timed systems. In particular, Kaisa Sere and her co-authors [54] extended the well-known formal modeling, verification, and refinement framework Event-B [6] with a notion of time, resulting in a formal transformational design approach where the proof-obligations resulting from the timing part in the refinement steps are captured by timed automata and verified by the UPPAAL tool [52].

The Java modeling language JML [70] is an interface specification language for Java which was used as the basis for the interface specification of service contracts in our paper. Extensions of JML have been proposed to capture timed properties and to support component-based reasoning about temporal properties [207, 208]. These extensions have been used to modularly verify so-called performance correctness [307, 306]). For this purpose, JML's interface specification language is extended with a special **duration**-clause, to express timing constraints. The JML-based treatment of time is abstract insofar as it formalizes the temporal behavior of programs in terms of abstract "JVM cycles". Targeting specifically safety critical systems programmed in SCJ (Safety-critical Java), SafeJML [150] re-interprets the **duration**-clause to mean the worst-case execution time of methods concretely in terms of absolute time units. For a specific hardware implementation for the JVM for real-time applications, [296] presents a different WCET analysis [271] for Java. The approach does not use full-fledged logical reasoning or theorem proving, but is a static analysis based on integer linear programming and works at the byte-code level. We are not aware of work relating real-time proof systems to virtualized software, as addressed in this paper.

7.7　Discussion

Cloud computing provides an elastic but metered execution environment for virtualized services. Services pay for the resources they lease on the cloud, and new resources can be dynamically added as required to offer the service to a varying number of end users at an appropriate service quality. In order to make use of the elasticity of the cloud, the services need to be *scalable*. A service which does not scale well may require a complete redesign of its business code. A *virtualized* service is able to adapt to the elasticity provided by the cloud. We believe that the deployment strategy of virtualized services and the assessment of their scalability should form an integral part of the service design phase, and not be assessed a posteriori after the development of the business code as it is done today. The design of virtualized services provides new challenges for software engineering and formal methods.

Virtualization empowers the designer by providing far-reaching control over the resource parameters of the execution environment. By incorporating a resource management strategy which fully exploits the elasticity of the cloud into the service, *resource-aware* virtualized services are able to balance the service contracts that they offer to their end users, to the metered cost of deploying the services. For resource-aware virtualized services, the integration of resource management policies in the design of the service at an early development stage seems even more important.

This paper pursues a line of research addressing the formal verification of service contracts for virtualized services. We have considered a very simple setting with an interface language which specifies services, including their service contracts in the form of response time guarantees, and a simple object-oriented language for realizing these services. To support non-functional behavior, the language is based on a real-time semantics and associates deadlines with method calls. Virtualization is captured by the fact that objects are dynamically created with associated execution capacities. Thus, the time required to execute a method activation depends not only on the actual parameters to the method call, but also on the execution capacity of the called object. This execution capacity reflects the processing power of virtual machine instances, which are created from within the service itself.

The objective of the proof system proposed in this paper is to apply deductive verification techniques to ensure that *all local deadlines are met during the execution of a virtualized service*. This proof system builds on previous work for real-time systems, and recasts the deductive verification of timing properties to a setting of virtualized programs. The extension of service interfaces with response-time guarantees, as proposed in this paper, allows a compositional design-by-contract approach to service contracts for virtualized systems.

Whereas our work goes in the direction of worst-case cost analysis, it would also be interesting to consider soft real-time requirements as typically encountered in service-oriented computing. This could in principle be done by incorporating probabilistic information about response times and execution cost into the models. As such, the approach taken in this paper could be complemented by simulations and statistical techniques (e.g., Monte Carlo simulations have been applied in the context of ABS [193]). However, as cloud computing is increasingly used for critical services in domains such as health and banking, we believe that there is a need for analysis techniques for hard deadlines also in the context of virtualized services.

Several challenges to the proposed approach are left for future work, in particular the extension to concurrency and asynchronous method calls. Currently we plan to address this challenge in terms of proof rules which make use of explicit assumptions about the size of the queues for concurrent ABS objects. The size of the queues can be statically detected; e.g., the size may be approximated by techniques such as may-happen-in-parallel analysis [133]. Furthermore, in the concurrent setting, it is interesting to work with reflection; e.g., an object may query the current load of its virtual machine and use this as a basis for resource management. In this paper, we have considered the explicit allocation of resources for each object. In future work, it would be advantageous to lift the bounded queue length from individual objects to groups of objects, and work with the deployment of such groups. Another challenge is the incorporation of code which reflects the actual computations (replacing the job-statements of this paper). In this case, the abstraction to job-statements could be done by incorporating a worst-case cost analysis [11] into the proof system. Another interesting challenge, which remains to be investigated, is how to incorporate the global requirements which we find in many service-level agreements into a compositional proof system, such as the maximum number of end users.

Chapter 8

Event-B and Linear Temporal Logic

Steve Schneider, Helen Treharne

University of Surrey

David M. Williams

VU University Amsterdam

Abstract. In this chapter we consider when temporal logic properties can be carried through Event-B refinement chains. We also identify conditions on temporal logic properties that make them suitable for use in a refinement chain, since not all properties are preserved by Event-B refinement. In particular we identify the particular notion of β-dependence for an LTL property: that only the events in the set β need to be checked to determine whether an execution meets the property. Such properties are not affected by the introduction of new events in a refinement step, which means they will carry through a refinement chain and will hold for any resulting refinement which is deadlock-free and does not contain anticipated events. The chapter presents a Lift example refinement chain to illustrate the results and how they are applied.

8.1 Introduction

Event-B [6] is a step-wise development method with excellent tools: the Rodin platform [2] providing proof support and ProB [226] providing model checking. As Hoang and Abrial [168] clearly state, the focus of verification within Event-B has been on the safety properties of a system to ensure that "something (bad) never happens". Typically, this has been done via the discharging of proof obligations. Nonetheless, the use of linear temporal logic (LTL) to specify temporal liveness properties has also been prevalant, for example in its application within the ProB tool [227]. The challenge is to identify more natural ways of integrating Event-B and LTL, so that LTL properties can be preserved by Event-B refinement, which is not currently the case in general.

Event-B describes systems in terms of *machines* with state, and *events* which are used to update the state. Events also have *guards*, which are conditions for the event to be enabled. One (abstract) machine may be refined by another (concrete) machine, using a *refinement step*. A *linking invariant* captures how the abstract and concrete states are related, and each abstract event must be refined by one or more concrete events whose state transformations match the abstract one in the sense of preserving the linking invariant. Refinement is transitive, so a sequence of refinement steps, known as a *refinement chain*, will result in a concrete machine which is a refinement of the original abstract one.

A particular feature provided by Event-B is the introduction of *new* events in a refinement step—events which do not refine any abstract event. This allows for refinements to add finer levels of granularity and concretisation as the design develops; there are many examples in [6]. These new events are invisible at the abstract level (they correspond to the abstract state not changing), and we generally need to verify that they cannot occur forever. Event-B makes use of *labels* to keep track of the status of events as a refinement chain progresses. Event-B labels are *anticipated, convergent* and *ordinary*. The labelling of events in Event-B form part of the core of a system description but their inclusion is primarily to support the proof of safety properties and ensuring that events cannot occur forever: convergent events must decrease a variant and anticipated events cannot increase it. The result described in this chapter is applicable when all newly introduced events are convergent or anticipated, and all anticipated events become convergent at some stage in the refinement chain. As an initial example, consider a machine $Lift_0$ (Figure 8.1) with two events *top* and *ground*, representing movement to the top and to the ground floor. This can be refined by a machine $Lift_1$ (Figure 8.2) introducing a new anticipated event *mid* corresponding to movement to one of the intermediate floors.

Linear temporal logic provides a specification language for capturing properties of executions of systems and is appropriate for reasoning about liveness and fairness. For example, we might verify for $Lift_0$ that whenever *top* occurs, then eventually *ground* will occur. However, this is not guaranteed for its refinement $Lift_1$: it may be that the intermediate floors are visited repeatedly forever following the *top* event, thus never reaching the next *ground* event. Alternatively it may be that some refinement system deadlocks, again preventing *ground* from occurring. Hence we see that LTL properties are not automatically preserved by Event-B refinement.

In this chapter we consider when temporal logic properties can be carried through Event-B refinement chains. We also identify conditions on temporal logic properties that make them suitable for use in a refinement chain, since some properties are not preserved by Event-B refinement (for example, the property "*mid* never occurs" holds for $Lift_0$ but not for its refinement $Lift_1$). The approach is underpinned by our process algebra understanding of the Event-B semantics, in particular the traces, divergences and infinite traces semantics used for CSP (Communicating Sequential Processes) and applied to Event-B in [294]. We do not present the proof of the main result here; it is provided in [295]. In this chapter we give the intuition of why it holds and in order to aid the reader to develop an understanding of why the conditions of the result are required.

> **machine** $Lift_0$
> **variables** flr_0
> **invariant** $flr_0 \in \{0, 102\}$
> **events**
>   init $\hat{=} \ flr_0 := 0$
>   top $\hat{=}$ **status**: ordinary
>   **when** $flr_0 < 102$ **then** $flr_0 := 102$ **end**
>   ground $\hat{=}$ **status**: ordinary
>   **when** $true$ **then** $flr_0 := 0$ **end**
> **end**

Figure 8.1: $Lift_0$

8.2 Event-B

8.2.1 Machines

An Event-B development is defined using *machines*. A machine M contains a vector of variables and a set of events. The *alphabet* of M, αM, is the set of events defined in M. Each event evt_i has the general form $evt_i \hat{=}$ **any** x **where** $G_i(x, v)$ **then** $v :| \ BA_i(v, x, v')$ **end**, where x represents the parameters of the event, and the guard $G_i(x, v)$ is the condition for the event to be enabled. The body is given by $v :| \ BA_i(v, x, v')$ whose execution assigns to v any value v' which makes the *before-after* predicate $BA_i(v, x, v')$ true. This simplifies to $evt_i \hat{=}$ **when** $G_i(v)$ **then** $v :| \ BA_i(v, v')$ **end** when there are no parameters, since the guard and the *before-after* predicate do not refer to the parameters x. Such state transformations can also be expressed in terms of abstract assignments using the Generalised Substitution Language.

Variables of a machine are initialised in an initialisation event *init* and are constrained by an invariant $I(v)$. The Event-B approach to semantics is to associate proof obligations with machines. The key proof obligation, INV, is that all events must preserve the invariant. There is also an optional proof obligation on a machine with respect to deadlock freedom which means that a guard of at least one event in M is always enabled. When this obligation holds M is *deadlock free*.

Figure 8.1 gives an example machine $Lift_0$ which describes the behaviour of a lift that can move to the top floor or the ground floor. It contains one variable, flr_0, which can take the values 0 or 102. The initialisation *init* starts the machine at floor 0. The event *top* has a guard $flr_0 < 102$ such that it is only enabled, and hence can only occur, when the lift is not at the top floor. This event will be blocked when the lift is at the top floor. The body of the event, $floor_0 := 102$, sets the variable to the value 102. On the other hand, the event *ground* has a guard of *true*, so it is always enabled. Its effect is to set the variable to the value 0. This event is not blocked when the lift is already at the ground.

```
machine Lift₁
refines Lift₀
variables flr₁
invariant flr₁ = flr₀ ∨ flr₁ ∈ {1..101}
variant 0
events
   init ≙ flr₁: = 0
   top ≙ status: ordinary
       when flr₁ = 101  then flr₁: = 102 end
   ground ≙ status: ordinary
       when flr₁>0  then flr₁: = 0 end
   mid ≙ status: anticipated
       when true  then flr₁: ∈ 1..101 end
end
```

Figure 8.2: $Lift_1$

8.2.2   Refinement

An Event-B development is a sequence of B machines $M_0, \ldots, M_i, \ldots, M_n$ each related to the next by a refinement relationship, and $n>0$.

A refinement step can introduce new events and split and merge existing events[1]. In any particular step, new events are considered as refinements of *skip*: the event that is always enabled and makes no change to the state. Their introduction is to provide more detail to a specification.

A machine M_i is considered to be refined by M_{i+1} if the given *linking invariant* J_{i+1} on the variables between the two machines is established by their initialisation, and preserved by all events. Formally, we denote the refinement relation between two machines, written $M_i \preccurlyeq M_{i+1}$, when all of the following proof obligations hold: (1) feasibility, (2) guard strengthening and (3) simulation. Feasibility of an event is the property that, if the event is enabled (i.e., the guard is true), then there is some after-state. Guard strengthening requires that when a concrete event is enabled, then so is the abstract one. Finally, simulation requires that the occurrence of events in the concrete machine can be matched in the abstract one (including the initialization event). Formal details of these proof obligations can be found in [6].

There are three kinds of labelling of events in Event-B: *anticipated* (a), *convergent* (c) and *ordinary* (o), where convergent events are those which must not execute forever, whereas *anticipated* events provide a means of deferring consideration of divergence-freedom until later refinement steps. The proof obligation which deals with divergences requires that the proposed variant v of a refinement machine satisfies the appropriate properties: that it is a natural number that decreases on occurrence of any convergent event, and that does not increase on occurrence of any anticipated event. We augment the previous refinement relation to include this additional requirement, and write $M_i \preccurlyeq_W M_{i+1}$ when it holds between M_i and

---

[1]For simplicity we omit the treatment of splitting and merging events in this chapter.

```
machine Lift₂
refines Lift₁
variables flr₂
invariant flr₂ = flr₁
variant 101 − flr₂
events
    init ≙ flr₂: = 0
    top ≙ status: ordinary
        when flr₂ = 101  then flr₂: = 102 end
    ground ≙ status: ordinary
        when flr₂ = 102  then flr₂: = 0 end
    mid ≙ status: convergent
        when flr₂<101  then flr₂: = flr₂ + 1 end
end
```

Figure 8.3: $Lift_2$

M_{i+1}. Ordinary events can occur forever and therefore this requirement is not mandatory for such events.

As an example, Figure 8.2 gives $Lift_1$, a refinement of $Lift_0$. It introduces a new variable flr_1 and gives a linking invariant which relates flr_1 to the state flr_0 of $Lift_0$. Any occurrence of an event in $Lift_1$ must be matched by some performance of the event in $Lift_0$, in the sense that it must be enabled in $Lift_0$ and the linking invariant must be preserved. The new event *mid* must refine *skip*, thus the linking invariant must be preserved. Observe that *mid* is nondeterministic, and can result in flr_1 being set to any value between 1 and 101. Observe also that the guard of *ground* is stronger in $Lift_1$, and that it will be blocked in $Lift_1$ when the lift is already on the ground. The variant is simply a constant, since anticipated events are only required not to increase the variant, and there are no convergent events.

Figure 8.3 gives $Lift_2$, a refinement of $Lift_1$. Observe that the nondeterminism in *mid* has been resolved so that it moves the lift up one floor. Furthermore, *mid* is now convergent, and the variant has been set so that it is strictly decreased by *mid*.

8.2.3   Development Strategy

Event-B has a strong but flexible refinement strategy which is described in [153]. In [294], we also discussed different Event-B refinement strategies and characterised them with respect to the approaches documented by Abrial in [6] and supported by the Rodin tool. In this chapter, we focus on the simplest strategy, and the one most commonly used. The strategy has the following set of restrictions on a refinement chain $M_0 \preccurlyeq_W M_1 \preccurlyeq_W \ldots \preccurlyeq_W M_n$:

1. each new event in M_i is either anticipated or convergent, where $i>0$;

2. refinements of anticipated event of M_i are either convergent or anticipated in M_{i+1};

3. refinements of convergent or ordinary events of M_i are ordinary in M_{i+1};

4. no anticipated events remain in the final machine M_n.

| | *top* (o) | *top* (o) | *top* (o) |
| | *ground* (o) | | |

text>*ground* (o)
text>*ground* (o)

text>*mid* (a)
text>*mid* (c)

Lift$_0$

text>*Lift*$_1$
text>*Lift*$_2$

Figure 8.4: Events and their annotations in the Lift development

Figure 8.4 illustrates the treatment of events in this development strategy for the refinement sequence $Lift_0 \preccurlyeq_W Lift_1 \preccurlyeq_W Lift_2$. The event *mid* is introduced by $Lift_1$ and hence must be anticipated or convergent—in this case it is anticipated. It will need to refine *skip* within $Lift_0$. In the step from $Lift_1$ to $Lift_2$, it must again be anticipated or convergent—in this case it is convergent. Finally, we observe that there are no anticipated events in the final machine $Lift_2$.

This strategy ensures that all new events introduced along a refinement chain must at some stage be convergent. This means that no execution of M_n can end with an infinite sequence of new events not already in M_0.

8.2.4   Semantics

We define a trace of M to be either an infinite sequence of events (a,c or o), i.e., $\langle e_0, e_1, \ldots \rangle$ or a finite sequence of events, i.e., $\langle e_0, \ldots, e_{k-1} \rangle$ where the machine M deadlocks after the occurrence of the final event. Traces correspond to maximal executions of machines. Plagge and Leuschel in [265] provided a definition of an infinite or finite path π of M in terms of a sequence of events and their intermediate states. In order to distinguish notation, we use u to represent a trace without the intermediate states. We need not consider the particular states within a trace in our reasoning which is based on infinite traces. When a machine M is deadlock free all of its traces are infinite. We use the functions of concatenation ($\frown$) and projection ($\upharpoonright$).

8.3   LTL Notation

We use the grammar for the LTL operators presented by Plagge and Leuschel [265]:

$$\phi \quad ::= \quad true \mid [x] \mid \neg\phi \mid \phi_1 \vee \phi_2 \mid \phi_1 \ U \ \phi_2$$

A machine M satifies ϕ, denoted $M \models \phi$, if all traces u of M satisfy ϕ. The definition for u to satisfy ϕ is defined by induction over ϕ as follows:

$$u \models true$$
$$u \models [x] \quad \Leftrightarrow \quad u = \langle x \rangle ^\frown u^1$$
$$u \models \neg \phi \quad \Leftrightarrow \quad \text{it is not the case that } u \models \phi$$
$$u \models \phi_1 \vee \phi_2 \quad \Leftrightarrow \quad u \models \phi_1 \text{ or } u \models \phi_2$$
$$u \models \phi_1 U \phi_2 \quad \Leftrightarrow \quad \exists\, k \geq 0.\, \forall\, i < k.\, u^i \models \phi_1 \text{ and } u^k \models \phi_2$$

where u^n is u with the first n elements removed, i.e., $u = \langle x_0, \ldots, x_{n-1} \rangle ^\frown u^n$.

From these operators Plagge and Leuschel derived several additional operators, including: conjunction ($\phi_1 \wedge \phi_2$), finally (or eventually) ($F\phi$), and globally (or always) ($G\phi$), in the usual way; for explicitness we also provide direct definitions for them:

$$u \models \phi_1 \wedge \phi_2 \quad \Leftrightarrow \quad u \models \phi_1 \text{ and } u \models \phi_2$$
$$u \models F\phi \quad \Leftrightarrow \quad \exists\, i \geq 0.\, u^i \models \phi$$
$$u \models G\phi \quad \Leftrightarrow \quad \forall\, i \geq 0.\, u^i \models \phi$$

For example, the informal specification for the $Lift_0$ given in Section 8.1, that whenever *top* happens then eventually *ground* will happen, could be written as

$$\phi_0 \;=\; G([top] \Rightarrow F[ground])$$

and so

$$Lift_0 \models \phi_0$$

Conversely, we have that $Lift_0 \not\models G([ground] \Rightarrow F[top])$, since $Lift_0$ has the execution $\langle ground, ground, ground \ldots \rangle$ for which the predicate $G([ground] \Rightarrow F[top])$ does not hold.

It will also be useful to identify the events mentioned explicitly in an LTL formula ϕ. This set is called the alphabet of ϕ, and is written $\alpha(\phi)$, similar to the use of αM for the alphabet of machine M. For example, we have $\alpha(\phi_0) = \{top, ground\}$.

8.4 Preserving LTL Properties in Event-B Refinement Chains

The main result when temporal logic properties are preserved by refinement chains is given in Lemma 8.1 below. We have already observed that new events can be introduced during a refinement, e.g., *mid*. We aim for such properties to be preserved even in the presence of new anticipated and convergent events.

We first identify a key property used in Lemma 8.1 below, which enables us to gain insights into the kinds of temporal properties that are appropriate to be proposed and have the potential of being preserved through a refinement chain. Definition 8.1 describes a maximal execution satisfying a property ϕ. The execution may include some events which do not have an impact on whether the property holds or not; therefore we can restrict the maximal execution to include only those events that impact on the property.

Definition 8.1. *Let β be a set of events. Then ϕ is β-dependent if $\alpha(\phi) \subseteq \beta$ and $u \models \phi \Leftrightarrow (u \upharpoonright \beta) \models \phi$.*

An example of a β-dependent property is $GF[ground]$ with $\beta = \{ground\}$. For any execution u, $u \models GF[ground] \Leftrightarrow u \upharpoonright \{ground\} \models GF[ground]$. Conversely, $\neg G(ground)$ is not $\{ground\}$-dependent. For example, if $u = \langle ground, mid, top, ground, ground, \ldots \rangle$ then $u \models \neg G(ground)$ but $u \upharpoonright \{ground\} \not\models \neg G(ground)$.

As another example, $G(top \vee ground)$ is not $\{top, ground\}$-dependent. This is exemplified by any trace u which contains events other than those in $\{top, ground\}$. In this case $u \upharpoonright \{top, ground\} \models G(top \vee ground)$ but $u \not\models G(top \vee ground)$. Observe that this property holds for $Lift_0$ but not for $Lift_2$: it is not preserved by refinement. Since it is not $\{top, ground\}$-dependent Lemma 8.1 below is not applicable for this property.

Lemma 8.1 identifies conditions under which an LTL property ϕ will be preserved in a refinement chain. The conditions are as follows:

- by the end of the refinement chain there should be no outstanding anticipated events (and so all newly introduced events have been shown to be convergent), as given by restriction 4 of the development strategy;

- the final machine in the refinement chain must be deadlock-free; and

- all of the events that have an effect on whether or not ϕ is true are already present in M_0 (ϕ is β-dependent for some $\beta \subseteq \alpha M_0$).

These conditions are enough to ensure that ϕ is preserved through refinement chains. This means that M_0 can be checked for ϕ, and we can be sure that the resulting system M_n will also satisfy it.

The lemma is formally expressed as follows:

Lemma 8.1. *If $M_0 \models \phi$ and $M_0 \preccurlyeq_W \ldots \preccurlyeq_W M_n$ and:*

1. *M_n is deadlock free;*

2. *M_n does not contain any anticipated events; and*

3. *ϕ is β-dependent for some $\beta \subseteq \alpha M_0$*

then $M_n \models \phi$.

The proof of this lemma is given in [295] and we do not repeat it here. Instead it gives more insight to consider the role played by each of the conditions. One would normally hope that $M_0 \preccurlyeq_W M_n$ and $M_0 \models \phi$ would imply $M_n \models \phi$ without the need for additional conditions, however in the context of Event-B machines and LTL properties we do have the need for additional conditions. Thus, consideration of how the lemma fails in the absence of the conditions provides insight into why they are necessary.

8.4.1   Example: Deadlock-Freedom

Our first example illustrates the need for deadlock-freedom:

Observe that $Lift_0 \models F[ground]$. We now consider $Example_0$ which is a refinement of $Lift_0$:

> **machine** $Example_0$
> **refines** $Lift_0$
> **variables** flr
> **invariant** $flr = flr_0 \vee flr \in \{1..101\}$
> **events**
>   init $\widehat{=}$ $flr:= 0$
>   top $\widehat{=}$ **status**: ordinary
>   **when** $flr = 101$ **then** $flr:= 102$ **end**
>   ground $\widehat{=}$ **status**: ordinary
>   **when** $flr = 102$ **then** $flr_0:= 0$ **end**
> **end**

This is a refinement because the effects of the events remain the same, but they now have stronger guards, thus whenever an event from $Example_0$ is enabled then it can be matched by the same event in $Lift_0$. Furthermore, $Example_0$ has no anticipated events, and $F[ground]$ is $\{ground\}$-dependent. Hence, the only condition of Lemma 8.1 which does not hold is requirement (1): for $Example_0$ to be deadlock-free. $Example_0$ can in fact deadlock, because the guards have been strengthened to the point where none of the events can occur.

We also have that $Example_0 \not\models F[ground]$. The execution consisting of initialisation followed by deadlock is a maximal execution, and it is not the case that $ground$ will eventually occur for that execution.

Thus we have $Lift_0 \preccurlyeq_W Example_0$ and $Lift_0 \models F[ground]$ but $Example_0 \not\models F[ground]$. The possibility of deadlock in the refinement machine means that the property is not preserved.

This arises because strengthening guards can introduce new deadlocks, and so new maximal executions can be introduced into the refinement machine that were not present in the abstract machine. Hence even if all maximal executions of the abstract machine model the property, it need not be the case that all maximal executions of the refinement machine will do so.

8.4.2 Example: Anticipated Events

Our second example illustrates the need for there to be no anticipated events:

Consider the refinement relationship $Lift_0 \preccurlyeq_W Lift_1$, and the property $\phi = F[ground]$. We have that $Lift_1$ is deadlock-free. We also have that $Lift_0 \models \phi$, and that ϕ is $\{ground\}$-dependent. Hence the only condition of Lemma 8.1 which does not hold is requirement (2): for $Lift_1$ to contain no anticipated events.

One possible execution of $Lift_1$ is an infinite sequence of mid events: $u = \langle mid, mid, mid, \ldots \rangle$. For this execution, there is no occurrence of $ground$, and hence $u \not\models F[ground]$. This means that $Lift_1 \not\models F[ground]$.

Thus we have $Lift_0 \preccurlyeq_W Lift_1$ and $Lift_0 \models F[ground]$ but $Lift_1 \not\models F[ground]$. The presence of anticipated events in the refinement machine means that the property is not preserved.

This arises because anticipated events can occur repeatedly, and hence can provide a maximal execution that corresponds to a finite sequence of events from the abstract machine, which is not necessarily a maximal execution in the abstract machine, but only a finite prefix of one. Even if the maximal executions of the abstract machine all satisfy an LTL formula

ϕ, this does not mean that all their finite prefixes will do so. The performance of anticipated events in a refinement might prevent the progress specified by ϕ.

8.4.3   Example: β-Dependence

Finally, we consider an example which illustrates the need for β-dependence:

We have that $Lift_0 \preccurlyeq_W Lift_1 \preccurlyeq_W Lift_2$ and $Lift_0 \models G([ground] \lor [top])$ but $Lift_2 \not\models G([ground] \lor [top])$. This is because the refinement step has introduced a new event mid, and the possibility of this event invalidates the LTL specification. All of the conditions hold except for requirement (3): that the predicate $G([ground] \lor [top])$ be $\{ground, top\}$-dependent. The absence of $\{ground, top\}$-dependence means that the introduction of new events such as mid into the execution may not preserve the predicate.

8.5   Discussion and Related Work

One of the few papers to discuss LTL preservation in Event-B refinement is Groslambert [146]. The LTL properties were defined in terms of predicates on system states rather than our paper's formulation in terms of the occurrence of events. His paper focused only on the introduction of new convergent events. It did not include a treatment of anticipated events but this is unsurprising since the paper was published before their inclusion in Event-B. Our results are more general in two ways. Firstly, the results support the treatment of anticipated events. Secondly, we allow more flexibility in the development methodology. A condition of Groslambert's results was that all the machines in the refinement chain needed to be deadlock free. The main lemma, Lemma 8.1, does not require each machine in a refinement chain to be deadlock free, only the final machine. It is irrelevant if intermediate M_is deadlock as long as the deadlock is eventually refined away.

Groslambert deals with new events via stuttering and leaves them as visible events in a trace. This is why the LTL operators used by the author do not include the next operator (X). As new events may happen this may violate the X property to be checked. Plagge and Leuschel in [265] permit the use of the X operator since they treat the inclusion of new events as internal events which are not visible, and hence do not appear in the trace. Since we deal with new events as visible events we also lose the ability to reason about a temporal property using the typical X operator. Our reasoning is simpler than both Groslambert and Plagge and Leuschel since we only focus on events but this means we cannot have atomic propositions in our LTL, whereas they can.

The notion of verification of temporal properties of both classical and Event-B systems using proof obligations has been considered in many research papers. Abrial and Mussat in an early paper, [4], introduced proof obligations to deal with dynamic constraints in classical B. In [168], Hoang and Abrial have also proposed new proof obligations for dealing with liveness properties in Event-B. They focus on three classes of properties: existence, progress and persistence, with a view to implementing them in Rodin. A recent paper by Hudon and Hoang [175] also introduces new proof obligations to deal with liveness properties in Event-B based on Unity. They propose a new language referred to as Unity-B so that refinement chains are guided by both safety and liveness requirements. It is a more comprehensive treatment than [168] since an event explicitly contains guards and fine and coarse grained scheduling assumptions. They demonstrate the preservation of these scheduling assumptions through refinement chains. This is an exciting development since their future work will focus on decomposition and composition of liveness properties. Bicarregui *et al.* in [55] introduced a temporal concept into events using the guard in the *when* clause and the additional labels of *within* and *next* so that the enabling conditions are captured clearly and separately. However, these concepts are not aligned with the standard Event-B labelling.

The interest of LTL preservation through refinement is wider than simply Event-B. Derrick and Smith [108] discuss the preservation of LTL properties in the context of Z refinement but the authors extend their results to other logics such as CTL and the μ-calculus. They focus on discussing the restrictions that are needed on temporal-logic properties and retrieve relations to enable the model checking of such properties. Their refinements are restricted to data refinement and do not permit the introduction of new events in the refinement steps. Our approach does permit new events to be introduced during refinement steps; the contribution is in identifying conditions for LTL properties to hold even in the context of such new events.

8.6 Conclusion

In this chapter we have considered when temporal logic properties can be carried through Event-B refinement chains. We have also identified conditions on temporal logic properties that make them suitable for use in a refinement chain, since not all properties are preserved by Event-B refinement. In particular we identified the notion of β-dependence for an LTL property: that only the events in the set β need to be checked to determine whether an execution meets the property. Such properties are not affected by the introduction of new events in a refinement step, which means they will carry through a refinement chain and will hold for any resulting refinement which is deadlock-free and does not contain anticipated events. This main result is expressed in Lemma 8.1. The chapter presented a *Lift* example refinement chain to illustrate the results and how they are applied.

Chapter 9

A Provably Correct Resilience Mediator Pattern

Mats Neovius

Åbo Akademi University, Turku, Finland

Mauno Rönkkö

Department of Environmental Science, University of Eastern Finland

Marina Waldén

Åbo Akademi University, Turku, Finland

Abstract. The computational processes that manifest as systems are getting ever more complex. Interconnecting several similar autonomous systems to a system of systems is frequent, e.g. manufacturers' autonomous products are integrated to a home automation system. Traditionally, the approach for this problem has been abstraction of the details and thus, chiseling and gluing the bits and pieces together. This has been done by declaring interfaces or by using formal methods. The former declares accessibility whereas the latter may be used to gain a rigorous mathematical-logical view on the complexity and for the ability to reason on this with a set of logical rules. On top of these views, the mediator pattern is defined to provide a reusable solution for a recurring general problem. The mediator pattern encapsulates interaction between a set of autonomous systems with the intension to ease maintenance and refactoring. In this chapter, we formally integrate the mediator pattern in a correct-by-construction manner in the Action Systems formalism. The contribution is in introducing the mediator on an abstract level to a contemporary distributed system as a correctness preserving refinement step. In this setting, the mediator may then be further refined to provide an isolated placeholder for introduction of domain specific intelligent resilience addressing possible issues of inconsistency.

9.1   Introduction

Software engineering is a rather new engineering discipline. It differs from the more traditional disciplines in nearly all aspects; as a bridge is built to serve a function in a specific environment, the contemporary software is expected to operate in a magnitude of possibly changing deployment environments. This requires the software to adapt its functionality defined by the input correctness. As a result, the development of contemporary software is not only complex, but often also complicated. To structure the development and to have a toolbox of solutions for recurring problems, design patterns are defined; and to make this in a rigorous manner a formal description of it may be devised. The design pattern that we formally integrate to a specification is the mediator pattern that may be used to provide resilience.

The mediator pattern describes the well-known producer-consumer relationship, also known as publisher-subscriber or sender-receiver. The purpose is for a group of autonomous agents to communicate with each other through a mediator. The frequent example is a mailbox to which a set of producers (senders) may put a message intended only for a designated subscriber (receiver) that receives this message the next time they connect. As this is a good explanatory example, we note that introducing simple logic to the mediator is possible, e.g. majority voting in terms of fault tolerance. Related to fault tolerance, resilience introduces a sense of intelligence [216, 321] in case of unexpected changes. "Resilience thus deals with conditions that are outside the design envelope whereas other dependability metrics deal with conditions within the design envelope" [330]. Hence, environmental issues and interaction in multi-agent systems are most definitely subject to resilience where optimally, in addition to correcting predefined faulty inputs, a sense of learning is implemented. This learning including means of adaption is outside the envelope and constitutes a critical functionality of the system at hand. However, as the learning and response are implementation and deployment specifc issues, we limit ourselves to introduce the mediator as a placeholder to which this domain specific knowledge may be added.

Evidently, the critical parts of a system are those on which applying formal methods with the purpose of proving correctness is motivated. Formal languages for specifying design patterns include BPSL [324] and LePUS [138]. As these languages do provide the basis for formalizing the pattern itself, they do not treat the pattern in context of some formal method that would provide the tools of formal correctness to the engineer's toolbox. Moreover, as the design envelope of the formal method is the requirements manifested as a specification, a mediator pattern featuring resilience in this context is outside the envelope. This outlines the contribution of this chapter; we show how to introduce a mediator as a correctness preserving refinement step in the Action Systems framework [27].

Action Systems enable a rigorous stepwise development that follows the rules of refinement calculus by Back [24]. We show how to introduce the mediator pattern to a specification consisting of sets of producer and consumer systems while preserving correctness without changing the producers or consumers. Moreover, we consider the producers and consumers to be autonomous, i.e. their service logic and interfaces are subject to change without notice. We do not restrict the mediator pattern's functionality by any means, hence allowing its further development according to the domain specific requirements. This implies that reuse of proof of introducing the mediator is reusable, but reuse of domain specific resilience logic is impossible or very difficult; an issue not considered in this chapter.

The structure of this chapter is as follows. Section 9.2 outlines the Action Systems framework and the related refinement calculus framework. Section 9.3 discusses previously published work on the resilience mediator pattern. Section 9.4 discusses how a resilience mediator pattern is introduced in a correct manner using the Action Systems formalism. In Section 9.5, we conclude and discuss the application of the resilience mediator pattern and its further development.

9.2 Provably Correct Stepwise Development with Action Systems

The Action Systems formalism [27, 33] is a rich formal language that is well suited for modelling critical systems in a correct-by-construction manner. It was further developed for specifying distributed systems by Sere [31, 298]. The Action Systems formalism relies on Hoare logic [170] and the weakest precondition predicate transformer (wp) semantics of Dijkstra [112]. The abstract action systems can be developed via stepwise refinement [23, 249] adding more details to the specification while preserving its mathematical correctness within the refinement calculus framework [38].

9.2.1 Weakest Precondition Predicate Transformers and the Action Systems Framework

The weakest precondition predicate transformer semantics is defined on the language of guarded commands [112, 113] for assigning meanings to programs. A wp is defined on a statement s and postcondition q as wp(s, q). It is a Boolean function wp(s, q):$(\Sigma \to Bool) \to (\Gamma \to Bool)$ where Σ and Γ denote the before and after state space [38] on a set of states $\Sigma_{\mathrm{wp}(s,q)} \subseteq \Sigma$ for which executing s guarantees q, i.e. $\Gamma_q \subseteq \Gamma$. The wp semantics assume monotonicity, conjunctivity, continuity and that wp$(s, false) = false$, i.e. the law of the excluded miracle. The semantics allow nested loops.

For the Action Systems framework used in this chapter, the properties of the wp semantics have been revised. For example, the continuity condition is violated by non-determinism and the law of the excluded miracle by angelic behavior. The predicate transformer semantics used in this chapter is defined in Table 9.1.

Of these statements, magic always establishes q. Disallowed behaviour is captured by **abort** statement, whereas **skip** does nothing. If a evaluates to *false* for assumption $[a]$, the statement behaves miraculously while for assertion $\{a\}$, the statement aborts. In both cases, if a evaluates to *true* the statement behaves as **skip**. We say that a statement is enabled iff executing the statement establishes q, i.e. defines the guard predicate gd on statement s as $gd(s) = \neg \mathrm{wp}(s, false)$. Thus, statements **abort**, **skip**, $x := E$ and $\{a\}$ are always enabled. On these fundamentals, we define the Action Systems framework.

Table 9.1: Semantics of conventional actions

action	notation	wp(action, q)
Miraculous statement	**magic**	*true*
Aborting statement	**abort**	*false*
Stuttering statement	**skip**	q
Multiple assignment	$x := E$	$q[E/x]$
Non-deterministic assignment	$X :\in S$	$\forall\, x'.x' \in S \Rightarrow q[x'/x]$
Sequential composition	$sA;sB$	$\mathrm{wp}(sA, \mathrm{wp}(sB, q))$
Non-deterministic choice	$sA \,[\!]\, sB$	$\mathrm{wp}(sA, q) \wedge \mathrm{wp}(sB, q)$
Assumption	$[a]$	$a \Rightarrow q$
Assertion	$\{a\}$	$a \wedge q$

9.2.2   The Action Systems Framework

The semantics of the Action Systems framework is based on the wp semantics with a basic building block of an action. An action is defined as a guarded statement denoted by $gA \rightarrow sA$ with the meaning of $[gA];sA$. For action A the gA is commonly called the guard whereas the action's body is sA. Thus, the enabledness $gd(A)$ of an action is defined $gd(A) = gA \wedge \neg \mathrm{wp}(sA, \textit{false})$. Consequently, the wp of an action on some q is $\mathrm{wp}([gA];sA, q) = gA \Rightarrow \mathrm{wp}(sA, q)$. Moreover, the actions are finitely conjunctive implying demonic non-determinism, i.e. $\mathrm{wp}(A, q \wedge r) \Rightarrow \mathrm{wp}(A, q) \wedge \mathrm{wp}(A, r)$. This also implies monotonicity, i.e. that $(q \Rightarrow r) \Rightarrow (\mathrm{wp}(A, q) \Rightarrow \mathrm{wp}(A, r))$. On these actions, the repetitive construct may be defined: $\mathrm{wp}(\textbf{do}\, A\, \textbf{od}, q) = (\forall\, n.\mathrm{wp}(A^n, gA \vee q)) \wedge (\exists\, n.\neg gA^n)$ where $A^0 = \textsc{skip}$ and $A^{n+1} = A^n;A$. Thus, the repetitive construct defines that after each action some other action is enabled or the postcondition q is satisfied, that the number of actions is finite and that there is a termination state (total correctness).

An action system $\mathcal{A}$ is a composition of an initialization statement a_0 and a repetitive construct of actions separated by non-deterministic choice. The outline is thereby as follows:

$$\mathcal{A} = |\,[\,\textbf{var}\ x, y^*\ ;\ a_0;\textbf{do}\, A_1\,[\!]\,...\,[\!]\, A_n\, \textbf{od}\,]\,|:z$$

In $\mathcal{A}$ the variables are x, y and z. More specifically, x are local variables, y are exported variables denoted by asterisk * and z are imported variables declared in some other action system. Statement a_0 initialises x and y followed sequentially by the repetitive construct of actions A_i.

The execution model of an action system does not provide fairness. The action system terminates when no action is enabled in the repetitive construct, i.e. when q of $\mathrm{wp}(\textbf{do}\, A\, \textbf{od}, q)$ is established. Moreover, parallel execution of actions is possible when the actions operate on disjoint sets of variables [31, 298].

Features of the Action Systems framework include parallel composition and prioritised execution. Parallel composition of action systems is denoted by " $\|$ " [25]. Thus, for action systems

$$\mathcal{A} = |\,[\,\textbf{var}\ x, y^*\ ;\ a_0;\textbf{do}\, A_1\,[\!]\,...\,[\!]\, A_n\, \textbf{od}\,]\,|:z$$

and

$$\mathcal{B} = |\,[\,\textbf{var}\ u, v^*\ ;\ b_0;\textbf{do}\, B_1\,[\!]\,...\,[\!]\, B_n\, \textbf{od}\,]\,|:w$$

the parallel composition is

$$\mathcal{A} \parallel \mathcal{B} = \mid [\, \mathbf{var}\ x, u, y^*, v^*\ ; a_0; b_0; \\ \mathbf{do}\ A_1 \parallel ... \parallel A_n \parallel B_1 \parallel ... \parallel B_n\ \mathbf{od} \\ \,] \mid : (z \cup w) - (v \cup y)$$

Overlapping variable names are solved by renaming before composition. Composition is associative and commutative, but irreversible. Therefore, it is used frequently only for analysis purposes.

9.2.3 Refinement

Refinement of Action Systems is based on work by Back et al. [38]. Fundamentally, refinement is a collection of conditions for unidirectional stepwise concretisation from a more abstract set of actions to a more concrete set while preserving the mathematical properties. Typically, refinement reduces non-determinism or adds functionality. Refinement is denoted by $\sqsubseteq$ and if $A \sqsubseteq A'$ we say that A' refines A. A statement S is refined by S' if and only if $\forall\ q{:}\mathrm{wp}(S, q) \Rightarrow \mathrm{wp}(S', q)$, i.e. if a postcondition is established by S then the same condition should be satisfied by S' as well. By transitivity it holds that if $S \sqsubseteq S'$ and $S' \sqsubseteq S''$ we have that $S \sqsubseteq S''$. As an action is a sequential composition of two statements, refinement of an action follows the same structure. The conditions to prove are (i) $\{gA\};sA \sqsubseteq sA'$ and (ii) $gA' \Rightarrow gA$, i.e. to show that if gA holds then sA is refined by sA' and that A is enabled whenever A' is enabled. An abstraction relation R mapping variables from A to A' is defined for data refinement. In relation $R(x, x', u)$, x denotes the abstract variables, x' the concrete variables and u the global variables. The refinement considering R is then denoted by $\sqsubseteq_R$, and the conditions to prove are (i) $\{gA\};sA \sqsubseteq_R sA'$ and (ii) $R \wedge gA' \Rightarrow gA$. Intuitively, this means that (i) the action A' behaves in the same way as A and (ii) that the guard of A' is strengthened.

A powerful kind of data refinement is superposition refinement [35]. Fundamentally, superposition introduces new functionality to the action system. The new functionality must not interfere with the old one. Consider

$$\mathcal{A} = \mid [\, \mathbf{var}\ x, y^*\ ; a_0; \mathbf{do}\ A\ \mathbf{od}\,] \mid : z$$

and

$$\mathcal{A}' = \mid [\, \mathbf{var}\ x', y'^*\ ; a_0; \mathbf{do}\ A' \parallel W'\ \mathbf{od}\,] \mid : z'$$

Then $\mathcal{A} \sqsubseteq_R \mathcal{A}'$ holds only if the following conditions hold:

1. *Initialisation*:

   (a) a_0' has the same effect on the old variables as a_0 when $R(x, x', u)$ holds;
   (b) a_0' establishes $R(x_0, x_0', u)$

2. *Old actions*:

   (a) Each old action A' has the same effect on the old variables as A when $R(x, x', u)$ holds;
   (b) The old actions A' will preserve $R(x, x', u)$;

(c) The guards of the old actions are strengthened, $gA' \Rightarrow gA$, whenever $R(x, x', u)$ holds

3. *Auxiliary actions*:

   (a) W' has no effect on x and u;

   (b) Each W' preserves $R(x, x', u)$

4. *Termination of auxiliary actions*: The auxiliary actions W' must not compute forever if $R(x, x', u)$ holds, i.e. $R \Rightarrow$ wp(**do** W' **od**, *true*)

5. *Exit condition*: $R \wedge \neg gA' \wedge \neg gW' \Rightarrow \neg gA$

6. *Non-interference*: Execution of an external action belonging to an environment action system, $\mathcal{E}$, where $R(x, x', u)$ holds, preserves $R(x, x', u)$

The three first conditions guarantee that the initialisation and the actions are correct refinements. The auxiliary actions W' must, for example, only assign the new variables x' such that relation R holds. The fourth condition stipulates that each auxiliary action must terminate if $R(x, x', u)$ holds. Finally, the fifth condition states that whenever $R(x, x', u)$ holds and no action is enabled in the refined system $\mathcal{A}'$, then no action is enabled in the old system $\mathcal{A}$. As we do not specify the environment here, we simply assume that condition 6 holds throughout the rest of this chapter.

9.2.4   Tool Support

We have chosen the Action Systems formalism in this chapter, since it is a very powerful and flexible modelling framework. However, in order to be able to develop complex systems that are correct-by-construction we need to be able to rely on tool support for the refinement proofs. The Action Systems formalism lacks direct tool support. However, via the Event-B formalism developed by Abrial [6] we can consider to have tool support also for Action Systems. Event-B originates from the B-Method [5], but has been heavily inspired by Action Systems. Moreover, the correspondence between Action Systems and Event-B is underpinned by the B-Action Systems formalism [341] that can represent Action Systems using the B-Method. System development in Event-B is based on set theory and atomicity for the events in the same way as for Action Systems. Moreover, abstract models are developed into concrete ones by stepwise refinement in both formalisms.

The Rodin Platform[1] is an Eclipse based tool that provides tool support for stepwise development of Event-B models. The tool generates proof obligations for the abstract as well as the concrete model to be able to show that they are consistent and correct with respect to the abstract model. These proof obligations correspond to the ones shown in Section 9.2.3. The tool tries to prove the proof obligation automatically. The ones that cannot be proved automatically can be discharged interactively by the user. The interactive proving process is facilitated by a number of Eclipse plug-ins. Also the development process is well supported by tools. Notable in the context of this chapter are ProB[2] that help to visualise the development and UML-B [310] translating UML diagrams[3] to Event-B models. A later

---

[1]http://rodin.cs.ncl.ac.uk/
[2]http://www.stups.uni-duesseldorf.de/ProB/index.php5/Main_Page
[3]http://www.uml.org/

version of the tool iUML-B[4] even allows a more integrated development between UML and Event-B.

In this chapter we rely on UML statechart diagrams when modeling our systems. Due to the tight correspondence between Action Systems and Event-B we can rely on the principle for translation of UML to Event-B when creating action systems from our statechart diagrams. State changes in a UML diagram are represented by a variable *state* in the action system. The transitions in UML are modeled by actions. The guards on the transitions have a one-to-one correspondence to the guards in the action systems. Moreover, the initial transition in UML is modeled by the initialisation statement in the action systems and the abort transition by the abort statement.

9.3 Resilience Mediator

As previously explained, our environment has become populated with autonomous systems, many of which are provided as "black boxes". Therefore, improving the resilience of an aggregated system is by no means trivial, as the sub-systems and services cannot be enhanced or refined in any way. Furthermore, the service interfaces and the service logic may change without notice, leading to faulty messaging and unpredictable system's dynamics. Also, as the number of connected components change, it becomes harder to track the cause of a fault. A fault may then propagate over communication from one component to another, causing the connected components to enter faulty states one by one. This cascading effect has been studied, for instance, by Zhu et al. [353].

With this problem setting in mind, a method for designing resilience mediators was proposed in an earlier article [283]. A resilience mediator is a data mediator [345]. It is placed in between the set of producer and consumer components of a system relaying their messages, as depicted in Figure 9.1. Based on the messages, the mediator detects and stores events. Stored events are inspected by a monitor that notifies the mediator about state changes. Hence, the resilience mediator can maintain state awareness [276] and correct or refine a message between the producers and consumer components accordingly. In this way, propagation of faults becomes restricted while requiring no changes to the original components. Consequently, a resilience mediator can address faulty dynamics that can be identified from the communication and that can also be mediated by enhancing the communication between the connected components. From a maintenance point of view, the use of resilience mediators ensures that all resilience improvements are localized to the mediators that coordinate the resilience actions, improving the management of later resilience extensions.

An earlier article [283] proposed an analysis and design process for the resilience mediators to uncover desired resilience actions. The process includes four steps: 1) behavioral analysis with sequence diagrams is used to uncover critical messaging; 2) deviation analysis with HAZOP tables [274] is used to identify all potential deviations; 3) mediation analysis is used to identify deviations that can be identified in communication and corrected during messaging; 4) fault state analysis is used to verify improved resilience and significance of remaining fault propagation.

Although the analysis and design process proposed in [283] is needed to understand the

---

[4]http://wiki.event-b.org/index.php/IUML-B

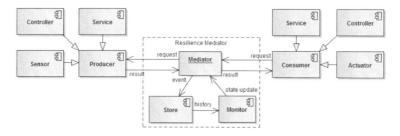

Figure 9.1: The resilience mediator as an architectural component.

potential deviations and their effect on the system as a whole, the process does not address the implementation or its correctness. In particular, the design process does not address the implementation phase at all. Because of this, the design process does not provide a provably correct design pattern that can be applied to improve the resilience of an aggregated system in a reliable manner. For this reason, we will show below how to introduce a resilience mediator to an aggregated system in a provable correct manner by using the Action Systems formalism.

9.4   Formal Development of the Resilience Mediator Pattern

With devices being connected all the time everywhere providing possibly faulty information to some consumers, the mediator acts as a relay of this information. Because of this, a mediator may introduce intelligent inspection of the information by applying filtering and reacting on controversial inputs. Such a mediator is then an instance of a resilience mediator. The purpose of the resilience mediator is to prevent the escalation of a faulty input to a system wide faulty state.

The abstract sequence diagram before applying the mediator pattern is depicted in Figure 9.2. The producer first produces data, and then signals the consumer to start consuming the data. After consuming the data, the consumer signals about the end of the consumption to the producer. When the resilience mediator pattern has been applied, the mediator acts as a relay, as depicted in Figure 9.2. The difference with the mediator is that it brings in intelligence via an additional preprocessing phase that allows to determine potential faults and anomalies in the signaling, thus giving the opportunity for the mediator to intervene and correct the signaling before passing it on. We shall now show how such a resilience mediator is introduced in a stepwise, provably correct manner in the Action Systems framework.

9.4.1   The First Model - Abstract System

Using Action Systems, we start the development with the most abstract system description that exhibits provably correct and desired dynamics. Such a system typically has only a few states and a predefined sequence of state transitions. All the details are then introduced to the abstract system step by step. During each step, the correctness is preserved by proving the refinement proof obligations as stated earlier.

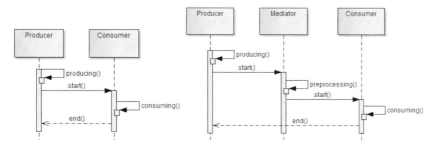

Figure 9.2: Traditional producer-consumer signaling sequence on the left and signaling with a mediator on the right.

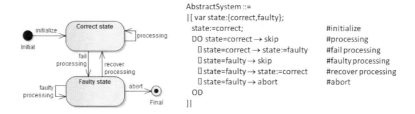

Figure 9.3: The abstract system as a UML statechart on the left and as an action system, called *AbstractSystem*, on the right.

In our case, the abstract system captures overall systems dynamics, as depicted in Figure 9.3 as a UML statechart. The abstract system has two states: a correct state and a faulty state. The system is initialized to the correct state. The system can do processing in both states; however, the processing in the faulty state is faulty processing. The system may change between the correct and faulty state. In particular, if the processing fails in the correct state, the system state changes to the faulty state. Correspondingly, the system may recover from the faulty state back to the correct state. There is also a possibility that the faulty state processing is unmanageable. Then the system aborts. The corresponding action system is also presented in Figure 9.3. In the action system, there is only one variable, *state*, capturing the two states depicted in the UML statechart. As mentioned earlier, each transition in the statechart corresponds to an action in an action system. The correspondence is indicated by referring to the name of the transition in the corresponding action.

9.4.2 The Second Model - Introduction of the Producer-Consumer Pattern

In the first refinement step, we introduce the traditional producer-consumer pattern, as depicted in Figure 9.2, to the abstract system. In terms of the statechart, this is done by refining the self-referential transitions *"processing"* and *"faulty processing"*. For the transition *"processing"* this is done by:

1. splitting the state *"Correct state"* into two states *"Correct state produce"* and *"Correct state consume"*

2. adding transitions *"start"* and *"end"* between the two states

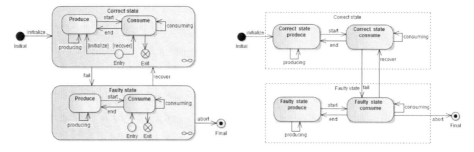

Figure 9.4: Statechart diagram of the system with producer-consumer states; hierarchical statechart on the left and flattened statechart on the right.

3. adding self-referential transition *"producing"* to the state *"Correct state produce"*

4. adding self-referential transition *"consuming"* to the state *"Correct state consume"*

5. assigning the transitions *"fail"* and *"recover"* from and to the state *"Correct state consume"*

The same is then repeated for the transition *"faulty processing"* with respect to the state *"Faulty state"*. The resulting statechart is shown in Figure 9.4. Note that the transitions *"fail"* and *"recover"* were purposefully assigned to the consumption states only. The reason for this is that we aim at introducing a resilience mediator that is able to detect and correct faults and anomalies occurring in the producer component. However, as the resilience mediator may not always be able to correct faulty dynamics, the failure may still take place on the consumer side. This is why the failure and recovery is considered only for the consumer states. Figure 9.4 also shows a statechart, where the hierarchical (inner) states are flattened into single level diagram. The resulting statechart also indicates which states belong to the producer and consumer components.

To prove that this refinement strategy is correct, we use the Action Systems formalism. This is done by using the superposition refinement. With respect to the abstract system shown above, we introduce two new variables *cState* and *fState* to describe the changes in the new statechart with four states. We then add new actions to capture how the values of these variables change with respect to the abstract state variable, *state*. Lastly, the existing actions are bound to the new variables by strengthening their guards and adding the appropriate assignments to the new variables. The resulting action system is shown in Figure 9.5. The names of the actions indicate which original action of the abstract system they refine. The names of the actions also indicate which transitions the actions capture in the new statechart.

The abstraction relation between *AbstractSystem* and *PCSystem* is as follows:

$$(state = correct \Rightarrow fState = consume) \land (state = faulty \Rightarrow cState = consume)$$

Hence, the communication between the correct and the faulty states takes place via state *consume*.

To prove that the refined system *PCSystem* is a correct refinement of *AbstractSystem*, we need to show that all the refinement proof obligations in Section 9.2.3 are satisfied. First, in *PCSystem*, the state is initialized to *correct* and is hence a correct simulation of *AbstractSystem*. The relation (invariant) is established since variable *fState* is assigned to value *consume*. Hence, proof obligation 1 is satisfied. In the refined actions *"fail"* and *"recover"* we strengthen the guards and only add new assignments to the substate variables

```
PCSystem ::=
|[ var state:{correct,faulty}
      cState:{produce,consume}
      fState:{produce,consume}
  state,cState,fState:=correct,produce,consume;                          #initialize
  DO state=correct ∧ cState=produce → cState:=consume                    #processing.1 / start
   ▯ state=correct ∧ cState=produce → skip                               #processing.2 / producing
   ▯ state=correct ∧ cState=consume → cState:=produce                    #processing.3 / end
   ▯ state=correct ∧ cState=consume → skip                               #processing.4 / consuming
   ▯ state=correct ∧ cState=consume → state:=faulty; fState:=consume     #fail processing / fail
   ▯ state=faulty ∧ fState=produce → fState:=consume                     #faulty processing.1 / start
   ▯ state=faulty ∧ fState=produce → skip                                #faulty processing.2 / producing
   ▯ state=faulty ∧ fState=consume → fState:=produce                     #faulty processing.3 / end
   ▯ state=faulty ∧ fState=consume → skip                                #faulty processing.4 / consuming
   ▯ state=faulty ∧ fState=consume → state:=correct; cState:=consume     #recover processing / recover
   ▯ state=faulty ∧ fState=consume → abort                               #abort
  OD
]|
```

Figure 9.5: An action system, called *PCSystem*, with producer and consumer actions introduced.

```
SystemIntegrated::=
|[ var pc_state: {c_produce,c_consume,f_produce,f_consume};
    pc_state:=c_produce;                                       #initialize
  DO pc_state=c_produce → pc_state:=c_consume                  #start
   ▯ pc_state=c_produce → skip                                 #producing
   ▯ pc_state=c_consume → pc_state:=c_produce                  #end
   ▯ pc_state=c_consume → skip                                 #consuming
   ▯ pc_state=c_consume → pc_state:=f_consume                  #fail
   ▯ pc_state=f_produce → pc_state:=f_consume                  #faulty start
   ▯ pc_state=f_produce → skip                                 #faulty producing
   ▯ pc_state=f_consume → pc_state:=f_produce                  #faulty end
   ▯ pc_state=f_consume → skip                                 #faulty consuming
   ▯ pc_state=f_consume → pc_state:=c_consume                  #recover
   ▯ pc_state=f_consume → abort                                #abort
  OD
]|
```

Figure 9.6: Action systems with merged state variable.

cState and *fState* according to the abstraction relation, respectively. In addition, the actions *"processing.2"* and *"processing.4"* refine the original stuttering action *"processing"*. Similarly, the actions *"faulty processing.2"* and *"faulty processing.4"* refine the original stuttering action *"faulty processing"*. This also holds for the actions *"processing.3"* and *"faulty processing.3"*. Hence, proof obligation 2 is fulfilled. The auxiliary actions *"processing.1"* and *"faulty processing.1"* assign the new variables *cState* and *fState*, respectively, in a manner that preserves the abstraction relation. Since they only assign the new variables and their guards are strengthened, proof obligation 3 is satisfied. The auxiliary actions disable themselves and hence they fulfill proof obligation 4. Both systems terminate only by abort, fulfilling proof obligation 5. Proof obligation 6 holds as there are no parallel composed environment action systems present.

For clarity, we shall next reorganize *PCSystem* by representing the original state-space involving three variables with a single variable state-space of all reachable states. As a result, we obtain an action system called *SystemIntegrated* as shown in Figure 9.6. As there

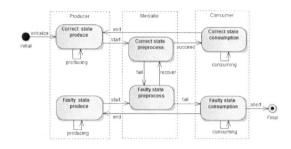

Figure 9.7: Flattened statechart diagram of the system with the resilience mediator states.

is a one-to-one correspondence between the actions of *PCSystem* and *SystemIntegrated*, the refinement invariant is the equivalence relation between the guards of the actions. Consequently, the refinement proof obligations are satisfied.

9.4.3   The Third Model - Introduction of the Resilience Mediator

In the second refinement step, we introduce to the action system *SystemIntegrated* the resilience mediator to the statechart with the producer-consumer pattern. We apply the exact same refinement strategy as in the first refinement step. This time, we refine the self-referential transition *"consuming"* for both of the states *"Correct state consume"* and *"Faulty state consume"* in Figure 9.4. For the transition *"consuming"* and the state *"Correct state consume"* this is done by:

1. splitting the state *"Correct state consume"* into two states *"Correct state preprocess"* and *"Correct state consumption"*

2. adding transitions *"succeed"* from the state *"Correct state preprocess"* to the state *"Correct state consumption"*

3. assigning the *"start"* transition to the state *"Correct state preprocess"*

4. assigning the *"end"* transition to the state *"Correct state consumption"*

5. assigning the transitions *"fail"* and *"recover"* to the state *"Correct state preprocess"* only

The same is then repeated for the transition *"consuming"* with respect to the state *"Faulty state consume"*. The resulting flattened statechart is shown in Figure 9.7. Note that the transitions *"fail"* and *"recover"* were purposefully assigned to the preprocessing states only. This is a design choice, whereby we delegate the responsibility of detecting a faulty operational state to the resilience mediator. Consequently, we may introduce the resilience mediator without changing the producer and consumer components. The correctness of the second refinement step is proven with Action Systems in the same way as for the first refinement step. By this refinement strategy, we arrive at an action system *MediatorSystemIntegrated* as shown in Figure 9.8. The resulting action system has, thus, a single variable for capturing the state change and transitions between the producer, the mediator and the consumer.

By reverse application of the parallel composition rule, we can decompose an action system into distributed sub-systems that operate in parallel by their shared *pmc_state*. By

```
MediatorSystemIntegrated::=
|[ var pmc_state: {c_produce,c_preprocess,c_consumption,
                   f_produce,f_preprocess,f_consumption};
      pmc_state:=c_produce;                                    #initialize
   DO pmc_state=c_produce → skip                               #producing
    ▯ pmc_state=c_produce → pmc_state:=c_preprocess            #start
    ▯ pmc_state=c_preprocess → pmc_state:=c_consumption        #succeed
    ▯ pmc_state=c_preprocess → pmc_state:=f_preprocess         #fail
    ▯ pmc_state=c_consumption → skip                           #consuming
    ▯ pmc_state=c_consumption → pmc_state:=c_produce           #end
    ▯ pmc_state=f_produce → skip                               #faulty producing
    ▯ pmc_state=f_produce → pmc_state:=f_preprocess            #faulty start
    ▯ pmc_state=f_preprocess → pmc_state=f_consumption         #faulty fail
    ▯ pmc_state=f_preprocess → pmc_state:=c_preprocess         #recover
    ▯ pmc_state=f_consumption → skip                           #faulty consuming
    ▯ pmc_state=f_consumption → pmc_state:=f_produce           #faulty end
    ▯ pmc_state=f_consumption → abort                          #abort
   OD
]|
```

Figure 9.8: An action system representation of the system with the resilience mediator states.

applying this technique to *MediatorSystemIntegrated*, we can obtain action systems that capture the behavior of the individual components, the producer, the mediator and the consumer. The resulting action systems are shown in Figure 9.9. For clarity, we have declared the global, shared variable *pmc_state* in an action system of its own to emphasize that the variable is a global state variable that is not owned by any of the components.

9.4.4 Example of a Resilience Mediator

As an example of adding a resilience mediator to a model, we can consider an adaptive house with components and activities [283]. More specifically, we focus on the adaptation of home automation to the occupants' vacation. During their vacation, the occupants are assumed to be away from home. Technically, the event starts when the occupants enter their vacation to a calendar. The adaptive house uses the entry for instance to plan heating, cooling and ventilation of an empty house. Its objective is to save energy during the vacation while maintaining nominal living conditions. The living conditions are restored at the end of the vacation, so that when the occupants arrive, the house has optimal living conditions. In order to do this, the adaptive house accesses not only the calendar, but also a weather forecast service as well as relevant pricing services regarding electricity used for HVAC.

For clarity, we focus here on the activity of scheduling the vacation and the timing of the home automation in the adaptive house example. Following the resilience mediator pattern introduced in this chapter we start with a very abstract model as in the pattern. Hence, we only state that we have a correct state and a faulty state. Then following the pattern, we introduce a producer and a consumer in the model. The producer corresponds to the calendar that takes care of scheduling the functions in the house, and the consumer corresponds to the adaptive controller that processes the scheduled data. The resilience mediator pattern allows us to add a mediator to the producer-consumer model as a refinement step. Here we introduce a mediator to handle the preprocessing of the scheduled data. The producer,

```
System::=
|[ var pmc_state: {c_produce,c_preprocess,c_consumption,
                   f_produce,f_preprocess,f_consumption};
   pmc_state:=c_produce;                                    #initialize
   DO false → skip OD
]|

Producer::=
|[ DO pmc_state=c_produce → skip                           #producing
   ▯ pmc_state=f_produce → skip                            #faulty producing
   ▯ pmc_state=c_consumption → pmc_state:=c_produce        #end
   ▯ pmc_state=f_consumption → pmc_state:=f_produce        #faulty end
OD
]| : pmc_state

Mediator::=
|[ DO pmc_state=c_produce → pmc_state:=c_preprocess        #start
   ▯ pmc_state=f_produce → pmc_state:=f_preprocess         #faulty start
   ▯ pmc_state=c_preprocess → pmc_state:=f_preprocess      #fail
   ▯ pmc_state=f_preprocess → pmc_state:=c_preprocess      #recover
   OD
]| : pmc_state

Consumer::=
|[ DO pmc_state=c_consumption → skip                       #consuming
   ▯ pmc_state=f_consumption → skip                        #faulty consuming
   ▯ pmc_state=c_preprocess → pmc_state:=c_consumption     #succeed
   ▯ pmc_state=f_preprocess → pmc_state=f_consumption      #faulty fail
   ▯ pmc_state=f_consumption → abort                       #abort
   OD
]| : pmc_state
```

Figure 9.9: The obtained action systems capturing the producer, mediator and consumer actions.

consumer and mediator can subsequently be refined to capture more details in scheduling the adaptation of the home automation.

As an example of the role of the mediator, we can consider the case for detecting when the end of a scheduled event was not correctly given in the system [283]. A mediator can detect the mismatch by having external contextual information, for instance, by monitoring use of water, energy and appliances or movement. Due to the calendar information, the adaptive control will start to adjust the house for an arriving occupant. After some time, however, the mediator can detect that the house is adjusted for occupants, but there is no-one there. Then, the mediator can contact the occupant for confirmation, and depending on occupants estimated arrival time, reschedule arrival or ask the occupant to reschedule the vacation duration. As a consequence, some intelligence has been brought to the adjustment via the mediator.

9.5   Discussion and Conclusion

Patterns in software design have an established meaning while pattern in formal methods are sometimes mixed with reuse. The reason is the objective. Patterns in formal methods have been used to facilitate proof reuse [136] as a model transformation that adds or modifies certain elements of the specification [232] given that the pattern specification matches the problem [169]. This view omits the generality of the design pattern as intended in software engineering, i.e. that a pattern is an unfinished product; which is the approach taken by this chapter on a system level.

Resilience and fault tolerance are sometimes used interchangeable. Fault tolerance on a priori known faulty behavior, sometimes called resilience, takes a slightly different approach as of the uniform recovery strategy. This fault tolerance has been extensively studied by Troubitsyna [332] and later, as resilience [262] However, with respect to the definition of resilience, a resilient entity is intelligent that we consider being more than sets of logical rules.

The resilience mediator as introduced in this chapter is valid regardless of the level of abstraction, whether this is a component or a system. The common denominator that the resilience mediator considers on systems, components, system of systems and multi-agent systems is the distance of the specification from reality. In the considered setting, the reality changes and thus, the distance varies. Hence, a valid formal approach cannot be fixed, but needs to adapt. This adaption is the point of introducing intelligence, i.e. the point of resilience acting as the motivation of this chapter. Thus, a formal specification with the mediator pattern implemented is always at least as close to the reality as without the mediator. We have shown in this chapter that the mediator pattern can be introduced as a refinement step to a formal producer-consumer specification in the Action Systems formal framework. This is the main contribution of this chapter.

Acknowledgments. This research is funded by the Academy of Finland project "FResCo: High-quality Measurement Infrastructure for Future Resilient Control Systems" (Grant numbers 264060 and 263925).

Part IV

Refinement

Chapter 10

Relational Concurrent Refinement - Partial and Total Frameworks

John Derrick

Department of Computer Science, University of Sheffield, Sheffield, S1 4DP, UK

Eerke Boiten

School of Computing, University of Kent, Canterbury, Kent, CT2 7NF, UK

10.1   Introduction

Data refinement as found in a state-based language such as Z or B, and refinement as defined in a process algebra may appear as very different notions. Data refinement is defined using a relational model in terms of the behaviour of abstract programs, whereas in models of concurrency, refinement is often defined in terms of sets of observations. The methodologies used to verify them are also different: downward and upward simulations for data-refinement, whereas calculation of traces, refusals etc. from the semantics is prevalent in a process algebra.

In this chapter we discuss work undertaken to reconcile the two approaches. Specifically, we show how one can embed concurrent semantics into a relational framework. This not only articulates the relationship between the two, but allows event-wise verification methods for concurrent refinement relations to be derived. We discuss how this can be achieved below, with particular emphasis on the type of relational model used - specifically whether one uses a model containing total relations, or one in which relations may be partial. Both are shown to have limits in expressiveness, and because of this we introduce a more general framework for simulations, called process data types.

10.2   Models of Refinement

This chapter starts out from a standard theory of refinement for abstract data types (ADTs) using relational models. The seminal paper on this topic was by He, Hoare and Sanders [159]. It uses what we would now call the *total relational model*: all operations are total relations, and it has been used in a number of contexts. For example, the standard refinement theory of Z [348, 105], is based on this version of the theory. However, it is equally feasible to use a model where the relations can be partial, and after the original framework was introduced the restriction to total relations was dropped (see [158]). The key result that makes the partial framework useable is that it is equally possible to prove soundness and joint completeness of the same set of simulation rules in the partial framework.

In this model, a program (defined here as a *sequence* of operations) is given a semantics as a relation over a global state G. The global state G is defined such that values in G, or more generally: relations on G form the observations that we wish to make of programs. The operations of the ADT, however, are concerned with a *local* state $State$. In order to make observations on global states from these, any sequence of operations is bracketed between an *initialisation* and a *finalisation* which translate between global and local state.

The following defines our notion of relational abstract data type.

Definition 10.1 (Data type; Total data type).
A (partial) *data type* is a tuple $(State, Init, \{Op_i\}_{i \in J}, Fin)$. In this tuple, $State$ is its local state, and operations $\{Op_i\}$, indexed by $i \in J$, are relations on this local state. The initialisation $Init$ is a total relation from G to $State$, and the finalisation Fin is a total relation from $State$ to G.

For a *total* data type, we impose the additional constraint that all operations are total on $State$.   □

For now we require that $Init$ and Fin are total; this is not a significant restriction of expressiveness, and it could be relaxed if needed. We use it as a view that it is always possible to start a program sequence, and that it is always possible to make an observation. We can then define programs, and hence data refinement as follows.

Definition 10.2 (Program; Data refinement).
Given a data type $D = (State, Init, \{Op_i\}_{i \in J}, Fin)$ a *program* is a sequence over J. For a program $p = \langle p_1, ..., p_n \rangle$ over D, its meaning is defined as $p_D = Init \mathbin{\S} Op_{p_1} \mathbin{\S} ... \mathbin{\S} Op_{p_n} \mathbin{\S} Fin$.

Given data types A and C, C is a *data refinement* of A, denoted $A \sqsubseteq_{data} C$, iff for every sequence p over J, $p_C \subseteq p_A$.   □

The standard way to verify a refinement is to use *simulations*, of which there are two varieties: *downward* and *upward*.

Definition 10.3 (Downward and upward simulations).
A relation $R \subseteq AState \times CState$ is a *downward simulation* between data types $A = (AState, AInit, \{AOp_i\}_{i \in J}, AFin)$ and $C = (CState, CInit, \{COp_i\}_{i \in J}, CFin)$ if and only if the following conditions hold:

$$CInit \subseteq AInit \mathbin{\S} R$$
$$R \mathbin{\S} CFin \subseteq AFin$$
$$\forall i{:}J \bullet R \mathbin{\S} COp_i \subseteq AOp_i \mathbin{\S} R$$

A relation $\mathsf{T} \subseteq \mathsf{CState} \times \mathsf{AState}$ is an *upward* simulation between A and C as above if and only if:

$$\mathsf{CInit} \mathbin{\substack{\circ\\\circ}} \mathsf{T} \subseteq \mathsf{AInit}$$
$$\mathsf{CFin} \subseteq \mathsf{T} \mathbin{\substack{\circ\\\circ}} \mathsf{AFin}$$
$$\forall\, i{:}J \bullet \mathsf{COp}_i \mathbin{\substack{\circ\\\circ}} \mathsf{T} \subseteq \mathsf{T} \mathbin{\substack{\circ\\\circ}} \mathsf{AOp}_i \qquad\qquad\qquad \square$$

They are sound (every simulation establishes data refinement) and jointly complete (every data refinement can be proved by a sequence of simulations) [159, 102].

An alternative view of refinement is that used for a behavioural notation such as a process algebra [171, 246, 53]. In a process algebra, such as CSP, CCS, ACP, LOTOS, refinement is defined over the semantics of the language, which is often given by labelled transition systems (LTS) and sets derived from these. For example, for CSP or CCS the language is modelled as an LTS where the state space is the set of terms in the language.

Refinement preorders and notions of equivalence are defined using the semantics with the same principle as in a relational model, that is, that two terms are identified whenever no observer can notice any difference between their external behaviours. Varying how the environment *interacts* with a process leads to a multitude of potential refinement relations, e.g., traces, failures, readiness, failure traces; an overview is provided by van Glabbeek in [338, 337]. Here we will just consider a small number of the possible relations. To begin we use the usual notation for labelled transition systems:

Definition 10.4 (Labelled Transition System).
A *labelled transition system* (LTS) is a tuple $(States, Act, T, Init)$ where $States$ is a non-empty set of states, such that $Init \subseteq States$ is the set of initial states, Act is a set of action(-label)s, and $T \subseteq States \times Act \times States$ is called the transition relation. $\qquad \square$

Additional notation needed includes the usual one for transitions: $p \xrightarrow{a} q$ for $(p, a, q) \in T$ and the extension of this to traces (written $p \xrightarrow{tr} q$) and the set of enabled actions of a process which is defined as: $next(p) = \{a \in Act \mid \exists\, q \bullet p \xrightarrow{a} q\}$. Further, $\sigma \in Act^*$ is a *trace* of a process p if $\exists\, q \bullet p \xrightarrow{\sigma} q$, which we also write as $p \xrightarrow{\sigma}$. The set of traces of p is denoted $\mathcal{T}(p)$. Finally, $(\sigma, X) \in Act^* \times \mathbb{P}(Act)$ is a *failure* of a process p if there is a process q such that $p \xrightarrow{\sigma} q$, and $next(q) \cap X = \varnothing$. The set of failures of p is denoted $\mathcal{F}(p)$.

10.3 Using a Partial Framework to Embed Concurrent Refinement Relations

Given two frameworks such as the partial relational model and the process algebraic semantics, and the variety of refinement relations that follow, the natural question to ask is *how do they relate?* To answer that, the work initiated in [104], and followed on in [60, 106], described a methodology to relate the two approaches. That work used the total and partial frameworks to embed each refinement relation and derive corresponding simulation rules. In this set-up, we consider original data types (e.g., using Z data types to specify LTSs in an obvious way) to which we give a concurrency refinement semantics – but we then aim to check this refinement relation within the relational framework. This works in a number of steps:

1. Define a relational data type with a state that contains sufficient information to encode the data type's subsequent behaviours (enabled operations), as well as any observations that may need to be made, through its finalisation operation. The choice of finalisation (and its target type, i.e., the global state) is taken so that we observe the characteristics of interest.

2. Describe how the relevant LTS observation is made directly from the original data type and its programs. For example, for trace refinement what denotes traces in the data type.

3. Prove that the induced notion of refinement from inclusion between sets of these observations is equivalent to data refinement on the relational embedding constructed in step 1.

4. Extract a characterisation of this particular refinement relation as simulation rules on the operations (etc.) of the data type.

The simulation conditions that are derived are, by construction, guaranteed to provide a sound proof method for the given notion of refinement; however, their joint completeness in general requires a separate proof – see [59] for a detailed discussion and an example.

In general there are two parameters that one can use to vary the embedding. First, there is the framework itself, does one use partial or total relations, and if the latter how are partial operations (e.g., a Z operation with a precondition that is not total) embedded as total relations? In particular, this gives rise to differing *encodings*. Second, what observations are made, and in a relational framework this is given by the choice of finalisation. In previous work we have considered both questions together, and encoded each process algebraic relation in an appropriate framework. Sometimes, however, multiple choices are appropriate, and here we are particularly interested in the tension between using a partial and total framework. We begin by illustrating the approach with two refinement relations, trace and failure preorder, in a setup that does not consider divergence.

10.3.1 Basic Relations without Divergence

The simplest refinement relation of interest is the trace preorder:

Definition 10.5 (Trace preorder). The trace preorder is defined by $p \sqsubseteq_{tr} q$ iff $\mathcal{T}(q) \subseteq \mathcal{T}(p)$. □

As observed in [104] the partial relations model records exactly trace information for the embedding with a trivial finalisation: possible traces lead to the single global value; impossible traces have no relational image. Thus in a partial model there is no need to record additional observations, and we can use the following embedding. Detail on Z data types is omitted here (see [105]), note that their initialisations are sets of initial states rather than relations, and they have no explicit finalisations. In the definition below, $\theta State$ returns the *bindings* of *State*, i.e., the elements of the relevant schema type, see [105].

Definition 10.6 (Trace embedding).
The *trace embedding* of a Z data type $(State, Init, \{Op_i\}_{i \in J})$ is given by $(\mathsf{State}, \mathsf{Init}, \{\mathsf{Op}_i\}_{i \in J}, \mathsf{Fin})$, where $\mathsf{G} == \{*\}$ and $\mathsf{State} == State$:

$$\mathsf{Init} == \mathsf{G} \times \{Init \bullet \theta State'\}$$
$$\mathsf{Op} == \{Op \bullet \theta State \mapsto \theta State'\}$$
$$\mathsf{Fin} == \mathsf{State} \times \mathsf{G}$$

We denote the trace embedding of a Z data type A by $\mathsf{A}\mid_{tr}$. □

The second step in the process is to define the traces of a Z abstract data type. This is simply each sequence of operations (each program) which does not have an empty semantics. Then we need to prove that inclusion on these corresponds exactly on data refinement on the trace embedding.

Definition 10.7. For a Z data type $A = (State, Init, \{Op_i\}_{i \in J})$ its set of *traces* $\mathcal{T}(A)$ is defined as all sequences $\langle i_1, \ldots, i_n \rangle$ such that $\exists\, State' \bullet Init \,\mathring{,}\, Op_{i_1} \,\mathring{,}\, \ldots \,\mathring{,}\, Op_{i_n}$. □

Theorem 10.1. For Z data types A and C, $\mathsf{A}\mid_{tr}\sqsubseteq_{data}\mathsf{C}\mid_{tr}$ iff $\mathcal{T}(C) \subseteq \mathcal{T}(A)$. □

Using this embedding one can then derive simulation rules that correspond to this refinement notion (here: traces), see [106]. We give them as they would appear written in the Z schema calculus. They are, in fact, the standard rules for Z refinement but without the constraint on applicability of operations. These are used also in Event-B [6], based on ideas from action systems [37].

Definition 10.8 (Trace simulations).
Given Z data types A and C, the relation R on $AState \wedge CState$ is a *trace downward simulation* from A to C if

$$\forall\, CState' \bullet CInit \Rightarrow \exists\, AState' \bullet AInit \wedge R'$$
$$\forall\, i \in J \bullet \forall\, AState; CState; CState' \bullet R \wedge COp_i \Rightarrow \exists\, AState' \bullet R' \wedge AOp_i$$

The total relation T on $AState \wedge CState$ is a *trace upward simulation* from A to C if

$$\forall\, AState'; CState' \bullet CInit \wedge T' \Rightarrow AInit$$
$$\forall\, i{:}J \bullet \forall\, AState'; CState; CState' \bullet$$
$$(COp_i \wedge T') \Rightarrow (\exists\, AState \bullet T \wedge AOp_i)$$
□

The trace embedding needed a trivial finalisation in the partial framework. For a refinement relation such as the failure preorder, one needs to observe more and hence needs a more complex finalisation. First the definition of the preorder itself:

Definition 10.9. The failures preorder, $\sqsubseteq_f$, is defined by $p \sqsubseteq_f q$ iff $\mathcal{F}(q) \subseteq \mathcal{F}(p)$. □

The embedding into the partial relations model is still fairly simple, we take the finalisation to observe a set E which is (any subset of) the set of all refused events $\{i{:}J \mid \neg\, \mathrm{pre}\, Op_i\}$:

Definition 10.10 (Failures embedding).
The *failures embedding* of a Z data type $(State, Init, \{Op_i\}_{i\in J})$, is $\mathsf{A}\mid_f = (\mathsf{State}, \mathsf{Init}, \{\mathsf{Op}_i\}_{i\in J},$ $\mathsf{Fin})$, where $\mathsf{G} == \mathbb{P}\, J$ and $\mathsf{State} == State$, and

$$\mathsf{Init} == \mathsf{G} \times \{Init \bullet \theta State'\}$$
$$\mathsf{Op} == \{Op \bullet \theta State \mapsto \theta State'\}$$
$$\mathsf{Fin} == \{State; E{:}\,\mathbb{P}\, J \mid (\forall\, i \in E \bullet \neg\, \mathrm{pre}\, Op_i) \bullet \theta State \mapsto E\}$$
□

In the relational embedding failures are pairs (tr, X), where tr is a trace, and there exists states $(\mathsf{State}, \mathsf{State}') \in tr$ (with State being initial) such that $\forall\, i{:}X \bullet \mathsf{State}' \notin \mathrm{dom}\,\mathsf{Op}_i$.

Theorem 10.2. With the failures embedding, data refinement corresponds to the failures preorder: $\mathsf{A}\mid_f \sqsubseteq \mathsf{C}\mid_f$ iff $\mathcal{F}(C) \subseteq \mathcal{F}(A)$. □

Given the failures embedding the changes to the simulation conditions are as follows (see [106]):

Downward simulations: $R \mathbin{\,\substack{\circ \\ \circ}\,} CFin \subseteq AFin$ is equivalent to

$$\forall\, i{:}J; AState; CState \bullet R \wedge \text{pre}\, AOp_i \Rightarrow \text{pre}\, COp_i$$

Upward simulations: $CFin \subseteq T \mathbin{\,\substack{\circ \\ \circ}\,} AFin$ is equivalent to

$$\forall\, CState \bullet \exists\, AState \bullet \forall\, i{:}J \bullet T \wedge (\text{pre}\, AOp_i \Rightarrow \text{pre}\, COp_i)$$

In the first work on relating concurrent and relational refinement relations [104, 60], this preorder was embedded into the total relational model. Lemma 3 in [60] suggested this was not necessary. Hence in later work a partial relations model was used.

10.3.2   Internal Events and Divergence

So far the partial relational framework has been sufficient to model our different concurrent refinement relations. However, when we extend the framework to model aspects of internal events, one naturally needs to consider divergence, and this is where totalisations (and other encodings) start to be necessary.

First, there is an easy extension of the trace refinement embedding to one that includes internal events, but does not consider infinitely enabled internal behaviour (livelock) a problem. We define a data type extended to include an internal operation, and give a first interpretation of that as a standard data type; refinement is then inherited through that interpretation.

Definition 10.11 (*i*-data type; embedding as a data type (no livelock))**.** An *i*-data type is an extension of the data type of Definition 10.1 with an additional component $i \subseteq State \times State$. The *i*-data type $D = (State, G, Init, \{Op_k\}_{k \in J}, i, Fin)$ is embedded as the data type $\widehat{D} = (State, G, \widehat{Init}, \{\widehat{Op_k}\}_{k \in J}, Fin)$ where $\widehat{Op} = Op \mathbin{\,\substack{\circ \\ \circ}\,} i^*$ and $\widehat{Init} = Init \mathbin{\,\substack{\circ \\ \circ}\,} i^*$.   □

The simulation rules deriving from this are those of Definition 10.3 with finite internal behaviour inserted after all occurrences of operations and initialisation. In the absence of (observed) divergence, joint completeness of the simulations follows from joint completeness of the partial relations simulations, plus the fact that the data type with internal operations is refinement equivalent to its embedding as in Definition 10.11, see also [105] for the latter point.

However, for a refinement relation that does not ignore divergence, the interpretation of the *i*-data type needs to ensure the correct observations are made when the final state records diverge. In the case of catastrophic interpretations (i.e., ones where divergence is propagated to all further behaviours) the embeddings need to generate arbitrary behaviour from the point of divergence onwards, and propagate this into all subsequent operations.

As usual in such encodings, we use an additional value, assumed not to be included in any local or global state space. For any set S, let $S_\omega = S \cup \{\omega\}$.

The trace refinement relation just given ignores divergence, and thus is different from trace inclusion in the CSP failures-divergences model. An interpretation of *i*-data types to encode CSP trace refinement is a simplification of the failures-divergences embedding [60], as follows.

Definition 10.12 (Embedding trace refinement (CSP model)).

An i-data type $D = (\mathsf{State}, \mathsf{G}, \mathsf{Init}, \{\mathsf{Op}_k\}_{k \in J}, i, \mathsf{Fin})$ is embedded as the data type $\widehat{D} = (\mathsf{State}_\omega, \mathsf{G}, \widehat{\mathsf{Init}}, \{\widehat{\mathsf{Op}_k}\}_{k \in J}, \widehat{\mathsf{Fin}})$ where

$$\widehat{\mathsf{Init}} = \mathsf{Init} \, ; \, i^* \ \cup \ \mathbf{if} \ \mathrm{divi}\,\mathsf{Init} \ \mathbf{then} \ \mathsf{G} \times \mathsf{State}_\omega$$

$$\widehat{\mathsf{Op}} = \mathsf{Op} \, ; \, i^* \ \cup \ \mathrm{divOp} \times \mathsf{State}_\omega$$

$$\widehat{\mathsf{Fin}} = \mathsf{Fin} \ \cup \ \{\omega\} \times \mathsf{G}$$

$$\mathrm{divOp} =_{def} \{s\!:\!\mathsf{State} \mid \exists\, s'\!:\!\mathsf{State} \bullet (s, s') \in \mathsf{Op} \wedge s' \xrightarrow{i^\infty}\}$$

$$\mathrm{divi}\,\mathsf{Init} =_{def} \exists\, s\!:\mathrm{ran}\,\mathsf{Init} \bullet s \xrightarrow{i^\infty} \qquad \qquad \square$$

Using this interpretation we can once again derive simulation rules, which do contain some modifications to account for potential divergence including at initialisation.

Definition 10.13 (Simulation for trace refinement (CSP model)).

The relation R between AState and CState is a downward simulation between i-data types A and C iff $\forall\, k\!:\!J$:

$$\mathbf{if} \ \mathrm{divi}\,\mathsf{CInit} \ \mathbf{then} \ \mathrm{divi}\,\mathsf{AInit} \ \mathbf{else} \ \mathsf{CInit} \, ; \, i_C^* \subseteq \mathsf{AInit} \, ; \, i_A^* \, ; \, R$$

$$R \, ; \, \mathsf{CFin} \subseteq \mathsf{AFin}$$

$$(\mathrm{divAOp}_k) \lhd R \, ; \, \mathsf{COp}_k \, ; \, i_C^* \subseteq \mathsf{AOp}_k \, ; \, i_A^* \, ; \, R$$

$$\mathrm{dom}(R \rhd \mathrm{divCOp}_k) \subseteq \mathrm{divAOp}_k \qquad \qquad \square$$

10.4 A Total Relational Framework

So far our relational framework has been one of partial relations. However, the last section is moving us towards a total framework, where additional values are used to represent aspects that one wants to additionally observe. So why not go the whole way? Indeed in a partial model the natural encoding of particular programmes being "impossible", e.g., leading to a deadlock, is through the empty relation. If non-deterministic choice is represented as union of relations, the choice ends up hiding the possibility of a deadlock. This may not be desirable, e.g., for a final implementation.

Of course the partial relational model can solve this by making more detailed observations at the end of a program, but it may be simpler to use a total relational framework. In this model error behaviour is now encoded *explicitly* in operations. Many such encodings increase operations' domains to become total, however sometimes we also use them just to augment partial relations.

When turning a partial model into a total one, there are two types of totalising encodings used: the *blocking* (or strict) totalisation maps error behaviour only to a "sink" state; the *non-blocking* (or non-strict, or chaotic) totalisation in addition maps it to *all* possible normal outcomes.

These totalisations use the distinguished value $\bot$ not in S, and the extended state is denoted $S_\bot$.

Definition 10.14 (Totalisation).

Let Op be a partial relation on State. Its non-blocking and blocking totalisations are total relations on $\mathsf{State}_\perp$, denoted $\widehat{\mathsf{Op}}^{nb}$ and $\widehat{\mathsf{Op}}^{b}$ respectively, where $\widehat{\mathsf{Op}}^{nb} == \mathsf{Op} \cup \{x, y : \mathsf{State}_\perp \mid x \notin \mathrm{dom}\,\mathsf{Op} \bullet (x, y)\}$, and $\widehat{\mathsf{Op}}^{b} == \mathsf{Op} \cup \{x : \mathsf{State}_\perp \mid x \notin \mathrm{dom}\,\mathsf{Op} \bullet (x, \perp)\}$. $\qquad\qquad\square$

With totalisations like these, a similar procedure to that described above can be used to extract simulation rules on the underlying data type. Indeed, although most embedding of concurrent refinement relations can be made directly into the partial relation model, historically, the work on embedding concurrent refinement relations began by using a framework of total relations.

At first sight, the total relational framework looks strictly more expressive than the partial model. However, the partial framework turns out to be necessary to embed concurrent refinement relations based on *ready sets*.

The readiness model, as proposed by Olderog and Hoare [255], does not record the failures of a process but its "ready sets" which contain events that the process is able to engage in. Thus (t, Y) is a ready set of a process P if t is a trace of P and $\exists\,Q \bullet P \xrightarrow{t} Q \wedge \forall\,e \in Y \bullet Q \xrightarrow{e}$. Readiness refinement is taken to be inclusion of the ready sets.

The readiness model can be coded in the relational context by taking the set of events observed at finalisation E to be the set of events that are enabled instead of refused. However, to do it is necessary to use the partial relational framework as the readiness model deals not with events that can be refused, as did the failures model, but with events that are enabled. Thus the aspects we wish to make observable in the finalisation must be concerned with just the enabled events. Taking E to be the set of events that are enabled instead of refused seems to achieve this. However, the totalisation of a partial relation (in the blocking model) maps the areas where operations were refused to $\perp$, and this gets percolated through to the finalisation. That is, the outcome of a program run will sometimes be $\perp$ which represents the observation of a certain amount of refusals.

This observation is sufficient to deny the correspondence between readiness refinement and the total relational embedding. It is worth noting that this is due to the artificial nature of the totalisation: it has included observations that are simply not observed via a finalisation that really only looks at enabled events. However, in a partial model there is no issue in setting the finalisation to be the set of events that are enabled, and [104] details the simulation rules that are derived.

So, should one always use the partial relational framework? Well, no, as we have seen the totalised framework is useful (and sometimes needed!) when incorporating divergence into the model. Although we extended the partial framework with divergence in Section 10.3.2, in some cases we need even more information in the model than that described above.

10.5 A General Framework for Simulations - Process Data Types

Indeed, the two approaches of partial or totalised relations are not exclusive. Languages such as B [5] have both preconditions and guards, and some specification languages include both situations that lead to *divergence* and situations that lead to *deadlock*. For example, in our treatment of internal operations in [107], and in process algebraic contexts, both occur – livelock causes divergence, and deadlocks may arise from synchronising actions that are not enabled.

Representing two kinds of exceptional situations cannot be done using only relations of the kinds described so far. Alternative approaches might be through predicate transformers, relations over sets, or by including extra sets alongside relations. Here we consider an approach, called *process data types* which essentially applies two encodings in sequence: one to account for deadlock, and another to account for livelock.

Such a model contains essential design decisions on the relative ordering of the different kinds of erroneous behaviours: what observations should be possible when the semantics leads to a non-deterministic choice between *normal, divergent*, and *blocking* behaviour?

If our encodings lead to potentially partial relations, a fairly natural choice in this contains three kinds of non-standard behaviour. One of them is a zero of non-deterministic choice, one of them is a unit of choice, and the third is neither. The zero represents an "error" whose possible occurrence hides all other behaviour; the unit represents an "error" whose possible occurrence is hidden by other behaviour, and only for the third type we can identify both *possible* and *certain* occurrence of it.

The "zero" is "chaotic" divergence as in CSP: it can also be viewed as representing the union of all possible behaviour with something even worse. The unit of non-deterministic choice is empty behaviour – this then also allows us to model choice as union of relations. The final kind of error is "deadlock" or "blocking", as an explicitly observed behaviour, consistent with CSP's observation of refusals. This set up allows us to record what happens to operations outside their domain in each of the three possible ways: blocking, non-blocking divergent, or absent behaviour.

Together this fixes the outcomes of choice between the different types of behaviour. The choice between divergence and anything else (including blocking) is divergence – as it is a zero. The choice between a behaviour and empty is that behaviour – as it is a unit. The choice between any type of behaviour and itself is that behaviour too. An extra category is then "possible deadlock".

These decisions are summarised in the table below; as before, $\bot$ represents blocking, and ω represents divergence. Choice is modelled as union of sets.

Choice		n	ω	$\bot$	p$\bot$	$\varnothing$	model
normal	n	n	ω	p$\bot$	p$\bot$	n	subsets of State
divergence	ω	ω	ω	ω	ω	ω	State $\cup \{\bot, \omega\}$
deadlock	$\bot$	p$\bot$	ω	$\bot$	p$\bot$	$\bot$	$\{\bot\}$
possible deadlock	p$\bot$	p$\bot$	ω	p$\bot$	p$\bot$	p$\bot$	subsets of State$_\bot$
empty	$\varnothing$	n	ω	$\bot$	p$\bot$	$\varnothing$	$\varnothing$

Process data types, discussed below, allow us to derive refinement rules directly for any formalism which gives rise to both kinds of errors once the areas of divergence and blocking have been made explicit. The area of empty behaviour is implicitly characterised.

Definition 10.15 (Process data type).
A *process data type* is a tuple (State, Inits, $\{Op_i\}_{i \in J}$, Fin) such that Inits is a subset of State; every operation $\{Op_i\}$ is a triple (N, B, D) such that dom N, D, and B are disjoint subsets of State; Fin is a relation from State to G.   □

Operations are split into three parts. When Op = (N, B, D) the normal effect of Op is

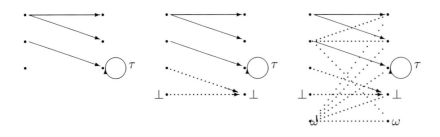

Figure 10.1: [60] The original Op, and a divergent after-state; with $\mathsf{B}_\perp \times \{\perp\}$ added; finally also with $\mathsf{D}_\omega \times \mathsf{State}_\omega \cup \{(\omega, \perp)\}$.

given by the relation N. The sets B and D describe areas from where the operation would lead to blocking and divergence, respectively. If the three sets form a partition, that excludes certain situations, such as miracles and possible (as opposed to certain) deadlock from a given state. This also ensures that (using $\perp$ and ω) it can be represented by a *total* data type. It is still possible for a *trace* to lead to possible deadlock, namely when some but not all of the states it leads to exhibit a deadlock.

This properly generalises the blocking and non-blocking totalisations, namely:

- the blocking operation Op is represented by $(\mathsf{Op}, \overline{\mathrm{dom}\,\mathsf{Op}}, \varnothing)$, i.e., it never diverges, and blocks in the complement of the operation's domain;

- the non-blocking operation Op is represented by $(\mathsf{Op}, \varnothing, \overline{\mathrm{dom}\,\mathsf{Op}})$, i.e., it diverges in the complement of the operation's domain, but never blocks.

In the embedding below, which translates process data types into total data types, we will use state spaces enhanced with special values as before – now also using a third value "no" to represent the impossibility of making an observation in a final state.

Definition 10.16 (Embedding process data types).
For any set State, let $\mathsf{State}_{\perp,\omega,\mathsf{no}} == \mathsf{State} \cup \{\perp, \omega, \mathsf{no}\}$, and similarly for sets subscripted with subsets of these special values.

A process data type $(\mathsf{State}, \mathsf{Inits}, \{(\mathsf{N}_i, \mathsf{B}_i, \mathsf{D}_i)\}_{i \in J}, \mathsf{Fin})$ with global state G is embedded into the total data type $(\mathsf{State}_{\perp,\omega}, \mathsf{Init}, \{[\![\mathsf{Op}]\!]_i\}_{i \in J}, [\![\mathsf{Fin}]\!])$ where

$$\mathsf{Init} == \mathsf{G}_{\perp,\omega,\mathsf{no}} \times \mathsf{Inits}$$
$$[\![(\mathsf{N}, \mathsf{B}, \mathsf{D})]\!] == \mathsf{N} \cup (\mathsf{B}_{\perp,\omega} \times \{\perp\}) \cup (\mathsf{D}_\omega \times \mathsf{State}_\omega)$$
$$[\![\mathsf{Fin}]\!] == \mathsf{Fin} \cup (\overline{\mathrm{dom}\,\mathsf{Fin}} \times \{\mathsf{no}\}) \cup \{(\perp, \perp)\} \cup \{\omega\} \times \mathsf{G}_{\omega,\mathsf{no}}$$

$\square$

The process data type has a set of initial states, each of which gets related to every global state. Operations are embedded in line with the explanation above: as a union of normal behaviour and behaviour representing blocking and divergence from the respective sets. Finalisation makes both blocking and divergence visible globally. Figure 10.1 [60] illustrates this operation embedding.

Now we can turn the handle on the refinement derivation machine again. Given two process data types, we construct their total data type embeddings, apply the simulation rules for total data types, and then simplify. In particular, any reference to the artificial values should be eliminated in this simplification step. This then leads to sound simulation rules for process data types.

Those for downward simulation on process data types come out as follows. The relation R between AState and CState is a downward simulation between the process data types $(\mathsf{AState}, \mathsf{AInits}, \{\mathsf{AOp}_i\}_{i \in J}, \mathsf{AFin})$ and $(\mathsf{CState}, \mathsf{CInits}, \{\mathsf{COp}_i\}_{i \in J}, \mathsf{CFin})$, iff

$$\mathsf{CInits} \subseteq \mathrm{ran}(\mathsf{AInits} \lhd \mathsf{R})$$
$$\mathsf{R} \,\mathbin{\mathrm{\mathring{9}}}\, \mathsf{CFin} \subseteq \mathsf{AFin}$$
$$(\mathrm{dom}\, \mathsf{AFin}) \lhd \mathsf{R} = \mathsf{R} \rhd (\mathrm{dom}\, \mathsf{CFin})$$

and $\forall\, i{:}J$, for $\mathsf{AOp}_i = (\mathsf{AN}, \mathsf{AB}, \mathsf{AD})$, $\mathsf{COp}_i = (\mathsf{CN}, \mathsf{CB}, \mathsf{CD})$

$$\mathsf{AD} \lhd \mathsf{R} \,\mathbin{\mathrm{\mathring{9}}}\, \mathsf{CN} \subseteq \mathsf{AN} \,\mathbin{\mathrm{\mathring{9}}}\, \mathsf{R}$$
$$\mathrm{dom}(\mathsf{R} \rhd \mathsf{CB}) \subseteq \mathsf{AB}$$
$$\mathrm{dom}(\mathsf{R} \rhd \mathsf{CD}) \subseteq \mathsf{AD}$$

As one might have expected, there are no changes in the initialisation and finalisation rules. For operation, "correctness" is modified only to constrain to non-divergent abstract states. The other two rules are reflections of "applicability", generalising it separately for blocking and non-blocking interpretations of partial operations.

Second, the conditions for upward simulation for process data types. The relation T is an upward simulation between the process data types $(\mathsf{AState}, \mathsf{AInits}, \{\mathsf{AOp}_i\}_{i \in J}, \mathsf{AFin})$ and $(\mathsf{CState}, \mathsf{CInits}, \{\mathsf{COp}_i\}_{i \in J}, \mathsf{CFin})$, iff

$$\mathrm{ran}(\mathsf{CInits} \lhd \mathsf{T}) \subseteq \mathsf{AInits}$$
$$\mathsf{CFin} \subseteq \mathsf{T} \,\mathbin{\mathrm{\mathring{9}}}\, \mathsf{AFin}$$
$$\overline{\mathrm{dom}\, \mathsf{CFin}} \subseteq \mathrm{dom}(\mathsf{T} \rhd\!\!\!\!\rhd \overline{\mathrm{dom}\, \mathsf{AFin}})$$

and $\forall\, i{:}J$, for $\mathsf{AOp}_i = (\mathsf{AN}, \mathsf{AB}, \mathsf{AD})$, $\mathsf{COp}_i = (\mathsf{CN}, \mathsf{CB}, \mathsf{CD})$

$$\mathrm{dom}(\mathsf{T} \rhd \mathsf{AD}) \lhd \mathsf{CN} \,\mathbin{\mathrm{\mathring{9}}}\, \mathsf{T} \subseteq \mathsf{T} \,\mathbin{\mathrm{\mathring{9}}}\, \mathsf{AN}$$
$$\mathsf{CB} \subseteq \mathrm{dom}(\mathsf{T} \rhd \mathsf{AB})$$
$$\mathsf{CD} \subseteq \mathrm{dom}(\mathsf{T} \rhd \mathsf{AD})$$

The comparison with "normal" rules is similar here to the downward simulation scenario. Initialisation is essentially the same, apart from describing sets rather than relations. Finalisation conditions account for partiality of finalisation. If we used process data types to embed a refusal semantics [60], correctness of finalisation ("are we making the correct observations at the end?") is the only rule significantly affected. The property (see [105]) that finalisation conditions resulting from an output embedding are trivially satisfied for downward simulation (and thus Z rules do not need to mention it) also transfers across.

10.6 Conclusions

As we have seen, it is possible to embed concurrent refinement relations into relational models in a natural fashion, and derive simulation rules for those refinement relations as a consequence. The two parameters that we varied in the embeddings used were the observations made by a finalisation, and the relational model – specifically additional values used in encodings.

The technical details of the definitions and results of some of the above work is given in more detail in [104, 60, 106]. Our focus here in this chapter though has been the relationship between the two relational models, and we have discussed how in general aspects of both are needed. This resulted in the definition of a process data type. Of particular relevance to our discussion here is the use of the partial model, and [106] contains further applications of this to, for example, notions of refinement in automata. A fuller discussion on modelling divergence is given in [58].

A notable absence in our discussion given above is the modelling of input and output. Surprisingly they (or more specifically the inclusion of outputs) add considerable complexity to the embeddings used in relational refinement modelling – and the full simulation rules are consequently complex depending on the exact model of outputs used. Further details of the subtleties involved are given in [60].

Chapter 11

Refinement of Behavioural Models for Variability Description

Alessandro Fantechi

Dipartimento di Ingegneria dell'Informazione, Università di Firenze, and ISTI–CNR

Stefania Gnesi

ISTI–CNR

Abstract. In Product Lines Engineering many studies are focused on the research of the best behavioural model useful to describe a product family and to reason about properties of the family itself. In addition the model must allow to describe in a simple way different types of variability needed to characterize several products of the family. Modal Transition System (MTS) is one of the more relevant behavioural models that has been broadly studied in literature and several extensions have been also described to more finely address behavioural variability. Furthermore MTS and its extensions define a concept of refinement which represents a step of design process, namely a step where some allowed requirements are discarded and other ones become necessary. In this chapter we present a bunch of variants of the classic MTS definition with the associated refinement definitions and a discussion on their expressive power.

11.1 Introduction

Product Lines or Families represent a new paradigm widely used to collectively describe products of a company that share similar functionality and requirements, in order to improve development efficiency and productivity [84, 267]. In this context many studies are focused on the search of the best behavioural models useful to describe a product family and to reason about properties of the family itself. In addition the model must allow to describe in a simple way different types of variability, needed to characterize several products of the family. One of these models are the Modal Transition Systems (MTSs) introduced by Larsen and Thomsen in [219]. They are an extension of the Labelled Transition Systems (LTSs) [200], and introduce two types of transitions useful to describe the necessary and allowed requirements, typical of product lines specifications, by distinguishing transitions that *must* (mandatory transitions) be done from those that *may* (optional transitions) be done. These models have been broadly studied as a possible means to express behavioural variability [132, 218, 130, 19, 20], and several extensions have been described. These extensions follow different approaches which entail the introduction of more complex and expressive requirements.

In this paper we review a bunch of variants of the classic MTSs definition, that have been proposed in the literature to more finely address behavioural variability discussing their expressive power. The discussed variants include *Disjunctive Modal Transition Systems* (DMTS) [220], where a must transition can be substituted by a must hypertransition, that is a set of transitions, and its semantics requires that at least one of these transitions has to be present in any element of the family. Another variant is *1-selecting modal transition system* (1MTS) defined in [131] which modifies the semantics of must hypertransition transforming the disjunctive choice in an exclusive choice: for every set of transitions related to a hypertransition one and only one transition has to be present in any element of the family. In [130] a generalization of DMTS and 1MTS was defined as *Generalized Extension Modal Transition System* (GEMTS), which introduces two new types of hypertransitions, described by means of $\diamond$ and $\square$, which allow to choose among "at most k of n" transitions for $\diamond$ and "at least k of n" transitions for $\square$.

Furthermore MTS and its extensions allow a concept of *refinement* to be defined. It represents a step of design process, namely a step where some allowed requirements are discarded and other ones become necessary. In this way when we say N is a refinement of M, we mean that N is a model derived by M removing some optional transitions and/or changing some optional transitions into mandatory ones. A refinement step therefore models the decision of fixing some optionality in the family description. Hence refinement can be used to restrict the variability exhibited by a family in order to derive from it subfamilies or products. When no more variability remains we call the model an *implementation*.

The paper is organized as follows. In Section 11.2 we introduce a running example and some preliminary definitions. In Section 11.3 we present the variants of MTSs together with the associated refinements relations, while in Section 11.4 we compare their expressiveness. Our conclusions are presented in Section 11.5.

11.2   Running Example and Background

We will use in this paper the running example from [19, 20]. It describes a family of (simplified) coffee machines through the following list of requirements:

1. A vending machine is activated by a coin. The only accepted coins are the one euro

coin for European products and the one dollar coin for Canadian products. Only one kind of coin is accepted.

2. After inserting a coin, the user has to choose whether (s)he wants sugar, after which (s)he may select a beverage;

3. The choice of beverage (coffee, tea, cappuccino) varies, but coffee must be offered by all products of the family;

4. Optionally, a ringtone may be rung after delivering a beverage;

5. After the beverage is taken, the machine returns idle.

As we can see, these requirements describe a set of possible, partially different products and we can divide them in two large categories: Canadian products and European products. Moreover each category has several possible products choosing the different alternatives.

11.2.1 Labelled Transition Systems

Definition 11.1 (LTS). *A Labelled Transition System (LTS) is a 4-tuple $(Q, A, \overline{q}, \delta)$, with set Q of states, set A of actions, initial state $\overline{q} \in Q$, and transition relation $\delta \subseteq Q \times A \times Q$. If $(q, a, q') \in \delta$, then we also write $q \xrightarrow{a} q'$.*

When modelling the behaviour of a product as an LTS, products of a family are considered to differ for the actions that they are able to perform at any given state; this means that the definition of a family has to accommodate all the possibilities desired for each derivable product, predicating on the choices that keep a product belonging to the family.

We can build an LTS representing all the possible behaviours conceived for the family of our running example as that represented in Figure 11.1.

We can notice that the given LTS cannot distinguish mandatory transitions from optional ones since variation points in the family definition are modeled as nondeterministic choices (i.e., alternative branches) independently from the type of variability.

The proposed example has indeed three *variation points* (namely, the choice of the coins, the choice among provided drinks, and the optional alert tone) which are:
- the first, *alternative*, that is, the two coin transitions cannot be present together in any product, but one should be present;
- the second and third *optional*, that is, the transitions may or may not be present in any product.

Starting from the LTS describing the family, a set of LTSs representing possible products may be derived. The derivation of a product will amount to choose some alternative branches from any node of an LTS, while removing the others. Removing branches from an LTS L produces an LTS L' such that L *simulates* L', according to the following definition [245].

Definition 11.2. *Let $L = (Q, A, \overline{q}, \delta)$ and $L' = (Q', A, \overline{q'}, \delta')$ We say that $q \in Q$ simulates $q' \in Q'$ if there exists a simulation that relates q and q'.*
$\mathcal{R} \subseteq Q \times Q'$ is a simulation if $\forall (q, q') \in \mathcal{R}$

- *$(q, a, q_1) \in \delta$ implies $\exists q_1':(q', a, q_1') \in \delta'$ and $(q_1, q_1') \in \mathcal{R}$.*

The above definition is naturally extended to LTSs by considering their initial states: an LTS L simulates L' iff $\overline{q}$ simulates $\overline{q'}$.

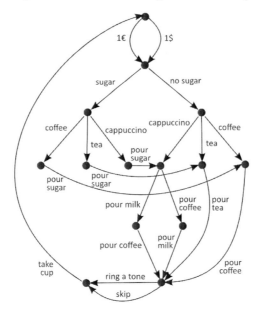

Figure 11.1: Modelling a product family with an LTS

Since an LTS cannot distinguish required transitions from optional ones not all LTS simulated by the one expressing the family are correct *products* of the family. Hence simulation between LTSs is a too coarse relation to express the refinement of a family in a subfamily.

11.3   Behavioural Models and Variability

In this section we introduce several frameworks developed to describe a specification with variability, more expressive than the LTSs seen in Section 11.2.1, so that they can properly distinguish mandatory from optional transitions. The models differ for their expressive power with respect to the type of variation points, that is, to their ability of expressing alternative transitions or more complex schemes of variability. Even though a categorization of these models has never been made, we can see easily some common characteristic in their definition:

1. every model is a particular type or an extension of transition systems.

2. some models adopt the modal operators □ "necessity" and ◇ "possibility": we could call the set of these models as *Modal Family*.

3. some models in the Modal Family introduce the hypertransition concept, that is a transition described by a pair (q, T) where q is a source state and T is a set of pairs (l, q') where l is a label and q' is a possible target state.

11.3.1   MTS: Modal Transition Systems

In [219] Larsen and Thomsen introduced a new formalism: the *Modal Transition System*. They noted that the LTS formalism is expressively too poor in order to provide a convenient specification. In effect, any specification defined through an LTS will limit the possible implementations to a single (behavioural) equivalence class. We would like to describe by a specification a wide collection of (possibly inequivalent) implementations and, exploiting some suitable techniques, this collection should be constantly reduced during the design process, in order to eventually determine a single implementation. It becomes clear that LTSs are not expressive enough for this task.

Definition 11.3 (MTS [219]). *A Modal Transition System (MTS) is a 5-tuple* $(Q, A, \overline{q}, \delta^{\diamond}, \delta^{\square})$, *with underlying LTS* $(Q, A, \overline{q}, \delta^{\diamond} \cup \delta^{\square})$. *An MTS has two distinct transition relations:* $\delta^{\diamond} \subseteq Q \times A \times Q$ *is the* may *transition relation, expressing* possible *transitions, while* $\delta^{\square} \subseteq Q \times A \times Q$ *is the* must *transition relation, expressing* mandatory *transitions. By definition,* $\delta^{\square} \subseteq \delta^{\diamond}$.
If $(q, a, q') \in \delta^{\diamond}$, *then we also write* $q \xrightarrow{a}_{\diamond} q'$ *and likewise we also write* $q \xrightarrow{a}_{\square} q'$ *for* $(q, a, q') \in \delta^{\square}$.

Note that an LTS is an MTS where the two transition relations $\delta^{\square}$ and $\delta^{\diamond}$ coincide.

An MTS provides an abstract description of a product family's set of products [132, 218, 130, 19, 20]. The products differ w.r.t. the actions they can perform in any given state: the MTS must allow all possibilities desired for each derivable valid product, affirming the choices that make a product belong to the family.

The MTS depicted in Figure 11.2, in which dashed arcs are used for the may transitions that are not must transitions $(\delta^{\diamond} \backslash \delta^{\square})$ and solid ones for must transitions $(\delta^{\square})$, is an attempt to model all possible behaviours conceived for the family of coffee machines described in Section 11.2.

Note that an MTS is thus able to model the constraints concerning *optional* and *mandatory* features (by means of may and must transitions). However, no MTS is able to model the constraint that 1€ and 1$ are exclusive nor that a cappuccino is not offered in Canadian products. We now make this claim somewhat clearer by defining how to generate products (LTSs) from a product family (MTS). This concept is formalized by the notion of refinement:

Definition 11.4 (Refinement [219]). *An MTS* $\mathcal{F}_p = (Q_p, A, \overline{q}_p, \delta_p^{\diamond}, \delta_p^{\square})$ *is a refinement of an MTS* $\mathcal{F} = (Q, A, \overline{q}, \delta^{\diamond}, \delta^{\square})$, *denoted by* $\mathcal{F} \preceq \mathcal{F}_p$, *if and only if there exists a refinement relation* $\mathcal{R} \subseteq Q \times Q_p$ *such that* $(\overline{q}, \overline{q}_p) \in \mathcal{R}$ *and for any* $a \in A$ *and for all* $(q, q_p) \in \mathcal{R}$, *the following holds:*

1. *whenever* $q \xrightarrow{a}_{\square} q'$, *for some* $q' \in Q$, *then* $\exists\, q_p' \in Q_p : q_p \xrightarrow{a}_{\square} q_p'$ *and* $(q', q_p') \in \mathcal{R}$, *and*

2. *whenever* $q_p \xrightarrow{a}_{\diamond} q_p'$, *for some* $q_p' \in Q_p$, *then* $\exists\, q' \in Q : q \xrightarrow{a}_{\diamond} q'$ *and* $(q', q_p') \in \mathcal{R}$.

Of course a straightforward generalization allows us to compare states from different MTSs.

Moreover if $\overline{q} \preceq \overline{q'}$, where $\overline{q}$ and $\overline{q'}$ are the initial states of L and M respectively and L is an LTS and M is an MTS then we will say that L is an *implementation* (or a product) of M.

From the MTS in Figure 11.2, the derivation of products (LTSs), including the European and Canadian coffee machines of Figures 11.3(a) and 11.3(b), is possible using Definition 11.4. Note, however, that also the derivation of the LTS of Figure 11.1 is possible by

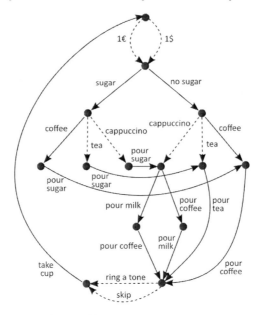

Figure 11.2: MTS modelling the family of coffee machines.

Definition 11.4, but the product it models violates the constraints of requirements 1 and 3 (cf. Section 11.2) by allowing the insertion of 1€ and 1$ and offering cappuccino.

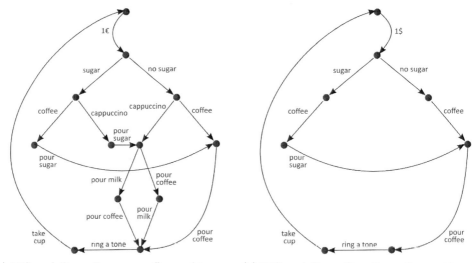

(a) LTS modelling a European coffee machine.   (b) LTS modelling a Canadian coffee machine.

Figure 11.3: (a)-(b) LTS modelling some correct products.

11.3.2 DMTS

Sometimes in the modelling of a product-line we would like to say "taken a set of possible features, at least one of them must be present in our products". Unfortunately, we cannot handle this situation using MTSs but the DMTS formalism resolves this problem. The DMTS, introduced in [220, 51], extends the MTS formalism: the type of must transition is modified from transition to *hypertransition*, whereas the may transitions are unchanged.

Definition 11.5 (DMTS [220]). *A Disjunctive Modal Transition System (DMTS) is a 5-tuple* $(Q, A, \overline{q}, \delta^\diamond, \delta^\square)$, *with set* Q *of states, set* A *of actions, initial state* $\overline{q} \in Q$, *and two distinct transition relations:* $\delta^\diamond \subseteq Q \times A \times Q$ *is the* may *transition relation and* $\delta^\square \subseteq Q \times \mathcal{P}(A \times Q)$ *is the* must *hypertransition relation.*

Intuitively, $q \to_\square U$, with $U \subseteq A \times Q$ may be understood as $\bigvee_{(a,q') \in U} q \xrightarrow{a}_\square q'$.
In contrast to MTS, DMTS allows us to define an inconsistent specification by the expression $q \to_\square \varnothing$. From now on we implicitly rule out this situation, by assuming that no transition of the form $q \to_\square \varnothing$ is allowed.

The next step is to introduce the refinement relation for DMTS:

Definition 11.6 (DMTS refinement). *A DMTS* $\mathcal{F}_p = (Q_p, A, \overline{q}_p, \delta^\diamond_p, \delta^\square_p)$ *is a refinement of a DMTS* $\mathcal{F} = (Q, A, \overline{q}, \delta^\diamond, \delta^\square)$, *denoted by* $\mathcal{F} \preceq \mathcal{F}_p$, *if and only if there exists a refinement relation* $\mathcal{R} \subseteq Q \times Q_p$ *such that* $(\overline{q}, \overline{q}_p) \in \mathcal{R}$ *and for any* $a \in A$ *and for all* $(q, q_p) \in \mathcal{R}$, *the following holds:*

1. *whenever* $q \to_\square V$, *for some* $V \subseteq A \times Q$, *then* $\exists U \subseteq A \times Q: \forall (a, q') \in V, \exists (a, q'_p) \in U$ *and* $(q', q'_p) \in \mathcal{R}$, *and*

2. *whenever* $q_p \xrightarrow{a}_\diamond q'_p$, *for some* $q'_p \in Q_p$, *then* $\exists q' \in Q: q \xrightarrow{a}_\diamond q'$ *and* $(q', q'_p) \in \mathcal{R}$.

As in the MTS case, a straightforward generalization allows us to compare states from different DMTSs, moreover if $\overline{q} \preceq \overline{q'}$, where $\overline{q}$ and $\overline{q'}$ are respectively the initial states of an LTS L and of a DMTS M, then we may say that L is an implementation of M.

Example 11.1. Suppose that our vending machine has this requirement: "The choice of drinks (coffee, tea, cappuccino) varies between the products. However, every product of the family delivers at least one different drink". Then we can model it using the DMTS as described in the Figure 11.4. Note that, for convenience, may transitions are not described. This is not an error since must hypertransitions guarantee us the presence of may transitions implicitly. Moreover, as we can see, the LTSs L, M and N are some of the possible implementations of our DMTS.

11.3.3 1MTS

As we said in the DMTS section, taken a set of possible features, the DMTS allows us to choose among these features in a disjunctive way, that is for every set we can make a disjunctive choice. A simple extension may be to change the choice type. Using 1MTS, introduced in [131], we can choose in an exclusive way and, from the modelling point of view, the exclusive choice is equivalent to say "taken a set of possible features, one and only one of them must be present in our products". The 1MTS takes advantage of the hypertransition concept and in addition it introduces a new concept: the choice function.

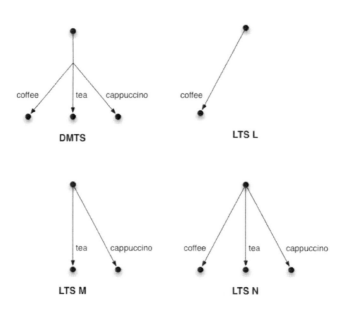

Figure 11.4: Modelling a product family with an LTS

Definition 11.7 (Choice function). *Let S be a set, $\mathcal{P}_S \subseteq \mathcal{P}(S)$ and $\gamma{:}\mathcal{P}_S \to S$. Then γ is a choice function if $\forall\, B \in \mathcal{P}_S, \gamma(B) \in B$. We denote the set of all choice functions on $\mathcal{P}_S$ by choice$(\mathcal{P}_S)$.*

In our context S will be the set of all possible transitions of the entire specification, $\mathcal{P}(S)$ will be the set of all possible hypertransitions that can be built over S, $\mathcal{P}_S$ will be the set of all possible hypertransitions present in our specification, T will be a particular hypertransition and a function γ, taken a hypertransition T, will return one and only one element of T, that is a transition. Moreover, in order to handle the hypertransition in a simpler way, we introduce the following notation:
Let $\to\, \subseteq S \times A \times S$ be a generic relation. If $s \in S$ we write:
$(s \xrightarrow{a})$ for $\{t \in S \mid (s, a, t) \in \to\}$ and
$(s \to)$ for $\{(a, t) \in A \times S \mid (s, a, t) \in \to\}$.

Definition 11.8 (1MTS [131]). *A 1-selecting Modal Transition System (1MTS) is a 5-tuple $(Q, A, \overline{q}, \delta^{\diamond}, \delta^{\square})$, with set Q of states, set A of actions, initial state $\overline{q} \in Q$, and two distinct transition relations: $\delta^{\diamond} \subset Q \times (\mathcal{P}(A \times Q)\backslash\varnothing)$ is the* may *hypertransition relation and $\delta^{\square} \subseteq Q \times (\mathcal{P}(A \times Q)\backslash\varnothing)$ is the* must *hypertransition relation.*
Moreover, $\delta^{\square} \subseteq \delta^{\diamond}$ (consistency requirement).

With respect to DMTSs, the may transition relation also uses hypertransitions and both may relation and must relation cannot include the "inconsistent" hypertransition, that is the hypertransition (s, T) where $T = \varnothing$. The reason for the introduction of may hypertransition is simple. Consider the system in Figure 11.5(a). It may be either interpreted as a DMTS (Figure 11.5(b)) or as a 1MTS (Figure 11.5(c)) and for a better understanding we draw the must hypertransition and the may transitions explicitly.

As we can see in Figure 11.6 LTSs L and I are two possible implementations of M, furthermore the DMTS N is a refinement of M. If we consider the system (b) in Figure 11.5 with the exclusive interpretation of must hypertransitions, we can easily note that this system fails, in effect the LTS I in Figure 11.6 is an implementation of this system but does not satisfy the exclusive interpretation. The 1MTS described in Figure 11.5(c) solves this problem.

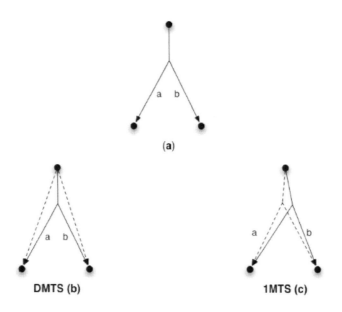

Figure 11.5: An example of DMTS and 1MTS

In order to define refinement over 1MTSs we need a novel concept: if $\mathcal{R} \subseteq \mathcal{S} \times \mathcal{S}$ is a generic relation between states, then the *extension* of $\mathcal{R}$ to $(A \times S) \times (A \times S)$ is: $(\theta, \theta_1) \in \mathcal{R} \Leftrightarrow a = a_1 \wedge (q', q'_1) \in \mathcal{R}$, where $(a, q'), (a_1, q'_1) \in A \times S$.

Definition 11.9 (1MTS refinement). *A 1MTS $\mathcal{F}_p = (Q_p, A, \overline{q}_p, \delta_p^\diamond, \delta_p^\square)$ is a refinement of a 1MTS $\mathcal{F} = (Q, A, \overline{q}, \delta^\diamond, \delta^\square)$, denoted by $\mathcal{F} \preceq \mathcal{F}_p$, if and only if there exists a refinement relation $\mathcal{R} \subseteq Q \times Q_p$ such that $(\overline{q}, \overline{q}_p) \in \mathcal{R}$ and for any $a \in A$ and for all $(q, q_p) \in \mathcal{R}$ and $\forall \gamma \in choice(q_p \to_\diamond). \exists \gamma' \in choice(q \to_\diamond)$ such that the following holds:*

1. *$\forall \Theta_p \in (q_p \to_\diamond). \exists \Theta_q \in (q \to_\diamond):(\gamma(\Theta_p), \gamma'(\Theta_q)) \in \mathcal{R}$*

2. *$\forall \Theta_q \in (q \to_\square). \exists \Theta_p \in (q_p \to_\square):(\gamma(\Theta_p), \gamma'(\Theta_q)) \in \mathcal{R}$*

Example 11.2. Suppose that our vending machine has this requirement: "The choice of drinks (coffee, tea, cappuccino) varies between the products. However, every product of the family delivers one and only one different drink". Then we can model this request using the 1MTS as described in the Figure 11.7. As we can see the LTSs L, M, and N are all possible implementations of our 1MTS.

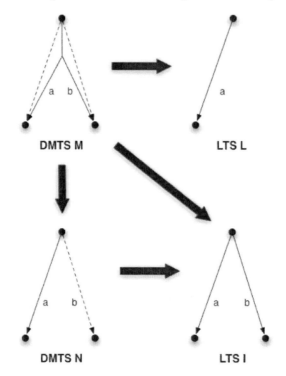

Figure 11.6: An example of DMTS and its problem with exclusive choices

11.3.4 Generalised Extended Modal Transition Systems

A more general notion is actually needed if we want to model *multiple optionality*, that is, the fact that a product may have (at least, at most, exactly) k of the n choices proposed by the family.

It is a matter of discussion whether multiple optionality is really useful in family engineering. Actually, it is rarely the case that a functional requirement on the products of a family gives bounds on the number of the possible features present in a product. However, nonfunctional requirements may do: for example, energy consumption or budget considerations may give an upper bound to the number of provided features, while marketing strategies may suggest a lower bound, under which the product loses its market. Also, upper and lower bounds may be used to define subfamilies on the basis of nonfunctional aspects. We will introduce the notion of multiple optionality in our running example by adding the requirement:

- "every product of the family delivers at least two different drinks".

This requirement could be dictated not by functional needs, but by marketing strategies.

In order to give a full range of behavioural models for variability types, the concept of Generalised Extended Modal Transition Systems, which is able to model multiple optionality has been introduced [130].

Definition 11.10 (GEMTS [129]). *A Generalised Extended Modal Transition System (GEMTS) is a quintuple $(Q, A, q_0, \Box, \Diamond)$, where Q is a set of states, A is a set of actions,*

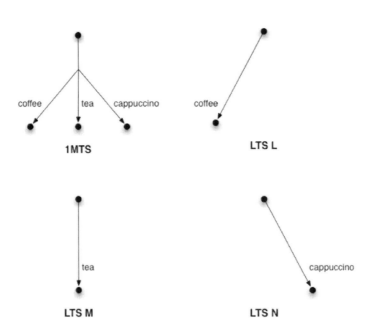

Figure 11.7: An example of 1MTS and its implementations

$q_0 \in Q$ *is the initial state,* $\square \subseteq Q \times 2^{A \times Q} \times N$ *is the* at least k-of-n *transition relation, and* $\diamond \subseteq Q \times 2^{A \times Q} \times N$ *is the* at most k-of-n *transition relation.*

We write respectively: $q \xrightarrow{a_1, a_2, \ldots, a_n}_{\square_k} q_1, q_2 \ldots, q_n$ and $q \xrightarrow{a_1, a_2, \ldots, a_n}_{\diamond_k} q_1, q_2 \ldots, q_n$ to denote elements of the two relations, meaning that in the first case any product of the family should have at least k of the n transitions $q \xrightarrow{a_i} q_i$, while in the second case any product of the family should have at most k of the n transitions (that is, it can also have no transition from this set). Again, the number of the actions on the arrow must coincide with that of target states, and order counts as well, since each action is paired to the corresponding state. Note that $0 < k \le n$ should always hold, otherwise the relation is meaningless.

The two defined transition relations have the following properties, that derive straight-forwardly from the interpretation of a GEMTS as model of a family of LTSs:

- $q \xrightarrow{a_1, a_2, \ldots, a_n}_{\square_k} q_1, q_2 \ldots, q_n \Rightarrow q \xrightarrow{a_2, \ldots, a_n}_{\square_{k-1}} q_2 \ldots, q_n$

- $q \xrightarrow{a_1, a_2, \ldots, a_n}_{\square_k} q_1, q_2 \ldots, q_n$ *and* $q \xrightarrow{a_1, a_2, \ldots, a_n}_{\diamond_k} q_1, q_2 \ldots, q_n$
  means: any product of the family should have *exactly* k of the n transitions $q \xrightarrow{a_i} q_i$; this defines the relation $\square \cap \diamond$ as the relation *exactly k of n*.

- If we let k to be equal to n, the may relation includes the must relation:
  $q \xrightarrow{a_1, a_2, \ldots, a_n}_{\square_n} q_1, q_2 \ldots, q_n \Rightarrow q \xrightarrow{a_1, a_2, \ldots, a_n}_{\diamond_n} q_1, q_2 \ldots, q_n$. Inclusion does not hold if $k \ne n$.

Note that the "at least 1-of-1" transitions coincide with the "exactly 1-of-1" ones and can be considered as the usual LTS transitions. For this reason, in the graphical notation we adopt for GEMTSs, in order to avoid notation overloading, we use the box and diamond symbols only in states corresponding to variation points: other states in the GEMTS with only "exactly 1-of-1" outgoing transitions are drawn as usual in LTS. This limited usage of the modality notation helps the reader to concentrate on variation points. For more simplicity, box and diamond symbols will be used without the number suffix when $n = 1$.

In our example, the added requirement that "every product of the family delivers at least two different drinks" can be modelled using the GEMTS shown in Figure 11.8.

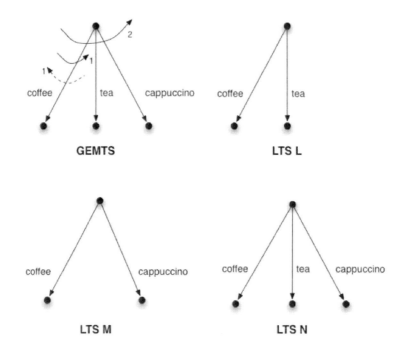

Figure 11.8: An example of GEMTS and its implementations

From the definition of GEMTS we can derive the definition of Extended Modal Transition Systems [129], in which at any state of the system, it can be defined whether to choose at least (or at most) one from a subset of the outgoing transitions.

Definition 11.11 (EMTS [129]). *An* Extended Modal Transition System *(EMTS) is a quintuple* $(Q, A, q_0, \Box, \Diamond)$, *where* Q *is a set of states,* A *is a set of actions,* $q_0 \in Q$ *is the initial state,* $\Box \subseteq Q \times 2^{A \times Q}$ *is the* at least 1-of-n *transition relation, and* $\Diamond \subseteq Q \times 2^{A \times Q}$ *is the* at most 1-of-n *transition relation.*

An EMTS can be defined as a GEMTS in which the relations $\Box$ and $\Diamond$ are restricted to the constant value $k = 1$: this is obvious considering that the two relations can be read "at least 1-of-n" and "at most 1-of-n". Consequently, an MTS can be defined as a GEMTS in which the relations $\Box$ and $\Diamond$ are restricted to the constant value $k = 1$, and where only singleton pairs from $A \times Q$ appear (that is, $n = 1$).

Note therefore that $s \xrightarrow{a}_{\Box_1} s' \Rightarrow s \xrightarrow{a}_{\Diamond_1} s'$ (if a transition is required is also possible, consistently with the definition of MTSs).

Defining a refinement relation on GEMTS is not immediate, due to the complex structure of hypertransitions. A convenient way is by recurring to the notion of product, that parallels on GEMTS the notion of implementation given for MTSs.

A GEMTS defines a family of LTSs according to the following definition:

Definition 11.12 (GEMTS products). *An LTS $P = (Q_P, A, p_0, \rightarrow)$ belongs to the family described by the GEMTS $F = (Q_F, A, f_0, \Box, \Diamond)$ (we say also P is a product of F, or P conforms to F, written $P \vdash F$) if and only if $p_0 \vdash f_0$, where:*

$p \vdash f$ *if and only if*

- $f \xrightarrow{a_1, a_2, \ldots, a_n}_{\Box_k} f_1, f_2 \ldots, f_n \Rightarrow$
  $\exists I \subseteq \{1, \ldots, n\}, k \leq | I | \leq n \colon \forall i \in I, p \xrightarrow{a_i} p_i$ *and* $p_i \vdash f_i$

- $f \xrightarrow{a_1, a_2, \ldots, a_n}_{\Diamond_k} f_1, f_2 \ldots, q_n \Rightarrow$
  $\nexists I \subseteq \{1, \ldots, n\}, k < | I | \leq n \colon \forall i \in I, p \xrightarrow{a_i} p_i$ *and* $p_i \vdash f_i$

- $p \xrightarrow{a} p' \Rightarrow \exists k, A' \subseteq A, Q \subseteq Q_F, f' \in Q_F \colon (a, f') \in A \times Q_F, (f, A' \times Q, k) \in \Box$ *or* $(f, A' \times Q, k) \in \Diamond$, *and* $p' \vdash f'$

We can read the previous clauses as saying: a product of a family has at least (at most) k of the n transitions specified in the family; moreover, if a product performs an action, this should be found among the n transitions specified for the family. The refinement relation defined in 11.3.4 is a restriction of a generic refinement relation, in effect it describes only the connection between a product (or LTS) and a specification (or GEMTS), all intermediate steps of the refinement process are ignored. In Figure 11.8, we can see that the LTSs L, M, and N are all possible implementations of the shown GEMTS.

According to the definition of *thorough refinement* proposed in [50], we give the following definition that is based on comparing the sets of products of two GEMTSs:

Definition 11.13 (GEMTS refinement). *Given two GEMTS F and G, we say that F refines G if $\{P_F \mid P_F \vdash F\} \subseteq \{P_G \mid P_G \vdash G\}$.*

The requirements given in Section 11.2 plus the one added above for the multiple optionality can be now fully expressed by the GEMTS shown in Figure 11.9.

The LTS of a product can be obtained from the family GEMTS by applying Definition 11.12. Hence, the set of products derivable are, again, a subset of those derivable by the family LTS (note that the family LTS can be obtained from the family GEMTS by just removing the modality notations in the graphical syntax). Therefore, any derivable product is still simulated by the family LTS.

The European vending machines of Figure 11.3a is derivable from the family GEMTS of Figure 11.9. Notice that now it is no more possible to derive the machine accepting both euro and dollar coins.

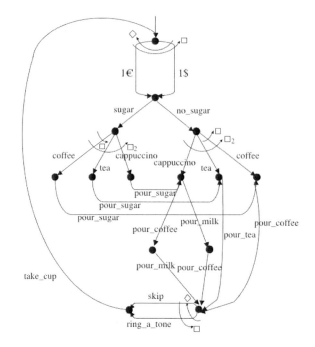

Figure 11.9: Vending machines family as a GEMTS

11.4   A Comparison on the Expressiveness

Table 11.1 shows the meaning of the modalities of the different variants of MTSs that have been proposed to model variability. DMTSs and 1-MTSs are analogous to GEMTSs for the fact that they all use hyper-transitions, that is, transition relations where the target is a set of states, in contrast with MTSs, in which the modal transition relations have only one target state. This allows modalities to predicate on a particular choice of a subset of transitions, rather than just on a single transition: note anyway that the possibility to have different (hyper)transitions from the same state is maintained in all the models.

Table 11.1: Meaning of modalities

modality	MTS	DMTS	1-MTS	EMTS	GEMTS
$\Diamond$ (may)	at most 1-of-1	at most n-of-n	at most 1-of-n	at most 1-of-n	at most k-of-n
$\Box$ (must)	at least 1-of-1	at least 1-of-n	exactly 1-of-n	at least 1-of-n	at least k-of-n

In [145] it has been shown that it is possible to determine a hierarchy comparing expressivity among all models of the Modal Family. The hierarchy, shown in Figure 11.10 has been defined under the assumption that all the considered models are *action-deterministic*, that is, a restriction is requested on the transition relation: the target state of any transition is univocally determined by its source state and its label.

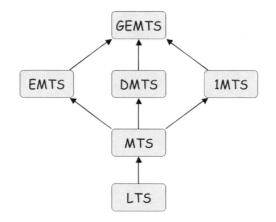

Figure 11.10: Hierarchy for the modal family

11.5 Conclusions

Refinement has traditionally been applied to the process of stepwise implementing an abstract specification through successive refinements adding implementation details. The seminal work of Kaisa Sere and her group has applied refinement to the action systems formalism according to this trend. [32]. We have indeed presented a different application of refinement, in which the abstract specification represents a family of different systems and each of them can be obtained by applying refinement steps, each of those fixes some variable aspect of the specification. We have reviewed some variants of Modal Transition Systems that are suitable to express behavioural variability in a family of systems. The concept of refinement in this context amounts to restricting the variability in order to define a subfamily of systems or a single system (*product*).

We have left out a few other more elaborate variants that have been proposed in the literature. In particular, other models, such as the OTS [48] (*Transition System with Obligations*) and the PMTS [49] (*Parametric Modal Transition System*), introduce the *obligation* concept, using logic formulae related to states. Every formula represents features requested. In [145] these models have been compared with the MTS variants discussed above and the appropriate definitions of refinement are given.

11.6 Acknowledgments

We wish to thank Christian Grioli for the work done on this subject during his master thesis [145].

Chapter 12

Integrating Refinement-Based Methods for Developing Timed Systems

Jüri Vain

Tallinn University of Technology

Leonidas Tsiopoulos

Åbo Akademi University

Pontus Boström

Åbo Akademi University

Abstract. Refinement-based development supported by Event-B has been extensively used in the domain of embedded and distributed systems design. For these domains timing analysis is of great importance. However, in its present form, Event-B does not have a built-in notion of time. The theory of refinement of timed transition systems has been studied, but a refinement-based design flow of these systems is weakly supported by industrial strength tools. In this paper, we focus on the refinement relation in the class of Uppaal Timed Automata and show how this relation is interrelated with the data refinement relation in Event-B. Using this interrelation we present how the Event-B and Uppaal tools can complement each other in a refinement-based design flow. The approach is demonstrated on a well-studied case-study, namely the IEEE 1394 tree identify protocol.

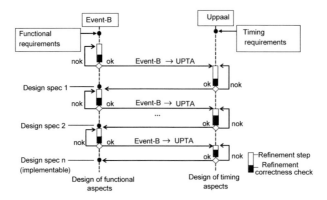

Figure 12.1: CCD workflow with data and timing refinement steps

12.1 Introduction

The Correct-by-Construction Design (CCD) workflow has proven itself in recent industrial practice. Peugeot Automobiles has developed a model of the car subsystems (lightings, airbags, engine, etc.) for Peugeot aftersales service; RATP (Paris Transportation) used to verify automatic doors for an existing metro line. Event-B [6] as one such CCD supporting formalism has proven its relevance in data intensive development while lacking sufficient support for timing analysis and refinement of timed specifications. Uppaal Timed Automata (UPTA) [47] address timing aspects of systems providing efficient data structures and algorithms for their representation and analysis but are less focused on supporting the refinement-based development, especially data refinement. The goal of this paper is to advocate the model-based design method where these two approaches are combined to mutually complement each other.

The design flow discussed in the paper comprises alternating steps of data and timing refinement (see Figure 12.1). The process starts with data refinement of requirements specification and proving refinement correctness within Event-B. The result of data refinement is input for timing refinement step performed using an UPTA model. The timing model is created as a result of mapping a refined Event-B model to an UPTA model that is then complemented with specific timing attributes. These attributes are initialized typically with values defined in the timing requirements specification. Further, the timing refinement step concerns the UPTA attributes that are not instantiated yet and/or those that are already instantiated but need to be modified to reduce timing non-determinism.

In the design workflow timing refinement steps are performed incrementally. If the UPTA model (the result of earlier transformation step), already exists, then it is reused for current Event-B to UPTA mapping. Only the new fragment of Event-B model that is introduced by current data refinement step needs to be mapped to UPTA and composed with the existing UPTA model. The resulting composition is subject to timing refinement.

Once specified, the timing refinement is verified using Uppaal [47] tool. The result of timing refinement step is also verified for consistency, i.e., against the properties such as deadlocks, non-Zenoness, connectedness, etc.

A first result for this approach was presented by Berthing et al. [54] where a part of an industrial safety critical system was abstractly specified and then refined to model the required redundancy. The refinement step was simple in the sense that it did not introduce new events to the refined model and used very simple data structures. This paper addresses introduction of new events and refinement of more complex data. The approach is applied to the IEEE 1394 case study, which is well-studied with both formalisms. The refinement cases concern incremental unfolding of details at each step with new variables and events. The tree identify process of IEEE 1394 is a leader election protocol which takes place after a bus reset in a network. The development steps are shown at first within Event-B and mapped then to UPTA for performing timing refinement and checking their correctness thereafter following the design workflow of Figure 12.1.

12.2   Related Work

An extensive study of automata models for timed systems is presented in [234] where a general model is defined for developing a variety of simulation proof techniques for timed systems. To improve model checking performance of timed systems timing constraint refinement methods such as the forward algorithm based on zones for checking reachability [63] and the counter example guided automatic timing refinement technique [111] have been studied. The refinement of timing has been addressed also as part of the specification technique in [97] where the constructs for refinement of Timed I/O specifications were defined for development of compositional design methodology. In [163] the authors propose a design approach where different forms of refinements are used to enhance model checking TCTL properties and preservation of these properties by refinements.

The motivation behind timing refinement in all these works has been rather model checking or automated design verification than refinement-based development. The way timing refinement steps are constructed in the course of practical design flow has deserved relatively little attention because the proof techniques rely rather on semantic models of refinement transformations than on syntactic ones the developer works with. The design transformations to be human comprehensible need to be specified in terms of high level modelling language and, thus, the refinement transformations need to be defined explicitly in terms of syntactic constructs of that language.

In [75, 215, 290] attempts have been made to incorporate discrete time directly into design description formalisms without native notion of time. Continuous time has been investigated also with a similar approach [322]. However, the clocks are not an integrated part, they are modelled as ordinary variables. Hence, continuous time specific problems such as Zeno behaviour cannot be addressed directly in, for example, the Event-B proof system. Modelling timed behaviour manually in Event-B is error prone and makes the model cluttered with timing properties, which also makes proofs harder to automate. Attempts to remedy this have been made with Hybrid Event-B [42]. An earlier attempt to integrate stepwise development in Event-B with model checking in Uppaal is given in [180] where events are grouped into more coarse grained processes with timing properties.

We aim to provide an integrated approach without deforming the underlying formalisms. Then the timing refinements can be addressed by reusing the model constructs of Event-B mapped to UPTA by keeping the event structure of the model as much as possible untouched. This allows verifying the data refinement steps also from the timing feasibility point of view.

The IEEE1394 example was studied earlier by following mostly a refinement and proof-based approach in [3] or by verifying the timing properties of the protocol for solving leader election contentions with timed automata specifications in [320]. Abrial et al. [3] presented an incremental specification of the protocol aiming at a generic specification without restrictions on the total number of nodes on the tree. The Event-B model was constructed in a usual, refinement-based manner where more details are unfolded at each refinement step with new variables and events. Probabilistic conflict resolving was applied for nodes choosing long or short waiting times. Thus, actual modeling of time was not concerned. On the other hand, Timed Automata applications to the protocol [320] reduced the timing correctness analysis of the protocol only on a pair of nodes contesting for network leadership. This approach helped in avoiding state explosion in the model checkers but ignored the aspects of data refinement outlined by Abrial et al.

For our proposed CCD methodology it is crucial to keep the Event-B and timing models in synchrony by mapping the Event-B specification to UPTA. Thus, it is of high interest to investigate to what extent UPTA can follow, as well as complement, the usual Event-B development and expose the timing related design faults that otherwise may remain undiscovered by the Event-B proof system.

12.3 Preliminaries

12.3.1 Preliminaries of Event-B

Consider an Event-B model M with variables v, invariant $I(v)$, and events $\mathcal{E}_1, \ldots, \mathcal{E}_m$. All events can be written in the form

$$\mathcal{E}_i = any\ x\ where\ G_i(v, x)\ then\ v :|\ S_i(v, x, v')\ end$$

where x are parameters, $G_i(v, x)$ is a predicate called the guard, and $v :|\ S_i(v, x, v')$ is a statement that describes a (nondeterministic) relationship between variable valuations before and after executing the event. The parameters can be omitted and the keyword *when* is then used instead of *where*. Event-B models do not have fixed semantics [6], but correctness of a model is defined by a set of proof obligations. We can use these proof obligations to prove correctness of many transition systems. In order to guarantee that $\mathcal{E}_i$ preserves invariant $I(v)$ we prove [6]:

$$I(v)\ \wedge\ G_i(v, x)\ \wedge\ S_i(v, x, v') \Rightarrow I(v')\ \text{(INV)}$$

In order to be able to relate Event-B with UPTA in the following sections we interpret an Event-B model as a Labelled Transition System (LTS) $(\Sigma, init, T, i)$, where Σ is the set of states, $init$ is the set of initial states, $T \subseteq \Sigma \times \Sigma$ is the set of transitions, and i is the set of legal states $i \subseteq \Sigma$. The set of states g_i where the guard of a transition σ_i holds is given as $g_i = \{v \mid \exists x \cdot G_i(v, x)\}$ and the relation s_i describing the before-after relation for states corresponding to the update statement $v :|\ S_i(v, x, v')$ is given as $s_i = \{v \mapsto v' \mid \exists x \cdot S_i(v, x, v')\}$. The relation describing the before-after states for each transition T_i is then given as[1] $g_i \lhd s_i$.

---

[1]The domain restriction operator $\lhd$ is defined as: $g \lhd s = \{\sigma \mapsto \sigma' \in s \mid \sigma \in g\}$.

We can now describe the Event-B model as a transition system $(\Sigma, init, T, i)$ where the state space Σ is formed from the variables $v_1, \ldots, v_n$, $\Sigma = \Sigma_1 \times \ldots \times \Sigma_n$, Σ_i is the type of v_i. The initial states are formed as $init = \{v \mid Init(v)\}$. The transitions T_i are given as $T_i = g_1 \lhd s_1 \cup \ldots \cup g_m \lhd s_m$. The set of legal states are the ones where the invariant $i = \{v \mid I(v)\}$ holds.

12.3.2 Preliminaries of UPTA

An UPTA is given as the tuple $(L, E, V, CL, Init, Inv, T_L)$, where L is a finite set of locations, E is the set of edges defined by $E \subseteq L \times G(CL, V) \times Sync \times Act \times L$, where $G(CL, V)$ is the set of constraints allowed in guards. *Sync* is a set of synchronisation actions over channels. An action *send* over a channel h is denoted by $h!$ and its co-action *receive* is denoted by $h?$. *Act* is a set of sequences of assignment actions as well as with clock resets. V denotes the set of data variables. CL denotes the set of real-valued clocks ($CL \cap V = \varnothing$). *Init* $\subseteq Act$ is a set of assignments that assigns the initial values to variables and clocks. $Inv: L \to I(CL, V)$ is a function that assigns an invariant to each location, $I(CL, V)$ is the set of invariants over clocks CL and variables V. $T_L: L \to \{ordinary, urgent, committed\}$ is the function that assigns the type to each location of the automaton.

We introduce the semantics of UPTA as defined in [47]. A clock valuation is a function $val_{cl}: CL \to \mathbb{R}_{\geq 0}$ from the set of clocks to the non-negative reals. A variable valuation is a function $val_v: V \to D$ from the set of variables to values. Let $\mathbb{R}^{CL}$ and D^V be the sets of all clock and variable valuations, respectively. The semantics of an UPTA is defined as an LTS $(\Sigma, init, \to)$, where $\Sigma \subseteq L \times \mathbb{R}^{CL} \times D^V$ is the set of states, the initial state $init = Init(cl, v)$ for all $cl \in CL$ and for all $v \in V$, with $cl = 0$, and $\to \subseteq \Sigma \times \{\mathbb{R}_{\geq 0} \cup Act\} \times \Sigma$ is the transition relation such that:

1. $(l, val_{cl}, val_v) \xrightarrow{\mathsf{d}} (l, val_{cl} + \mathsf{d}, val_v)$ if $\forall \mathsf{d}':0 \leq \mathsf{d}' \leq \mathsf{d} \Rightarrow val_{cl} + \mathsf{d}' \models Inv(l)$,

2. $(l, val_{cl}, val_v) \xrightarrow{act} (l', val'_{cl}, val'_v)$ if $\exists\, \mathsf{e} = (l, act, G(cl, v), r, l') \in E$ s.t. $val_{cl}, val_v \models G(cl, v)$, $val'_{cl} = [re \mapsto 0]val_{cl}$, and $val'_{cl}, val'_v \models Inv(l')$,

where for delay $\mathsf{d} \in \mathbb{R}_{\geq 0}$, $val_{cl} + \mathsf{d}$ maps each clock cl in CL to the value $val_{cl} + \mathsf{d}$, and $[re \mapsto 0]val_{cl}$ denotes the clock valuation which maps (resets) each clock in re to 0 and agrees with val_{cl} over $CL \backslash re$.

12.4 Mapping from Event-B Models to UPTA

As demonstrated in Sections 12.3.1 and 12.3.2 both Event-B and UPTA have common semantic ground defined as LTS. Though, for efficient mapping of Event-B to UPTA the mapping rules need to be defined on a syntactic level because the semantic mapping via LTS may cause huge models when having parallel composition and many data variables. Also mapping from LTS to UPTA may produce syntactically different (though semantically equivalent) results that are expensive to compare. Therefore, the goal in this paper is to define a syntactic level mapping between Event-B and UPTA and in that way to extend the Event-B based stepwise CCD to timed systems. Due to the incrementality of CCD the

Event-B to UPTA mappings depicted in Figure 12.1 preserve the locality of model changes introduced by Event-B refinements, i.e., only those model fragments that are introduced by Event-B refinements need to be mapped to the corresponding UPTA fragments. The rest of the UPTA model remains untouched by the Event-B refinement step.

In the following we progressively arrive to the mapping of models after we first discuss the syntax correspondence between Event-B and UPTA and the mapping of events using as a running example the development of IEEE 1394 tree identify protocol.

Mapping of functions and predicates. Integers and enumerated types in Event-B become integers, while finite sets and relations in Event-B are mapped to (multidimensional) arrays in UPTA. We can then implement the set and relational operators as C-functions in UPTA. For instance, guards of events stating that an element belongs to the domain of a function or a pair of mapped elements is present in a function. An example is the guard send_req_grd in UPTA shown in Figure 12.5a that is specified as a boolean function:

$$bool\ send\_req\_grd(id\_t\ x,\ id\_t\ y)\ \{$$
$$return\ forall\ (i : int\ [0, N])\ (REQ\ [x]\ [i] == 0)\ ;\}$$

returning false if the element $REQ\ (x, i)$ of two-dimensional boolean array REQ is 0 and true otherwise. In other words, when this guard is true it means that node x has not yet sent a parent request to node y. Generally, the functions are used in UPTA for encoding complex predicates and calculations to avoid visual overhead in the model graphical representation.

Mapping of events. Let

$$\mathcal{E}_i = when\ G_i(v)\ then\ v :|\ S_i(v, v')\ end$$

be an event specification in Event-B, then UPTA model template shown in Figure 12.2a is direct mapping of the event in UPTA where the occurrence of event $\mathcal{E}_i$ is simulated by executing the edge that has attributes: i) when_Gi(V) is the mapping of the event guard G_i, ii) V′ = then_Si(V, U′) is assignment statement that corresponds to the deterministic assignments of $v :|\ S_i(v, v')$, iii) U′:dom_U is an assignment statement that corresponds to non-deterministic assignments of $v :|\ S_i(v, v')$, and iv) location PrePost_Ei models pre- and post-control states of event $\mathcal{E}_i$.

Meaning of skip in Event-B. Assuming the skip-event in Event-B: i) does not have effect on the system state, ii) can be executed in any state at any time moment, iii) can interleave with all other events, iv) the number of skip events is unbounded, and v) it is not specified explicitly, we interpret the skip-event in UPTA as an abstraction of the event template depicted in Figure 12.2a and we present its model in Figure 12.2b. Due to i) and ii) the guard in skip-template is always true and the update statement keeps the variables unchanged. Like in Event-B the skip-events need not be specified explicitly in UPTA.

Timing of events. When adding in UPTA explicit timing constraint to events it is assumed that the occurrence of an event is instantaneous. An event takes place within some time interval $[lb,\ ub]$ provided it is enabled by guard G_i. For specification of these constraints new variables, namely set CL of clocks is added to the template of Figure 12.2a. The subject of weakest timing specification is the skip-event (Figure 12.2c) where the event occurrence is not restricted in time (time interval is $[inf\ dom(cl),\ supr\ dom(cl)]$). We usually assume the continuous interval $[lb,\ ub]$ where $lb = inf\ dom(cl) = 0$ and $ub = supr\ dom(cl) =$

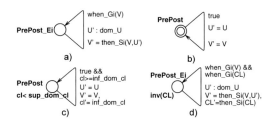

Figure 12.2: UPTA model template of Event-B a) un-timed event, b) skip-event, c) timed skip-event, and d) timed event.

$+\infty$. In UPTA syntactic notation the upper bound ub of the skip-event occurrence interval (see Figure 12.2c) is specified by the invariant of location PrePost, i.e., inv (PrePost) $\equiv$ $cl<supr\ dom\ (cl)$ and lower bound lb by the guard of edge $\langle$PrePost, PrePost$\rangle$. The same unbounded interval is assumed by default if the invariant of the location and the guard of the edge are left unspecified. Note that having $inf\ dom(cl) = 0$ and guard $cl \geq inf$ may introduce Zeno computations if there is a loop in the model that has maximum of lower bounds equal to 0.

In general, the timing specification of events, as illustrated in Figure 12.2d, introduces bounded intervals of occurrence that are specified as location invariant $inv(CL) \equiv \wedge_i cl_i \leq ub_i$ and the guard when_$Gi(CL) \equiv \wedge_j cl_j \geq lb_j$ of edge $\langle$PrePost, PrePost$\rangle$ over a set of clocks $cl_i \in CL$.

Modelling of system of events and ANY-construct. The translation of a set of Event-B events such as the ones in Figure 12.3 results in a parallel composition of UPTA processes that are instances of the template of Figure 12.2a above. The template may have parameters of the type bounded integer. That allows modelling of ANY-construct of Event-B, where the choice is finite. The parameter of template specified by its type defines the instances (processes) of the template, one for each value in the parameter type.

12.5   IEEE 1394 Case Study

The IEEE 1394 tree identify protocol case study will act as a running example throughout the paper. Before we proceed with the actual Event-B system model and its mapping to UPTA we first introduce shortly the idea of the protocol. The tree identify process of IEEE 1394 is a leader election protocol which takes place after a bus reset in the network. Immediately after a bus reset all nodes in the network have equal status, and know only to which nodes they are directly connected. A leader (root of the tree) needs to be elected as the manager of the bus. The protocol is designed for use in connected networks and will correctly elect a leader if the network is acyclic. Specifically, each node has two phases based on the number of children and the number of neighbours. If there are more than one neighbour, the node waits for requests from its neighbours to become their parent. If there is only one neighbour (and this neighbour is not a child), then the node sends a request to the

```
MACHINE  Ref2_IEEE1394
    ...
VARIABLES  req, ack, cnt
INVARIANT  req ∈ ND ⇸ ND ∧ ack ⊆ req ∧ ack ∩ ack~ = ∅ ∧ cnt ⊆ req ∧ ack ∩ cnt = ∅
EVENTS
    send_req = ANY x, y WHERE x↦y ∈ g ∧ y↦x ∉ ack ∧ x ∉ dom(req) ∧ g[{x}] = tr~[{x}] ∪ {y}
                      THEN req := req ∪ {x↦y} END;
    send_ack = ANY x, y WHERE x↦y ∈ req ∧ x↦y ∉ ack ∧ y ∉ dom(req) THEN ack := ack ∪ {x↦y} END;
    discover_cnt = ANY x, y WHERE x↦y ∈ req \ ack ∧ y ∈ dom(req) THEN cnt := cnt ∪ {x↦y} END;
    solve_cnt = ANY x, y WHERE x↦y ∈ cnt ∧ y↦x ∈ cnt THEN req, cnt := req \ cnt, ∅ END;
    progress = ANY x, y WHERE x↦y ∈ ack ∧ x ∉ dom(tr) THEN tr := tr ∪ {x↦y} END;
    elect = ANY x WHERE x ∈ ND ∧ g[{x}] = tr~[{x}] THEN ld, sp := x, tr END
END
```

Figure 12.3: Intermediate Event-B refinement of IEEE 1394 protocol

neighbour to become its parent. This implies that leaf nodes are the first to communicate with their neighbours, and that the spanning tree is built from the leaves.

Furthermore, the protocol may not proceed this straightforwardly because the parent requests are not atomic and contention may arise (two nodes simultaneously send the parent requests to each other). Since only one node can be the leader, contention must be resolved. This is achieved by timing. The standard specifies that each node chooses randomly whether to wait for a long or short time. If, after the wait period is over, there is a parent request from the other node, then the node becomes the root. If there is no such request, then the node resends its own parent request and contention may result again.

12.5.1 IEEE 1394 in Event-B

The deployment of our CCD methodology starts with a specification which captures an abstract behaviour of the whole system using the modelling approach by Abrial et al. [3]. For the purposes of this paper we do not present the very abstract specification and its first refinement since they do not contain essential timing information except the total time bound requirement for the protocol converging. The creation of the tree and the election of the leader node are supposed to be performed in "one shot" (abstract case of event *elect* in Figure 12.3). For the first refinement the authors gave the essence of the distributed algorithm with the tree being constructed in a step by step manner (addition of abstract case of event *progress* in Figure 12.3).

For the second refinement the communication mechanism has been introduced between the nodes. We start our case study from this point. The model is shown in Figure 12.3. In this refinement step the contention problem between two nodes is addressed. Let us first explain the main functionality regarding the construction of the tree. Two variables, *req* and *ack*, are used to handle the requests and the acknowledgements, respectively. When a pair $x \mapsto y$ belongs to *req* (guard of event *send_ack*) it means that node x has sent a request to node y asking it to become its parent (if this is the only one neighbour node not yet connected with node x - last guard of event *send_req*). Then node y can send an acknowledgement back to node x if node y has not already sent an acknowledgement to node x (second guard of event *send_ack*), nor has it sent a request to any node (guard $y \notin dom(req)$). Event *progress* is enabled when a node x receives an acknowledgement from node y ($x \mapsto y \in ack$) and it has not yet established any parent connection ($x \notin dom(tr)$). When these conditions are fullfilled the connection is established; the pair $x \mapsto y$ is added to *tr*, *tr* representing a sub-graph of graph g which gradually converges with event *progress* to the final tree.

The discovery and solving of contentions is handled by events *discover_cnt* and *solve_cnt* shown in Figure 12.3. If node y has also sent a request to node x at the same time, this means that event *send_ack* is not enabled because its third guard ($y \notin dom(req)$) does not hold while the first two guards hold. On the other hand, the enabling conditions for event *discover_cnt* state that node x has sent a parent request to node y and has not yet received an acknowledgement, while node y has also sent a request. As stated earlier, in the actual protocol specification the problem is solved by timers. In the Event-B case, for solving the contention problem, a new variable *cnt* for contention was introduced. The action of event *discover_cnt* adds the pair $x \mapsto y$ to *cnt*. Event *solve_cnt* is enabled when both pairs $x \mapsto y$ and $y \mapsto x$ are present in *cnt* and it removes these pairs from *req* and resets *cnt*, formalizing what happens after the short delay.

12.5.2 Mapping IEEE 1394 Event-B Model to UPTA

The resulting UPTA model corresponding to Event-B specification of Figure 12.3 is illustrated in Figure 12.5a. In order to arrive to this model several optimisations are performed allowing to reduce the global state-space by factor n and keep the structure of the model relatively compact. Specifically, the mapping of Event-B events, each to an individual UPTA process of template demonstrated in Figure 12.2d, produces a very high degree of parallelism of UPTA models. This is not feasible from the model checking point of view (the set of global states grows exponentially in the number of processes). Though there is not general minimization theory for the network of non-deterministic timed automata the deterministic fragments of the model can be minimized by applying bisimulation quotient method [165] and the classical Hopcroft minimization algorithm [174]. This allows merging the non-distinguishable (from trace semantics point of view) locations and edges in templates separately and in the product automaton of the parallel composition of templates.

The direct mapping of the specification of Figure 12.3 using template of Figure 12.2a provides a process network consisting of $m * n$ processes were m is the number of events and n the size of the template parameter domain. In this case study it is feasible to choose the id of a node as a template parameter since the set of events of nodes is uniform modulo the id of the node. In order to further reduce the size of the model we introduce a few more simplifications.

1. Non-interference of local events. We assume that once a guard component $G_i(V)$ of an event $\mathcal{E}_i$ becomes true it remains true until the guard component $G_i(CL)$ becomes true due to the progress of clocks and event $\mathcal{E}_i$ occurs within time interval satisfying inv(CL) $\wedge$ when_Gi(CL).

2. Mutually exclusive and non-blocking events. The events of a node are assumed to be consecutive, i.e., mutually exclusive, non-blocking, and conspiracy-free which implies that exactly one of these events is enabled at a time till the final state (after *progress* or *elect*) will be reached.

3. Terminal events. It is assumed that events *progress* and *elect* of a node may occur at most once after which no other event of the node can be enabled anymore, i.e., the protocol terminates in the node either after *progress* or *elect* event. So the self-loop pattern of these events (template in Figure 12.4a) can be substituted without side effects with the one of Figure 12.4b.

Simplification 1. above allows substituting the general event template of Figure 12.4a

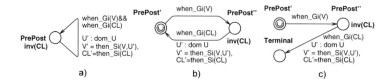

Figure 12.4: UPTA event template transformations: a) timed event, b) non-blocking event and c) terminal event

with the one depicted in Figure 12.4b. Here the location PrePost is split to sub-locations PrePost′ and PrePost″ connected with a new edge that has guard component $G_i(V)$. Urgent location PrePost′ represents the control state where the guard component $G_i(V)$ is supposed to become true and sub-location PrePost″ with invariant inv(CL) models the waiting state for $G_i(CL)$ becomes true. Thus, inv(CL) together with guard $G_i(CL)$ specify the time interval from the moment when G_i became true till some time instant that satisfies inv(CL)∧ when_$G_i(CL)$ and finally enables event $\mathcal{E}_i$.

Due to simplification 2. the parallel composition of event automata can be replaced with one automaton by merging locations PrePost′ of individual events of a node. It means that while one event is in the location PrePost″, there should not be any other event of that node progressing at the same time. The location merging of our IEEE1394 model events results in the model of Figure 12.5a.

12.6 Refinement of Timed Systems

Assuming the Event-B models are mapped to corresponding UPTA models the next goal in the CCD workflow is to apply timing refinement and ensure its correctness. Let us first introduce the very generic refinement definition for timed systems presented by Lynch and Vaandrager [234]. We adapt this definition in order to correspond to the UPTA semantics of Section 12.3.2. Let cl_c and cl_a be the concrete and abstract clocks of refinement and specification models N and M respectively.

Definition 12.1. *A specification M is refined by a specification N, written $M \sqsubseteq N$, iff there exists a binary relation $\mathsf{R} \subseteq \Sigma^N \times \Sigma^M$ such that for each pair of states $(n, m) \in \mathsf{R}$ we have:*

1. *whenever $n_{(l_{ref}, val_{cl_c}, val_w)} \xrightarrow{act^N} n'_{(l'_{ref}, val'_{cl_c}, val'_w)}$ for some $n' \in \Sigma^N$ then $m_{(l_{abs}, val_{cl_a}, val_v)} \xrightarrow{act^M} m'_{(l'_{abs}, val'_{cl_a}, val'_v)}$ and $(n', m') \in \mathsf{R}$ for some $m' \in \Sigma^M$*

2. *whenever $n_{(l_{ref}, val_{cl_c}, val_w)} \xrightarrow{\mathsf{d}^N} n'_{(l_{ref}, val_{cl_c}+\mathsf{d}, val_w)}$ for $\mathsf{d} \in \mathbb{R}_{\geq 0}$ then $m_{(l_{abs}, val_{cl_a}, val_v)} \xrightarrow{\mathsf{d}^M} m'_{(l_{abs}, val_{cl_a}+\mathsf{d}, val_v)}$ and $(n', m') \in \mathsf{R}$ for some $m' \in \Sigma^M$*

The conditions in Definition 12.1 state that all concrete transitions are also possible in the abstract model. Refinement of timing (condition 2), is here simplified by the restriction that all abstract clocks are also present in the concrete model, only new clocks are added.

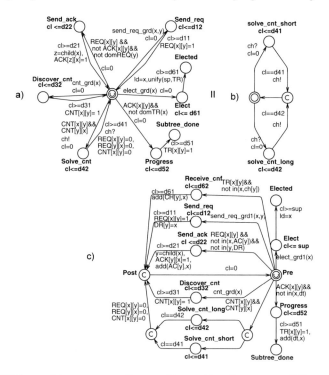

Figure 12.5: a) UPTA model corresponding to Event-B refinement of Figure 12.3, b) timing refinement, and c) UPTA model corresponding to final Event-B refinement of Figure 12.7

Hence, old clock guards and invariants can be strengthened. Clock resets must be preserved. All data invariants can be checked by the proof rules for refinement in Event-B. We have shown in our previous work [54] how to extend the Event-B proof obligations to check the timing properties of all conditions of Definition 12.1. The extended proof obligations could be added to the proof system for Event-B if development of timed systems is to be addressed solely within Event-B. The reader is referred to [54] for a detailed view. Here we focus on verification of refinement of timing by model checking in Uppaal after the corresponding data refinement step has been verified in Event-B.

In UPTA, we use an approach where the refinements are added to the abstract model incrementally to support compositional verification of refinement steps. To keep the clear correspondence between syntactic units of the abstract model and their refinement we define the refinement transformations syntactically by elements that are refined, i.e., by locations and edges of UPTA calling them *location* and *edge* refinements respectively. The model fragments introduced by refinement are composed with the abstract model by means of a wrapping construct "context frame". Technically, it means that without changing the semantics of the abstract model we add the refinement of the syntactic element el as new automaton M^{el} that is composed with the original model M via synchronized parallel composition $\|_{sync}$, s.t. $M \sqsubseteq M \|_{sync} M^{el}$. This synchronization is needed to preserve the contract of el with its context after refinement. It requires decorating automaton M with auxiliary channel labels to synchronize the entry and leave points to/from element el of M.

A representation of the model fragments that schematically represent location and edge refinement is depicted in Figure 12.6. The context frame for constructing $\sqsubseteq_l$ consists of the

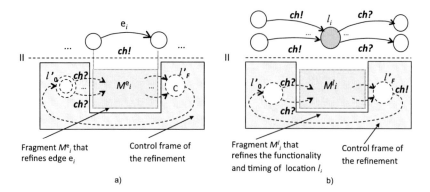

Figure 12.6: (a) Edge refinement, (b) location refinement.

following elements: a) synchronizing channel ch is needed to synchronize the executions of entering to and departing from location l_i transitions with those entering and departing to and from refining model M^{l_i} and b) auxiliary initial location l'_0 and final location l'_F of M^{l_i} are introduced to model waiting before the synchronization via channel ch arrives and after the execution of M^{l_i} terminates respectively. Location refinement is applied when refinement M^{l_i} specifies non-instantaneous time bounded behaviours that are represented in abstract model by location l_i. The *event guard* and *update refinement* introduced in Event-B [6] can be considered as un-timed case of edge refinement where the guards of edges are strengthened and new updates are added respectively in the refining model M^e consisting of exactly one edge.

Purely timing refinement without introducing new model elements means reducing the non-determinism of time intervals between the occurrence of events (events themselves are instantaneous). The bounds of event occurrence interval $[lb, ub]$ are specified according to the timed event template (Figure 12.2d) by clock constraints when_$G_i(CL)$ and inv(CL) respectively. Thus increasing the time constants in when_$G_i(CL)$ and reducing constants in inv(CL) are the simplest forms of respective edge and location refinement.

In general, edge and location refinements introduce new structural elements that require stating their correctness conditions formally:

Edge refinement. We say that a synchronous parallel composition of automata M and M^{e_i} is an edge refinement for edge e_i of M, $(M \sqsubseteq_e M \parallel M^{e_i})$ iff $e_i \in E(M)$, and there exists M^{e_i} s.t. $P_1 \wedge P_2 \wedge P_3 \wedge P_4$, where:

- P_1(*interference free new updates*): No variable of M is updated in M^{e_i}.

- P_2(*guard splitting*): Let $\langle l'_0, l'_F \rangle$ denote a set of all feasible paths from the initial location l'_0 to final location l'_F in M^{e_i} and $\langle l'_0, l'_F \rangle_k \in \langle l'_0, l'_F \rangle$ be k^{th} path in that set. Then,

  $$- \forall k \in [1, |\langle l'_0, l'_F \rangle|]. \vee_{j \in [1, Length(k)]} G(e'_j) \Rightarrow G(e_i),$$

i.e., the disjunction of edge guards of any path in $\langle l'_0, l'_F \rangle$ is not weaker than the guard of edge e_i.

- P_3(*0-duration unwinding*): $\forall l'_i \in (L_{M^{e_i}} \setminus l'_0). \mathrm{T}(l'_i) \in \{committed, urgent\}$,

i.e., all locations in the refinement M^{e_i} must be either *urgent* or *committed*.

- P_4(*non-divergency*): $G(\mathsf{e}_i) \Rightarrow [M^{\mathsf{e}}, \, l'_0 \models \mathsf{A}\Diamond l'_F]$,

i.e., validity of $G(\mathsf{e}_i)$ implies the existence of a feasible path in M^{e_i}.

Location refinement. We say that a synchronous parallel composition of automata M and M^{l_i} is a location refinement for location l_i of M, $(M \sqsubseteq_l M \parallel M^{l_i})$ iff $l_i \in L(M)$, and exists M^{l_i} s.t. $P_1 \wedge P_2 \wedge P_3$, where:

- P_1 (*interference free new updates*): No variable of M is updated in M^{l_i}.

- P_2 (*preservation of non-blocking invariant*):
  $[(M \parallel M^{l_i}),(l_0, l'_0) \models \mathsf{E}\Diamond deadlock] \Rightarrow [M, \, l_0 \models \mathsf{E}\Diamond deadlock]$.

- P_3 (*non-divergency*): $inv(l_i) \equiv x \leq \mathsf{d}$ for $x \in CL_M$, $\mathsf{d} < \infty \Rightarrow [M^{l_i}, \, l'_0 \models l'_0 \leadsto_{\mathsf{d}} l'_F]$,

where "$\leadsto_{\mathsf{d}}$" denotes bounded 'leads to' operator with non-negative integer time bound, CL_M is the set of clocks of M, locations l'_0 and l'_F denote respectively auxiliary pre- and post-locations in the context frame of the refinement.

P_2 and P_3 are specified as model checking queries expressed in TCTL. "$deadlock$" denotes a standard predicate in Uppaal about the existence of deadlocks in the model. P_3 requires that the invariant of l_i is not violated due to accumulated delays of M^{l_i} runs.

The practical use of location refinement followed with later minimization steps is demonstrated on event *solve_cnt* in our running example. Design requirements for solving the contention situation state that the contention is resolved by choosing different waiting times after the contention situation has been detected. The minimum and maximum of non-deterministic waiting time is specified in the UPTA model in Figure 12.5a as d41 and d42 respectively. In the given level of abstraction we leave this timing specification loose intentionally not to restrict the forthcoming design decisions that can determine exact time delays due to practical engineering considerations. Since the choice of waiting time is left to be arbitrary within the interval [d41, d42] it can be proved by model checking that both nodes can choose the delay that differs too little and leads to the next contention. In Figure 12.5b this condition is refined by applying location refinement. Location Solve_cnt is refined to solve_cnt_short and solve_cnt_long that results in two exact waiting times d41 and d42 that excludes intermediate values as required by the protocol specification.

12.6.1 Event-B and UPTA Final Refinement of IEEE 1394

The final Event-B refinement of our running example is shown in Figure 12.7. It is concerned with data refinement for the *localisation* of the global constants and variables. Variables nb (for neighbours), ch (for children), ac (for acknowledged), dr (for domain of req), and dt (for domain of tr) are declared and their connection to the variables of the previous models is shown in the invariant clause in Figure 12.7. For a node x, the sets for neighbours ($nb(x)$), children ($ch(x)$), and acknowledged nodes ($ac(x)$) are supposed to be stored locally within the node. The definition of variable ch above is not given in terms of an equality, rather in terms of an inclusion due to the fact that the set $ch(y)$ cannot be updated while event *progress* takes place. This is because this event can only act on its local data. A new event *receive_cnf* for receiving confirmation is thus necessary to update the set $ch(y)$. This new event refines *skip*. Events *discover_cnt* and *solve_cnt* remain unchanged. Events *send_req*, *send_ack*, *progress*, and *elect* are refined with the new variables.

Applying the mapping and optimization steps described in Section 12.4 to the final refinement of Figure 12.7 results in the UPTA model of Figure 12.5c. At first, the new

```
MACHINE  Ref3_IEEE1394

...
VARIABLES nb, ch, ac, dr, dt
INVARIANT nb ∈ ND → ℙ(ND) ∧ ch ∈ ND → ℙ(ND) ∧ ac ∈ ND → ℙ(ND) ∧ dr ⊆ ND ∧ dt ⊆ ND ∧
    ∀x · (x ∈ ND ⇒ nb(x) = g~[{x}]) ∧ ∀x · (x ∈ ND ⇒ ch(x) ⊆ tr~[{x}]) ∧ ∀x · (x ∈ ND ⇒ ac(x) = ack~[{x}]) ∧
    dr = dom(req) ∧ dt = dom(tr)
EVENTS
    send_req = ANY x, y WHERE  x ∈ ND \ dr ∧ y ∈ ND \ ac(x) ∧ nb(x) = ch(x) ∪ {y}
                        THEN req, dr := req ∪ {x↦y}, dr ∪ {x} END;
    send_ack = ANY x, y WHERE x↦y ∈ req ∧ x ∉ ac(y) ∧ y ∉ dr THEN ack, ac(y) := ack ∪ {x↦y}, ac(y) ∪ {x} END;
    receive_cnf = ANY x, y WHERE x↦y ∈ tr ∧ x ∉ ch(y) THEN ch(y) := ch(y) ∪ {x} END;
    discover_cnt = ANY x, y WHERE x↦y ∈ req \ ack ∧ y ∈ dom(req) THEN cnt := cnt ∪ {x↦y} END;
    solve_cnt = ANY x, y WHERE x↦y ∈ cnt ∧ y↦x ∈ cnt THEN req, cnt := req \ cnt, ∅ END;
    progress = ANY x, y WHERE x↦y ∈ ack ∧ x ∉ dt  THEN tr, dt := tr ∪ {x↦y}, dt ∪ {x} END;
    elect = ANY x WHERE x ∈ ND ∧ nb(x) = ch(x)THEN ld := x END
END
```

Figure 12.7: Event-B final refinement of IEEE 1394 tree identify protocol

event *receive_cnf* is introduced by instantiating the template of Figure 12.4b. By applying the minimization step 2 described in Section 12.5.2, the PrePost' location of *receive_cnf* is merged with the one of the model in Figure 12.5a. Thereafter, the ⟨PrePost'', PrePost'⟩ edges of all events are refined so that the clock resets and data variable assignments remain on consecutive edges separated by committed location Post. Finally, since ⟨Post, Pre⟩ edges of all nonterminating events have identical clock resets and destination location, they are merged into one ⟨Post, Pre⟩ edge as shown in Figure 12.5c.

Though the final refinement by Abrial et al. refers to timing constraint related to contention resolving and message transmission delays, practical implementation of the protocol may introduce much more complex timing constraints and that is not only for single events in isolation but also for those that depend on the ordering and timing of other events. For instance buffering time may depend on the arrival rate of messages sent by other nodes. This information cannot be introduced in un-timed models. Similarly, as our model checking experiments revealed, simple timing errors may remain unnoticed when introducing timing constraints for events *progress* and *elect* (the protocol total convergence time bound requirements were violated if the occurrence interval of events was specified too loose).

12.7 Conclusion and Future Work

We propose a correct-by-construction design workflow where model-based design transformations combine alternating data and timing constraints refinement steps. The goal is to benefit from mutually complementing formalisms Event-B and Uppaal automata and related verification techniques. For bridging the data and timing refinement steps the Event-B to UPTA map and its timing refinement transformations have been defined and their relevance discussed. As an advantage of integrating theses two it allows to verify the data refinement correctness also from its timing feasibility point of view. The approach is demonstrated on the development case study of the IEEE 1394 tree identify protocol. The automation of the proposed design transformations as plug-in either for Rodin or Uppaal tool remains for future work. Still, the results are encouraging from the usability point of view due to the simplicity of transformations and the locality of proof obligations to be discharged for timing refinement verification.

Part V

Applications

Chapter 13

Action Systems for Pharmacokinetic Modeling

M.M. Bonsangue

LIACS, Leiden University, The Netherlands

M. Helvensteijn

LIACS, Leiden University, The Netherlands

J.N. Kok

LIACS, Leiden University, The Netherlands

N. Kokash

LIACS, Leiden University, The Netherlands

Abstract. In system biology, models play a crucial role to understand, communicate, and analyze a system. Such models are often the result of complex interactions between continuous and discrete logics. In this chapter we give a short introduction to pharmacokinetics, the biological study of the process interactions between a drug and the human body. We show how action systems with continuous behaviour can be used as a precise model for pharmacokinetics processes. The support of action systems for stepwise refinement open the doors to new analysis and prediction tools for better understanding what happens between an administration of a drug dose and the body response.

Kaisa, in this paper we use "your" action systems in a field you would have never predicted: pharmacokinetics. We miss you!

13.1 Introduction

Pharmacokinetics is a branch of pharmacology that studies processes caused by substances (such as drugs, hormones, nutrients, or toxins) administered to a living organism, from the moment of substance administration to the point of its complete elimination from the body. Pharmacokinetics describes how the body affects a specific drug after administration through the mechanisms of absorption and distribution, as well as the chemical changes of the substance in the body, and the effects and routes of its excretion.

Compartment models are typically used as mathematical models to describe level changes of the drug concentration in our bodies [301]. Compartments are used to model drug concentration in body tissues by using differential equations for absorption, distribution, and elimination rates. Compartment models exist for many types of drugs, absorption methods (e.g. oral, intramuscular, intravenous), and elimination methods.

Compartment models are continuous-value models, based on systems of ordinary differential equations that can be analyzed by examining a graph with compartments as node and in-flow/out-flow rates labeling the arcs. However, pharmacokinetic models are not purely continuous. They also use a finite state logic to decide, for example, when absorption stops, when elimination starts, or when a process is triggered by cell death or division. As such they involve continuous states and dynamics, as well as some discrete logic corresponding to discrete states and dynamics. Because existing pharmacokinetic models often represent a tradeoff among accurately describing the data, having confidence in the results, and mathematical tractability, the discrete logic of the system is often ignored.

In this paper we use hybrid action systems as a model of pharmacokinetics. Action systems were originally proposed as a formalism for developing parallel and distributed systems [28]. They are based on a predicate transformer semantics for discrete computation, where parallel composition is described by interleaving of atomic actions. In hybrid action systems, computations may also continuously evolve over time [280, 281], and parallel composition is defined by interleaving discrete actions and combining continuous ones.

While it appears useful to be able to describe and reason about properties of pharmacokinetic models using hybrid systems, we are not aware of any existing consistent attempts to do so. In the last decade, there has been an increasing interest in quantitative and hybrid models of computation based on systems of ordinary differential equations. However, all models that we are aware of are concentrating on biological, bio-chemical, or bio-medical processes. Examples of successful approaches in these areas include several process algebraic languages [269, 166, 275, 95, 76, 83]. In this paper we take a novel alternative approach and use hybrid action systems for a compositional construction of pharmacokinetic models. We concentrate on the modelling aspects, but our framework allows properties of the systems to be proven formally within the refinement calculus [34, 281]. For example, in pharmacokinetics we deal with models for the study of a drug's pharmacological effect on the body. Understanding what happens between the administration of a drug and the body's response could help in recommending a proper dosing regimen for that drug (how much, how often, under what assumptions) with regard to a population, a subpopulation, or an individual patient. Drug concentrations must be kept high enough to produce a desirable response, but not so high that they produce toxic effects. The problem is that current models are so complex that it is difficult to extract useful predictions from them. However, since the magnitude of an effect is proportional to the concentration of the drug, our action system model could be used to prove that concentration always remain below the toxic level.

13.2 Actions and Action Systems

Let Var be a countable set of *variables* and assume that each variable in Var is associated with a nonempty set of *values*. A *state* is a function mapping variables to values in their respective associated set of values. We denote by *true* the predicate on Var which holds for every state, and by *false* the predicate on Var which holds for no state. Given a predicate $\mathcal{P}$ over Var, a list of variables x and a list of values v, we denote by $\mathcal{P}[x/v]$ the predicate that holds for those states s such that $\mathcal{P}$ holds for $s[x/v]$, where $s[x/v]$ is the state that maps the variables x to the values v, but otherwise behaves as s.

A *conjunctive predicate transformer* is a function π mapping predicates to predicates such that, for every nonempty index set I,

$$\pi\left(\bigwedge\{\mathcal{P}_i \mid i \in I\}\right) \;=\; \bigwedge\{\pi(\mathcal{P}_i) \mid i \in I\}.$$

Conjunctive predicate transformers form the semantic domains for a class of statements, called *actions*, interpreted by means of a *weakest precondition* semantics [113]. These should be considered specification statements rather than actual programs, as they need not be strict, disjunctive, or preserving of directed disjunctions [36].

More syntactically we consider *actions* denoting conjunctive predicate transformers. We define *ordinary actions* by the grammar

$$a \quad ::= \quad abort \mid skip \mid x := v \mid b \to a \mid p \mid a_1 \,;\, a_2 \mid [\![_I a_i \,.$$

Here, x is a list of variables, v is a list of values (possibly resulting from the evaluation of a list of expressions), b is a predicate over Var, p is a procedure name, and I is an index set ranged over by i. Intuitively, '*abort*' is the action which models unwanted or disallowed behaviour, '*skip*' is a stuttering action (i.e. not changing the state), '$x:=v$' represents multiple assignment, '$b \to a$' is a guarded action that executes a only when the guard b holds, 'p' is a procedure call, '$a_1 \,;\, a_2$' is the sequential composition of two actions 'a_1' and 'a_2', and '$[\![_I a_i$' is the nondeterministic choice among actions 'a_i' for $i \in I$.

A procedure declaration $p = P$ consists of a header p and an action P—the body of the procedure. Given a declaration for each procedure, we define the *weakest precondition* semantics of the above language in a standard way [113, 36], as the *least* conjunctive predicate transformer wp such that, for any predicate $\mathcal{P}$,

$$
\begin{aligned}
wp(abort, \mathcal{P}) &= false & wp(skip, \mathcal{P}) &= \mathcal{P} \\
wp(x := v, \mathcal{P}) &= \mathcal{P}[x/v] & wp(b \to a, \mathcal{P}) &= b \Rightarrow wp(a, \mathcal{P}) \\
wp(p, \mathcal{P}) &= wp(P, \mathcal{P}) & wp(a_1 \,;\, a_2, \mathcal{P}) &= wp(a_1, wp(a_2, \mathcal{P})) \\
wp([\![_I a_i, \mathcal{P}) &= \forall i \in I.wp(a_i, \mathcal{P}),
\end{aligned}
$$

where P is the action denoting the body of the procedure p. The *greatest* conjunctive predicate transformer satisfying the above is the *weakest liberal precondition* $wlp(a, \mathcal{P})$ of an action a with respect to the post-condition $\mathcal{P}$. Various form of iterations and other simple statements can be encoded in the above language. For example, assertions $\{b\}$ can be encoded as $b \rightarrow skip \| \neg b \rightarrow abort$: if the condition b does not hold in a given state, then the assertion aborts, otherwise has no effect. The details of the definition of the above functions is studied elsewhere [61].

Further on we will need the following notions. An action a is *enabled* in a given state if its *guard*

$$gd(a) \ = \ \neg\, wp(a, false)$$

holds in that state. For example, $gd(b \rightarrow x := v) = b$. An action a *terminates* in a given state if $t(a) = wp(a, true)$ holds in that state. An action a_1 *cannot enable* another action a_2 if

$$t(a_1) = true \quad \text{and} \quad \neg\, gd(a_2) \Rightarrow wp(a_1, \neg\, gd(a_2))\,.$$

Moreover, a_1 *cannot disable* a_2 whenever

$$t(a_1) = true \quad \text{and} \quad gd(a_2) \Rightarrow wp(a_1, gd(a_2))\,.$$

Actions that capture continuous-time dynamics are called *differential actions* [281]. They act not only on ordinary variables, but also on some *evolution variables* in $EVar$. Evolution variables describe the observation of an evolution but not its relation with respect to time. As such, they take values in the set of real numbers $\mathbb{R}$. A differential action, written as $e :\rightarrow d$, describes the evolution in time according to the predicate d of both evolution and ordinary variables from their initial value (as given by a state) to the values they reach when the guard e does not hold. The *evolution guard* e is a predicate over Var and a list $X = x_1 \cdots x_n$ of evolution variables. In this paper we consider the predicate d to be a partially defined system of differential equations of the form $\dot{X} = F(X)$, where $\dot{X}$ is a syntactic variant used to denote the component-wise first derivative of the variables in X.

A continuous function $f : \mathbb{R} \rightarrow \mathbb{R}^n$ with a continuous first derivative $\dot{f}$ is a *solution* for $e :\rightarrow d$, denoted by $SF(f, e :\rightarrow d)$, if $f(0) = X$ (the current value of the variables) and for every positive $r \in \mathbb{R}$, if the guard e is *true* then it satisfies the system of differential equations d. More formally

$$SF(f, e :\rightarrow d) \ \equiv \ f(0) = X \wedge \forall r \in [0, \infty).\ (e \Rightarrow d)[f(r)/X, \dot{f}(r)/\dot{X}]$$

The first point in time, if any, when the guard e does not hold, give the *duration* $\Delta(f, e)$ of a differential action. In other words, $\Delta(f, e) = inf\{r \in [0, \infty) \mid \neg e[f(r)/X]\}$.

For a postcondition $\mathcal{P}$ over X and Var, the *weakest precondition* of a differential action is the smallest set of states (assignments to X and Var) for which the evolution of $e :\rightarrow d$ is guaranteed to terminate in a state satisfying $\mathcal{P}$. More formally,

$$wp(e :\rightarrow d, \mathcal{P}) \equiv \forall f.(SF(f, e :\rightarrow d) \wedge \Delta(f, e) > 0 \Rightarrow$$
$$\Delta(f, e) < \infty \wedge \mathcal{P}[f(\Delta(f, e))/X]\,)$$

As for ordinary action, we say that a differential action $e :\rightarrow d$ is *enabled* if $gd(e :\rightarrow d) = \neg\, wp(e :\rightarrow d, false)$ holds, and that it *terminates* if $t(e :\rightarrow d) = wp(e :\rightarrow d, true)$ holds. Enabledness holds in states where there is evolution (i.e. $gd(e :\rightarrow d) \Rightarrow e$) and termination holds exactly when all evolutions terminate. Since $wp(e :\rightarrow d, \mathcal{P}) = wp(gd(e :\rightarrow d) :\rightarrow d, \mathcal{P})$ [281], we will only consider *stutter free* differential actions [280], that is, differential actions $e :\rightarrow d$ that are enabled exactly when there is evolution (i.e. $gd(e :\rightarrow d) = e$). Finally, the weakest precondition of a differential action is a conjunctive predicate transformer. These and other interesting results on differential actions are studied in, e.g. [281].

13.2.1 Action Systems

Action systems are a formalism for developing parallel and distributed systems originally proposed in [28]. We will consider here *hybrid action systems* [281]. They consist of a set of ordinary and differential actions operating on local and global (evolution) variables. First, the variables are created and initialized. Then, repeatedly, enabled actions are chosen and executed. Actions operating on disjoint sets of variables can be executed in any order. The execution terminates if no action is enabled, otherwise it continues infinitely.

Syntactically, an action system $\mathcal{A}$ is a statement of the form

$$\mathcal{A} \;=\; \|[\quad \textbf{var} \quad Y^* := V \; ; \; X := U$$
$$\textbf{proc} \quad p_1 = P_1 \; ; \ldots ; \; p_m = P_m$$
$$\textbf{do } A_1 \; \| \ldots \| \; A_n \textbf{ od}$$
$$]| \quad : Z$$

An action system provides a declaration section for variables and one for procedures. Here Y is a list of *global (evolution) variables*, marked with an asterisk $*$, that can be used locally by $\mathcal{A}$ and also by other action systems when put in parallel with $\mathcal{A}$. The (evolution) variables in the list X are *local* to $\mathcal{A}$. The global variables Y get initial values componentwise from the list V and the local variables x get initial values componentwise from the list U. Finally, Z is the list of imported variables, i.e. global variables declared in action systems that can be put in parallel with $\mathcal{A}$. We assume that X, Y, and Z are disjoint.

A *procedure* declared as $p_i = P_i$ is *local* and can be called only by the ordinary actions of $\mathcal{A}$ which can thus enable or disable the body P_i. *Actions* of the action system $\mathcal{A}$ are the actions $A_1, \ldots A_n$ and the bodies of all procedures declared in $\mathcal{A}$. Each action A_i can be either an ordinary action a as defined above, or a differential action $e :\to d$. All actions of $\mathcal{A}$ can refer to (evolving) variables which are in X, Y, or Z. Actions are atomic, meaning that if an action A_k is chosen for execution, then it is executed to completion.

Action systems are models of reactive systems. The *behaviour* of an action system is described by the set of all its computations. A *computation* here is a finite or infinite sequence of states (i.e. maps of *global* variables to values), without finite repetition (i.e. no finite stuttering), possibly terminating with a special symbol to denote abortion [25].

To model the dynamics of a system consisting of several reactive components, action systems are equipped with a parallel composition operation: global variables are merged together, local variables are kept separate, and the imported variables will consist of all variables imported by one component that are not declared as global in the other. Procedures are local, and thus kept separate, whereas actions are combined non-deterministically, thus modelling parallelism by interleaving [25].

13.2.2 Hybrid Action Systems

While parallel composition of action systems works fine with ordinary actions that have no duration in time, interleaving of differential actions does not model true concurrent evolution. Parallel composition of differential actions can be defined as the linear composition '$\oplus$' of two partially defined functions [280]: their values are added together on their common domain, and remain unchanged on the remaining domains.

$$e_1 :\to \dot{X} = F(X) \;\oplus\; e_2 :\to \dot{X} = G(X) \;=\; \begin{array}{l} e_1 \wedge e_2 :\to \dot{X} = F(X) + G(X) \\ \| \quad e_1 \wedge \neg e_2 :\to \dot{X} = F(X) \\ \| \quad \neg e_1 \wedge e_2 :\to \dot{X} = G(X) \end{array}$$

We want the dynamics of a hybrid action system to be an alternation between ordinary actions and differential ones, so that in a parallel composition we can linearly compose the continuous-time differential actions and interleave the discrete-time ordinary actions. A hybrid action system $\mathcal{A}$ is a statement of the form

$$\mathcal{A} \;=\; [\![\;\; \textbf{var} \quad Y^* := V \;;\; X := U$$
$$\textbf{proc} \quad p_1 = P_1 \;;\; \cdots \;;\; p_n = P_n$$
$$\textbf{alt } A \textbf{ with } D$$
$$]\!] \;\; : Z$$

where A is a non-deterministic composition of ordinary actions $a_1 \,[\!]\, \ldots \,[\!]\, a_n$, D is the non-deterministic composition of differential actions $e_1 :\to d_1 \,[\!]\, \ldots \,[\!]\, e_n :\to d_n$, and $\textbf{alt } A \textbf{ with } D$ is their alternation, defined as

$$\textbf{do } A \,[\!]\, \neg gd(A) \to D \textbf{ od}$$

When $gd(D)$ is $false$ we just write '$\textbf{alt } A$', and when $gd(A)$ is $false$ we just write '$\textbf{with } A$'. A computation of a hybrid action system is, a finite or infinite sequences of evolution states (mapping $evolution$ variables to $\mathbb{R}$), interleaved with ordinary states (mapping of $evolution$ variables to values).

We conclude by recalling the definition of parallel composition of hybrid action systems [280, 281]. Consider the action systems $\mathcal{A}_i$ for $i = 1, 2$:

$$\mathcal{A}_i \;=\; [\![\;\; \textbf{var} \quad Y_i^* := V_i \;;\; X_i := U_i$$
$$\textbf{proc} \quad p_{i,1} = P_{i,1} \;;\; \ldots \;;\; p_{i,n_i} = P_{i,n_i}$$
$$\textbf{alt } A_i \textbf{ with } D_i$$
$$]\!] \;\; : Z_i$$

where the global variables, local variables, and the local procedure headers declared in each action system $\mathcal{A}_i$ are required to be distinct. The *parallel composition* $\mathcal{A}_1 \,\|\, \mathcal{A}_2$ is defined as the action system

$$[\![\;\; \textbf{var} \quad (Y_1^* := V_1 \;;\; Y_2^* := V_2 \;;\; X_1 := U_1 \;;\; X_2 := U_2)$$
$$\textbf{proc} \quad p_{1,1} = P_{1,1} \;;\; \ldots \;;\; p_{1,n_1} = P_{1,n_1} \;;\; p_{2,1} = P_{2,1} \;;\; \ldots \;;\; p_{2,n_2} = P_{2,n_2}$$
$$\textbf{alt } A_1 \,\|\, A_2 \textbf{ with } D_1 \oplus D_2$$
$$]\!] \;\; : (Z_1 \cup Z_2) \backslash (Y_1 \cup Y_2)$$

13.3   Pharmacokinetic Modeling

The pharmacokinetics of a drug is a very complex biological process and there are no exact mathematical models for describing the concentration of a drug at any time in any part of the body. Pharmacometricians approximate the physiological process of the interactions between an organism and a drug by a compartmental model. Compartments represent the fluids and tissues of the human body, and each of them is assigned with input and output rates modeling the process of absorption and removal of the drug with regard to the tissues. This method is known as the ADME modeling scheme [223]:

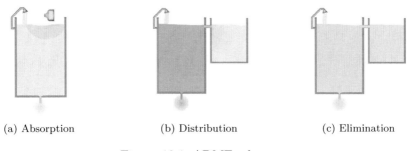

<table>
<tr><td>(a) Absorption</td><td>(b) Distribution</td><td>(c) Elimination</td></tr>
</table>

Figure 13.1: ADME scheme

- *Absorption* - the process of a substance movement to the blood stream.

- *Distribution* - the dispersion or dissemination of substances throughout the fluids and tissues of the body.

- *Metabolization* - the recognition by the organism that a foreign substance is present and the irreversible transformation of parent compounds into daughter metabolites.

- *Excretion* (or elimination) - the removal of the drug from the body. In rare cases, some drugs irreversibly accumulate in body tissue.

A drug can be given to a patient in several ways, e.g. intravenously, orally, subcutaneously, intramuscularly, with a skin patch, or by inhalation. Intravenous administration does not involve absorption and, hence, there is no loss of the drug. On the other hand, with oral administration, several factors affect the absorption of a drug: solubility and permeability, gastrointestinal pH, gastric emptying time, small intestinal transit time, etc. [311]

The fraction of an administered dose of unchanged drug that reaches the systemic circulation is known as *bioavailability* and is one of the principal pharmacokinetic properties of a drug. After absorption, most of the drug is distributed via the blood to the body tissues. Distribution describes the reversable transfer of the drug between blood, tissues, and organs. The drug is more easily distributed in highly perfused organs such as heart, brain, lungs, liver, and kidney than in poorly perfused tissues such as fat, skin, and muscles.

Metabolization refers to the bio-transformation of the drug to other molecules—called *metabolites*—inside the body. Metabolites are normally pharmacologically inactive, but they can be active or toxic. Some drugs remain inactive until metabolized. The principal organ involved in excretion of a drug is the kidney, which eliminates water soluble drugs with urine. Bio-flow from the liver is also an important route for the elimination of a drug. Drugs also leave the body via other natural routes: breath, tears, sweat, and saliva.

About half of the human body weight is water. Water is eliminated from the body in urine, with about 6 liters of water passing through the kidney filter per hour. Thus, we can model this process as a filled up tank with a tap and an open plug where the flow through the tap equals the flow through the plug hole. We can think of water flowing in the blood-stream through the kidneys as water in the tap. Now imagine adding some drug to the water. The speed with which the drug distributes through the tank can represent the process of drug absorption (Figure 13.1a). How the drug penetrates to the peripheral tanks can represent the distribution process (Figure 13.1b). Finally, the elimination process can be represented by the flow out of the tank through the plug hole (Figure 13.1c).

The pharmacokinetics compartment model assumes that the drug concentration is perfectly homogenous in each compartment of the body at all times. Thus, we also assume that the drug concentration is homogenous at each time in each tank. This is a strong assumption but it allows a quantitative description of the pharmacokinetics processes. The quantitative pharmacokinetic model describes how the drug amount varies in each compartment, or equivalently, how the drug amount varies in each water tank. Such a description mainly reduces to three main components:

- *Rate in* describes how the drug moves to the blood stream (or how the drug moves to the main tank)

- *Rate of distribution* describes the movement of a drug between the compartments (or the distribution of the drug between tanks)

- *Rate out* describes how the drug is eliminated from the body (or how the drug flows out of the main tank)

Absorption, distribution, and elimination are continuous processes limited by the physiological limitation of the tank, and the administered dose of the drug. Using the chemical balance law

$$\text{rate of change} = \text{input rates - output rates}$$

distribution can be described in terms of the other two components of each compartment. The pharmakinetics model of the system combines all these rates for each compartment into a single large system of differential equations. Next we describe in more detail the absorption and elimination processes, and give an example of their models as action systems.

13.3.1   Absorption

Absorption starts right from the beginning and stops when the amount of drug in the entire system is a bioavailability fraction $0 \leq f \leq 1$ of the administered dose (denoted *dose*). We use the local Boolean variable *abs* to record whether absorption takes place or not, and the global variable *dose* to record the amount of administered dose of the drug. The global evolution variable x describes the amount of drug present in the tissue modeled by a compartment (here, for example, blood), and it is dependent on time t.

Intravenous administration bypasses the absorption phase: the entire drug dose enters the general circulation (bioavailability fraction $f = 1$). We distinguish intravenous *bolus administration* and intravenous *infusion*. In the first case, the drug is administrated through the intravenous route in a negligible time and achieves instantaneous distribution throughout the central compartment. Thus, just before the administration, the amount of drug is 0, and just after administration the amount equals the administered dose (Figure 13.2a). In the case of intravenous infusion, the dose is administered with a constant rate during some time of infusion T (Figure 13.2b).

With oral administration of a drug (e.g. in tablet, capsule, or liquid form), the entire dose does not reach the systemic circulation ($f < 1$) due to factors such as breakdown in the intestine, poor absorption, and pre-systemic extraction. When swallowed, the drug enters the gastrointestinal tract and is then absorbed to the blood stream. The simple pharmacokinetic model considers the gut as a depot compartment that receives the dose. Describing the absorption process then reduces to describing how the dose is transferred from the depot

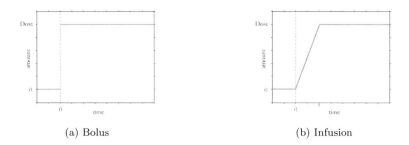

(a) Bolus (b) Infusion

Figure 13.2: Intravenous administration

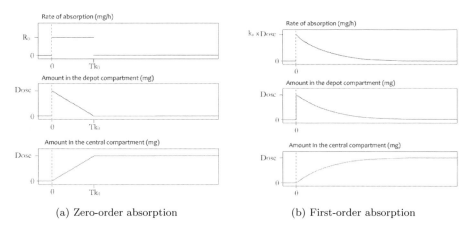

(a) Zero-order absorption (b) First-order absorption

Figure 13.3: Oral administration

compartment to the central blood compartment. In a zero-order absorption process (see Figure 13.3a), a drug is absorbed over time-period T with a constant rate $k_0 = dose/T$. In a first-order absorption process (see Figure 13.3b), the absorption rate is proportional to the amount of drug in the depot compartment. The proportionality constant is usually denoted k_a.

For example, let us consider oral administration of a drug, assuming the mono-exponential rate of transfer to the blood circulation given by $dx/dt = f \cdot dose \cdot k_a e^{-k_a \cdot t}$, where f is the fraction of bioavailability and k_a is the proportionality constant. Absorption into the blood can then be modeled as the following hybrid action system:

$$
\begin{aligned}
Absorption \quad = \quad & [\![\quad \textbf{var} \quad abs^* := true \\
& \quad \textbf{alt } abs \wedge x \geq f \cdot dose \rightarrow abs := false \\
& \quad \textbf{with } abs \wedge x < f \cdot dose :\rightarrow \dot{x} = f \cdot dose \cdot k_a e^{-k_a \cdot t} \\
&]\!] \quad : dose, x
\end{aligned}
$$

According to the above differential equation, the amount of drug in the blood stream at time t is given by $x(t) = e^{f \cdot dose \cdot k_a t}$. Here t ranges from 0 to the first point in time when $x(t)$ is not smaller than the bioavailability fraction of the administered dose.

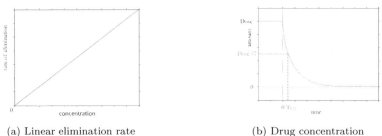

(a) Linear elimination rate   (b) Drug concentration

Figure 13.4: First-order elimination process

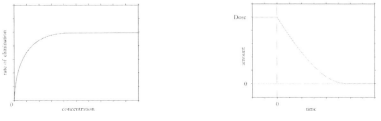

(a) Saturable elimination rate   (b) Drug concentration

Figure 13.5: Mixed elimination process

13.3.2   Elimination

As soon as drug concentration is above a certain threshold, the elimination process starts. Elimination is modeled by opening the plug of a water tank. We assume that the concentration of the drug that comes out of the tank at any time equals the concentration of the drug in the tank. There are several mathematical models to describe the elimination process. With first-order (linear) elimination, the elimination rate is directly proportional to the plasma concentration (Figure 13.4a). Thus, the amount of the drug in the tank decreases with a decreasing rate (Figure 13.4b).

With saturable elimination, the elimination rate increases with the increase in the concentration until a certain concentration is reached, after that the elimination rate stays constant (Figure 13.5a). Such capacity-limited elimination is known as a mixed-order process. The corresponding drug amount over time is shown in Figure 13.5b: the amount decreases with a rate which, at first, is almost constant and then slows down to resemble linear elimination.

For example, in a zero-order one-compartment model, let us assume that the drug elimination rate is $dx/dt = kx$, for some constant $k > 0$. An action system modeling the elimination processes is as follows:

$$Elimination \quad = \quad |[\quad \textbf{var} \quad x^* := 0$$
$$\textbf{with } x > min :\to \dot{x} := -k \cdot x$$
$$]| \quad :$$

where min is a threshold of minimal amount of drugs in the blood-stream before starting, so that the amount of drugs at time t is defined by the function $x(t) = x_0 e^{-kt}$ in terms of the initial amount x_0 at time 0.

13.3.3 One-Compartment Model

The overall rate of change of the amount of drugs in a single compartment over time is the combination of the rate *in* and the rate *out*. In this simplified model, distribution does not play a role.

The overall pharmacokinetic model is obtained through the parallel composition of the hybrid action systems modeling absorption and elimination of that compartment. Continuing our example we have

$$BloodTank = Absorption \parallel Elimination$$

Unfolding the definition we obtain the following one-compartment model of drug concentration in blood from an oral administration:

$$
\begin{aligned}
BloodTank = \|[\mathbf{var} \quad &abs^* := true, x^* := 0 \\
&\mathbf{alt}\ abs \wedge x \geq f \cdot dose \rightarrow abs := false \\
&\mathbf{with}\ abs \wedge min < x < f \cdot dose :\rightarrow \dot{x} = f \cdot dose \cdot k_a e^{-k_a \cdot t} - k \cdot x \\
&\| abs \wedge x \leq min \wedge x < f \cdot dose :\rightarrow \dot{x} = f \cdot dose \cdot k_a e^{-k_a \cdot t} \\
&\| (\neg abs \vee x \geq f \cdot dose) \wedge x > min :\rightarrow \dot{x} = -k \cdot x \\
&\|] : dose
\end{aligned}
$$

The first differential equation gives the concentration of drug in the blood-stream when the drug is absorbed and eliminated at the same time. Until the minimal concentration is reached, only absorption takes place (as we see in the second differential equation). The third differential action represents elimination when absorption is terminated.

13.3.4 Distribution

The above standard pharmacokinetics equations for one-compartment models are rather simple. Many categories of drugs, however, cannot be accurately characterized by one-compartment models. For example, the effects of anesthetic drugs greatly depend on distribution of the active substance into and out of peripheral tissues. To reflect the differences in perfusion of tissues, several compartments are used in the pharmacokinetic model. In this case, the construction of a certain pharmacokinetic model consists of choosing a model for absorption (or distribution) and elimination for each compartment from the current models pharmacometricians have developed (Figure 13.6a), and which have already been validated against observations in clinical studies (Figure 13.6). For example, the DDMoRe model repository[1] is a place where pharmacokinetic models can be stored, retrieved, and shared with the community.

A multi-compartment model can be translated into the parallel composition of hybrid action systems, each modeling a single compartment. Using one global evolution variable for each compartment, we can describe the amount of a drug present, assuming circular drug distribution among compartments, e.g. as in the case when drug metabolites are reabsorbed from kidneys along with water before they are excreted in the urine.

The overall rate of change of the amount of a drug in the body over time is the combination of the rate in, *in*, the rate of *distribution*, and the rate *out*.

---

[1]http://repository.ddmore.eu

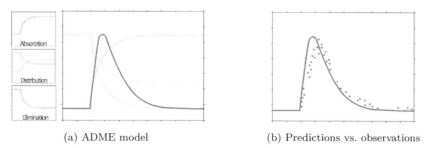

(a) ADME model   (b) Predictions vs. observations

Figure 13.6: Pharmacokinetic modeling

13.4   Conclusions and Future Work

In this paper, we propose action systems as precise models of the compositional semantics of pharmacokinetic processes. For simple compartmental models, we found no need to develop new extensions for action systems, as existing work on continuous and hybrid actions fit perfectly. However, we have not fully explored the potential of action systems for reasoning about pharmacokinetic (and pharmacodynamic) processes. In particular, it will be interesting to prove properties of such systems formally within the refinement calculus [34, 281].

We are planning to use hybrid action systems in the context of the ApiNATOMY project[2] to automatically combine and formally analyze quantitative descriptions of physiological processes across multiple scales. A method central to the ApiNATOMY effort is the visualization of ontology terms as tiles in a treemap, which allows us to automatically generate body plans from given ontologies, preserving spatial relations among selected components [100]. Metadata related to ontology terms is represented in the form of objects associated with tiles as well as visual connections between tiles and their associated objects. To be able to generate views which are of interest to a user, we need to employ compositional formal models of related processes. Action systems represent a promising framework for our goal. As future work, we will employ action systems to model and control heterogeneous sets of independently developed pharmacokinetic and pharmacodynamic models, e.g. as in the alcohol consumption process described by de Bono and Hunter [101]. Each of the individual models can be represented by a separate hybrid action system combining both discrete and continuous logic, which are then assembled into a complete process using sequential, parallel, or guarded choice composition operators defined for action systems.

---

[2]http://www.apinatomy.org

Chapter 14

Quantitative Model Refinement in Four Different Frameworks, with Applications to the Heat Shock Response

Diana-Elena Gratie

Turku Centre for Computer Science and Åbo Akademi University, Turku, Finland

Bogdan Iancu

Turku Centre for Computer Science and Åbo Akademi University, Turku, Finland

Sepinoud Azimi

Turku Centre for Computer Science and Åbo Akademi University, Turku, Finland

Ion Petre

Turku Centre for Computer Science and Åbo Akademi University, Turku, Finland

Abstract. Quantitative model refinement is an essential step in the model development cycle. Starting with a high level, abstract representation of a biological system, one often needs to add details to this representation to reflect changes in its constituent elements. Any such refinement step has two aspects: one structural and one quantitative. The structural aspect of the refinement defines an increase in the resolution of its representation, while the quantitative one specifies a numerical setup for the model that ensures its fit preservation at every refinement step. We discuss in this paper the implementation of quantitative model refinement in four extensively used biomodeling frameworks: ODE-based models, rule-based models, Petri net models, and guarded command language models, emphasizing the specificity for every model implementation. We argue that quantitative model refinement is framework-independent, being implementable in all chosen frameworks despite their different underlying modeling paradigms.

14.1 Introduction

Research in molecular biology has been traditionally conducted targeting individual molecular entities of a biological system. Discovery of the functional specificity of these molecular entities has brought about tremendous progress in the understanding of biological systems. However, the understanding of individual molecules alone does not suffice to tackle complex biological phenomena, such as those involved in diseases. Systems biology promotes precisely an integrative bottom-up approach for the analysis of such systems, which takes advantage of previous knowledge about individual proteins, but which is essentially concerned with a holistic analysis of the system: understanding the structure and the dynamics of the system, see [273, 201]. Moreover, in the past decades, we have witnessed a convergence of the fields of computer science and biology towards a new field that is expanding evermore, bioinformatics, see [270].

At the core of systems biology lies the concept of computational modeling, driven by the production of massive sets of experimental data which necessitate computer analysis, see [237]. In this context, formal frameworks prove to be essential in the synthesis and analysis of large biological models as an effort to predict the system-level behavior of such systems. This approach commences with an abstraction of the biological phenomena, which is ultimately converted into a model derived through an iterative process of model building involving system design, model analysis, hypothesis generation, hypothesis testing, experimental verification, and model refinement, see [273]. The model obtained as such very often needs to be refined to include more details regarding some of its biological processes, encapsulating presumably new experimental data over a number of new parameters. A reiteration of the whole process of quantitative model fitting and validation is, however, unfeasible for a large model. The alternative we discuss in this paper is that of quantitative model refinement.

Quantitative model refinement focuses on the step-wise construction of models, from small abstract models to large, detailed ones. Each refinement step consists of two parts. The structural part of the refinement consists of fixing the details to add to the model (e.g., new attributes in existing species, new species, new interactions, new modules, etc.) and identifying the new set of species and interactions yielded by adding these details. The quantitative part of the refinement consists in fixing the numerical setup of the refined model (kinetic parameters and initial values) in such a way that the quantitative behavior of the model, in particular its experimental fit, remains unchanged.

The goal of this paper is to give an overview of our approach to quantitative model refinement. We introduce first the main mathematical concepts we use in our framework: reaction-based model, refinement relation, structural refinement, and quantitative refinement. We then discuss the implementation of model refinement in four of the most widely used approaches in biomodeling: ODE-based models, rule-based models, Petri net models, and guarded command models. The structural part of the refinement has a different solution in each approach, in some cases leading to a compact representation of the refined models. The quantitative part of the refinement aims to avoid the computationally expensive procedure of parameter estimation (especially since the models get larger in each refinement step); instead, we apply in each approach the sufficient condition recently proposed in [143]. As a case study, we consider the heat shock response in eukaryotes. We refer to the model in [263] as the *basic heat shock response model* and to the model in [176] as its *refined model*. A short version of this paper was presented in [178].

The paper is organized as follows: we discuss the concept of fit-preserving refinement in Section 14.2. We succinctly describe in Section 14.3 the heat shock response and its underlying reaction-based model. We focus in Section 14.4 on the refinement of ODE-based models. In Section 14.5, we present rule-based modeling and the rule-based implementation of the heat shock response and of its corresponding refinement. In Section 14.6, we discuss Petri nets and their capabilities with regards to the implementation of the heat shock response and its refinement to include acetylation using colored Petri nets. Section 14.7 comprises a brief description of guarded command languages, focusing on PRISM, and a discussion regarding the implementation of the basic and the refined model for the heat shock response. We conclude the paper with a discussion in Section 14.8. All models developed in this chapter can be downloaded at [177].

14.2   Quantitative Model Refinement

Model refinement has been extensively investigated in the field of software engineering, especially in connection to parallel computing. For example, among other approaches, such studies brought about a logical framework for the construction of computer programs, called *refinement calculus*. This framework tackles the derivation of computer programs correct by construction and refinement of computer programs ensuring correctness preservation, see [38].

Quantitative model refinement is a concept introduced in systems biology as an approach to step-wise construction of biomodels. It focuses on preserving the quantitative behavior of such models, especially their model fit, while avoiding parameter estimation in the context of the combinatorial explosion in the size of the models. Quantitative model refinement was introduced for rule-based models in [251, 94] and for reaction-based models in [247, 176]. We introduce in the following the quantitative refinement of reaction-based models following the approach of [143] based on refinement relations.

A model M comprises *species* $\Sigma = \{A_1, \ldots, A_m\}$ and *reactions* $R = \{r_1, \ldots, r_n\}$, where reaction $r_j \in R$ can be expressed as a rewriting rule of the form:

$$r_j : s_{1,j}A_1 + \ldots + s_{m,j}A_m \xrightarrow{k_{r_j}} s'_{1,j}A_1 + \ldots + s'_{m,j}A_m, \qquad (14.1)$$

where $s_{1,j}, \ldots, s_{m,j}, s'_{1,j}, \ldots, s'_{m,j} \in \mathbb{N}$ are the *stoichiometric coefficients* of r_j and $k_{r_j} \geq 0$ is the *kinetic rate constant* of reaction r_j. We denote by $r_j^{(1)} = [s_{1,j}, \ldots, s_{m,j}]$ the vector of stoichiometric coefficients on the left hand side of reaction r_j and by $r_j^{(2)} = [s'_{1,j}, \ldots, s'_{m,j}]$ the vector of stoichiometric coefficients on its right hand side. We also denote reaction r_j as $r_j^{(1)} \xrightarrow{k_{r_j}} r_j^{(2)}$.

The goal of the refinement is to introduce details into the model, in the form of distinguishing several subspecies of a given species. The distinction between subspecies may represent post-translational modifications such as phosphorylation, acetylation, etc., but it could also account for different possible types of a particular trait (e.g., fur color of animals in a breeding experiment).

We consider that all species are refined at once. Thus, each species in some initial model M will be replaced by a non-empty set of species in its refined model M_R, according to a *species refinement relation* ρ. The refinement of a set of species to a new set of [sub]species is formalized in Definition 14.1.

Definition 14.1 ([143]). *Given two sets of species Σ and Σ', and a relation $\rho \subseteq \Sigma \times \Sigma'$, we say that ρ is a* species refinement relation *iff it satisfies the following conditions:*

1. *for each $A \in \Sigma$ there exists $A' \in \Sigma'$ such that $(A, A') \in \rho$;*

2. *for each $A' \in \Sigma'$ there exists* exactly one *$A \in \Sigma$ such that $(A, A') \in \rho$.*

We denote $\rho(A) = \{A' \in \Sigma' \mid (A, A') \in \rho\}$. We say that all species $A' \in \rho(A)$ are siblings.

Intuitively, each species $A \in \Sigma$ is refined to the set of species $\rho(A)$, and replaced in the refined model with its refinements. Each species must be refined to at least a singleton set (and in the singleton case one may say that the refinement is trivial and the species does not change, although it may be denoted by a different symbol in Σ'), and no two species in Σ can be refined to the same species $A' \in \Sigma'$.

We introduce in the following definition the refinement of a vector (of stoichiometric coefficients), of a reaction, and of a reaction-based model.

Definition 14.2 ([143]). *Let $\Sigma = \{A_1, \ldots, A_m\}$ and $\Sigma' = \{A'_1, \ldots, A'_p\}$ be two sets of species, and $\rho \subseteq \Sigma \times \Sigma'$ a species refinement relation.*

1. *Let $\alpha = (\alpha_1, \ldots, \alpha_m) \in \mathbb{N}^\Sigma$ and $\alpha' = (\alpha'_1, \ldots, \alpha'_p) \in \mathbb{N}^{\Sigma'}$. We say that α' is a ρ-refinement of α, denoted $\alpha' \in \rho(\alpha)$, if*

$$\sum_{\substack{1 \le j \le p \\ A'_j \in \rho(A_i)}} \alpha'_j = \alpha_i, \text{ for all } 1 \le i \le m.$$

2. *Let r and r' be two reactions over Σ and Σ', resp.:*

$$r : s_1 A_1 + \ldots + s_m A_m \xrightarrow{k_r} s'_1 A_1 + \ldots + s'_m A_m;$$

$$r' : t_1 A'_1 + \ldots + t_p A'_p \xrightarrow{k'_r} t'_1 A'_1 + \ldots + t'_p A'_p.$$

We say that r' is a ρ-refinement of r, denoted $r' \in \rho(r)$, if

$$r'^{(1)}_j \in \rho(r^{(1)}_j) \text{ and } r'^{(2)}_j \in \rho(r^{(2)}_j).$$

3. *Let $M = (\Sigma, R)$ and $M' = (\Sigma', R')$ be two reaction-based models, and $\rho \subseteq \Sigma \times \Sigma'$ a species refinement relation. We say that M' is a ρ-structural refinement of M, denoted $M' \in \rho(M)$, if*

$$R' \subseteq \bigcup_{r \in R} \rho(r) \text{ and } \rho(r) \cap R' \ne \varnothing, \ \forall\, r \in R.$$

In case $R' = \bigcup_{r \in R} \rho(r)$, we say M' is the full structural ρ-refinement *of M.*

Let A be a species with stoichiometry coefficient s in a reaction r, that is refined to a set of species $\rho(A)$. There are $\left(\!\!\left(\begin{array}{c} |\,\rho(A)\,| \\ s \end{array}\right)\!\!\right)$ ways of choosing s species (not necessarily distinct) from the refined set $\rho(A)$, where $\left(\!\!\left(\begin{array}{c} n \\ k \end{array}\right)\!\!\right) = \binom{n+k-1}{k}$ is the *multiset coefficient*, denoting the number of multisets of cardinality k taken from a set of cardinality n. It follows that the number of all possible ρ-refinements of a reaction r of the form (14.1) is

$$\prod_{i=1}^{m} \left(\!\!\left(\begin{array}{c} |\,\rho(A_i)\,| \\ s_i \end{array}\right)\!\!\right) \cdot \left(\!\!\left(\begin{array}{c} |\,\rho(A_i)\,| \\ s_i' \end{array}\right)\!\!\right). \tag{14.2}$$

We introduce in the following definition the notion of *quantitative refinement* of a model. We denote by $[A](t)$ the concentration of species A at time t. Associating time-dependent concentration functions to the variables of a model can be done either directly from the reaction model by choosing a kinetic law for each reaction, see, eg., [204], or by translating the reaction model to another modeling framework, such as Petri nets, rule-based model, or guarded command language, and using a suitable semantic for that translation.

Definition 14.3 ([176, 143]). *Given two reaction-based models $M = (\Sigma, R)$ and $M' = (\Sigma', R')$ and a species refinement relation $\rho \subseteq \Sigma \times \Sigma'$ such that M' is a ρ-refinement of M, we say that M' is a* quantitative ρ-refinement *of M if the following condition holds:*

$$[A](t) = \sum_{B \in \rho(A)} [B](t), \text{ for all } A \in \Sigma, t \geq 0. \tag{14.3}$$

A simple sufficient condition for a model M' to be a quantitative refinement of a model M is given in [143].

14.3   Case Study: The Heat Shock Response (HSR)

The heat shock response is a cellular defence mechanism against stress (high temperatures, toxins, bacterial infection, etc.) that is highly conserved among eukaryotes. We consider here the heat shock response model proposed in [263], consisting of the set of reactions listed in Table 14.1.

Upon exposure to stress, proteins misfold (reaction (10) in Table 14.1) or aggregate into multi-protein complexes that impair cellular functions up to cell death; the flux of the misfolding reaction depends exponentially on the temperature. To counter these proteotoxic effects of thermal stress the expression of a special family of molecular chaperones, called heat shock proteins (hsp's), increases. The chaperone role of hsp's is to bind to misfolded proteins and assist them in their correct refolding (reactions (11),(12) in Table 14.1) thus preventing multi-protein aggregation and cell death.

Table 14.1: The molecular model of the eukaryotic heat shock response proposed in [263].

No. Reaction	No. Reaction
(1) $2\,\mathsf{hsf} \rightleftarrows \mathsf{hsf}_2$	(7) $\mathsf{hsp} + \mathsf{hsf}_3 \rightarrow \mathsf{hsp{:}hsf} + 2\,\mathsf{hsf}$

Table 14.1: The molecular model of the eukaryotic heat shock response proposed in [263] - Continued

(2) $\mathsf{hsf} + \mathsf{hsf_2} \rightleftarrows \mathsf{hsf_3}$	(8) $\mathsf{hsp} + \mathsf{hsf_3}{:}\mathsf{hse} \rightarrow \mathsf{hsp}{:}\mathsf{hsf} + 2\,\mathsf{hsf} + \mathsf{hse}$
(3) $\mathsf{hsf_3} + \mathsf{hse} \rightleftarrows \mathsf{hsf_3}{:}\mathsf{hse}$	(9) $\mathsf{hsp} \rightarrow \varnothing$
(4) $\mathsf{hsf_3}{:}\mathsf{hse} \rightarrow \mathsf{hsf_3}{:}\mathsf{hse} + \mathsf{hsp}$	(10) $\mathsf{prot} \rightarrow \mathsf{mfp}$
(5) $\mathsf{hsp} + \mathsf{hsf} \rightleftarrows \mathsf{hsp}{:}\mathsf{hsf}$	(11) $\mathsf{hsp} + \mathsf{mfp} \rightleftarrows \mathsf{hsp}{:}\mathsf{mfp}$
(6) $\mathsf{hsp} + \mathsf{hsf_2} \rightarrow \mathsf{hsp}{:}\mathsf{hsf} + \mathsf{hsf}$	(12) $\mathsf{hsp}{:}\mathsf{mfp} \rightarrow \mathsf{hsp} + \mathsf{prot}$

The expression of hsp's is regulated by a family of proteins called heat shock transcription factors (hsf's). In a trimeric state (hsf$_3$) they bind to heat shock elements (hse's - the hsp-encoding gene promoter regions), forming hsf$_3$:hse complexes, and activate the transcription of hsp's, process modeled through reactions (1)-(4) in Table 14.1. The concentration levels of hsf$_3$:hse measure the DNA binding activity. Reaction (4) implies that with a higher level of DNA binding we get a faster transcription/synthesis of hsp. The hsp's downregulate their expression levels by binding to hsf$_3$:hse's, hsf$_3$'s, hsf$_2$'s, and hsf's and breaking down the complexes, thus stopping the expression activity (reactions (5)-(8)). The degradation of hsp molecules is modeled in reaction (9).

The hsf protein can undergo post-translational modifications (phosphorylation, acetylation, sumoylation), some of which influence hsf binding activity, see [10]. In particular, the acetylation of hsf's plays a role in the attenuation of the heat shock response. We consider in this paper the refinement of hsf molecules as described in [176]. The refinement considers the acetylation status (ON/OFF) of hsf proteins. The order of acetylated sites is not important in a compound with two or more hsf molecules, only their count. Thus,

- hsf is refined to $\{\mathsf{rhsf}^{(0)}, \mathsf{rhsf}^{(1)}\}$,

- a dimer molecule hsf$_2$ is refined to $\{\mathsf{rhsf_2}^{(0)}, \mathsf{rhsf_2}^{(1)}, \mathsf{rhsf_2}^{(2)}\}$,

- and a trimer molecule hsf$_3$ is refined to $\{\mathsf{rhsf_3}^{(0)}, \mathsf{rhsf_3}^{(1)}, \mathsf{rhsf_3}^{(2)}, \mathsf{rhsf_3}^{(3)}\}$,

where the superscript denotes the number of acetylated sites. This leads to an expansion of the model from 10 species and 17 irreversible reactions to 20 species and 55 irreversible reactions.

This refinement can be described via the following species refinement relation:

$$\begin{aligned}
\rho = \{ &(\mathsf{hse}, \mathsf{rhse}), (\mathsf{hsp}, \mathsf{rhsp}), (\mathsf{prot}, \mathsf{rprot}), (\mathsf{mfp}, \mathsf{rmfp}), (\mathsf{hsp}{:}\mathsf{mfp}, \mathsf{rhsp}{:}\mathsf{rmfp}), \\
&(\mathsf{hsf}, \mathsf{rhsf}^{(0)}), (\mathsf{hsf}, \mathsf{rhsf}^{(1)}), \\
&(\mathsf{hsf_2}, \mathsf{rhsf_2}^{(0)}), (\mathsf{hsf_2}, \mathsf{rhsf_2}^{(1)}), (\mathsf{hsf_2}, \mathsf{rhsf_2}^{(2)}), \\
&(\mathsf{hsf_3}, \mathsf{rhsf_3}^{(0)}), (\mathsf{hsf_3}, \mathsf{rhsf_3}^{(1)}), (\mathsf{hsf_3}, \mathsf{rhsf_3}^{(2)}), (\mathsf{hsf_3}, \mathsf{rhsf_3}^{(3)}), \\
&(\mathsf{hsp}{:}\mathsf{hsf}, \mathsf{rhsp}{:}\mathsf{rhsf}^{(0)}), (\mathsf{hsp}{:}\mathsf{hsf}, \mathsf{rhsp}{:}\mathsf{rhsf}^{(1)}), \\
&(\mathsf{hsf_3}{:}\mathsf{hse}, \mathsf{rhsf_3}^{(0)}{:}\mathsf{rhse}), (\mathsf{hsf_3}{:}\mathsf{hse}, \mathsf{rhsf_3}^{(1)}{:}\mathsf{rhse}), (\mathsf{hsf_3}{:}\mathsf{hse}, \mathsf{rhsf_3}^{(2)}{:}\mathsf{rhse}), \\
&(\mathsf{hsf_3}{:}\mathsf{hse}, \mathsf{rhsf_3}^{(3)}{:}\mathsf{rhse}) \}.
\end{aligned}$$

The full set of the refined reactions is given in Table 14.2.

Table 14.2: The list of reactions for the refined model that includes the acetylation status of hsf. A reaction (i.j) is a refinement of reaction (i) of the basic model, see Table 14.1

No.	Reaction
(1.1)	$2\,\text{rhsf}^{(0)} \rightleftarrows \text{rhsf}_2{}^{(0)}$
(1.2)	$\text{rhsf}^{(0)} + \text{rhsf}^{(1)} \rightleftarrows \text{rhsf}_2{}^{(1)}$
(1.3)	$2\,\text{rhsf}^{(1)} \rightleftarrows \text{rhsf}_2{}^{(2)}$
(2.1)	$\text{rhsf}^{(0)} + \text{rhsf}_2{}^{(0)} \rightleftarrows \text{rhsf}_3{}^{(0)}$
(2.2)	$\text{rhsf}^{(1)} + \text{rhsf}_2{}^{(0)} \rightleftarrows \text{rhsf}_3{}^{(1)}$
(2.3)	$\text{rhsf}^{(0)} + \text{rhsf}_2{}^{(1)} \rightleftarrows \text{rhsf}_3{}^{(1)}$
(2.4)	$\text{rhsf}^{(1)} + \text{rhsf}_2{}^{(1)} \rightleftarrows \text{rhsf}_3{}^{(2)}$
(2.5)	$\text{rhsf}^{(0)} + \text{rhsf}_2{}^{(2)} \rightleftarrows \text{rhsf}_3{}^{(2)}$
(2.6)	$\text{rhsf}^{(1)} + \text{rhsf}_2{}^{(2)} \rightleftarrows \text{rhsf}_3{}^{(3)}$
(3.1)	$\text{rhsf}_3{}^{(0)} + \text{rhse} \rightleftarrows \text{rhsf}_3^{(0)}{:}\text{rhse}$
(3.2)	$\text{rhsf}_3{}^{(1)} + \text{rhse} \rightleftarrows \text{rhsf}_3{}^{(1)}{:}\text{rhse}$
(3.3)	$\text{rhsf}_3{}^{(2)} + \text{rhse} \rightleftarrows \text{rhsf}_3{}^{(2)}{:}\text{rhse}$
(3.4)	$\text{rhsf}_3{}^{(3)} + \text{rhse} \rightleftarrows \text{rhsf}_3{}^{(3)}{:}\text{rhse}$
(4.1)	$\text{rhsf}_3^{(0)}{:}\text{rhse} \rightarrow \text{rhsf}_3^{(0)}{:}\text{rhse} + \text{rhsp}$
(4.2)	$\text{rhsf}_3{}^{(1)}{:}\text{rhse} \rightarrow \text{rhsf}_3{}^{(1)}{:}\text{rhse} + \text{rhsp}$
(4.3)	$\text{rhsf}_3{}^{(2)}{:}\text{rhse} \rightarrow \text{rhsf}_3{}^{(2)}{:}\text{rhse} + \text{rhsp}$
(4.4)	$\text{rhsf}_3{}^{(3)}{:}\text{rhse} \rightarrow \text{rhsf}_3{}^{(3)}{:}\text{rhse} + \text{rhsp}$
(5.1)	$\text{rhsp} + \text{rhsf}^{(0)} \rightleftarrows \text{rhsp:}\,\text{rhsf}^{(0)}$
(5.2)	$\text{rhsp} + \text{rhsf}^{(1)} \rightleftarrows \text{rhsp:}\,\text{rhsf}^{(1)}$
(6.1)	$\text{rhsp} + \text{rhsf}_2{}^{(0)} \rightarrow \text{rhsp:}\,\text{rhsf}^{(0)} + \text{rhsf}^{(0)}$
(6.2)	$\text{rhsp} + \text{rhsf}_2{}^{(1)} \rightarrow \text{rhsp:}\,\text{rhsf}^{(0)} + \text{rhsf}^{(1)}$
(6.3)	$\text{rhsp} + \text{rhsf}_2{}^{(1)} \rightarrow \text{rhsp:}\,\text{rhsf}^{(1)} + \text{rhsf}^{(0)}$
(6.4)	$\text{rhsp} + \text{rhsf}_2{}^{(2)} \rightarrow \text{rhsp:}\,\text{rhsf}^{(1)} + \text{rhsf}^{(1)}$
(7.1)	$\text{rhsp} + \text{rhsf}_3{}^{(0)} \rightarrow \text{rhsp:}\,\text{rhsf}^{(0)} + 2*\text{rhsf}^{(0)}$
(7.2)	$\text{rhsp} + \text{rhsf}_3{}^{(1)} \rightarrow \text{rhsp:}\,\text{rhsf}^{(0)} + \text{rhsf}^{(1)} + \text{rhsf}^{(0)}$
(7.3)	$\text{rhsp} + \text{rhsf}_3{}^{(1)} \rightarrow \text{rhsp:}\text{rhsf}^{(1)} + 2*\text{rhsf}^{(0)}$
(7.4)	$\text{rhsp} + \text{rhsf}_3{}^{(2)} \rightarrow \text{rhsp:}\,\text{rhsf}^{(0)} + 2\,\text{rhsf}^{(1)}$
(7.5)	$\text{rhsp} + \text{rhsf}_3{}^{(2)} \rightarrow \text{rhsp:}\text{rhsf}^{(1)} + \text{rhsf}^{(1)} + \text{rhsf}^{(0)}$
(7.6)	$\text{rhsp} + \text{rhsf}_3{}^{(3)} \rightarrow \text{rhsp:}\text{rhsf}^{(1)} + 2\,\text{rhsf}^{(1)}$
(8.1)	$\text{rhsp} + \text{rhsf}_3^{(0)}{:}\text{rhse} \rightarrow \text{rhsp:}\,\text{rhsf}^{(0)} + 2\,\text{rhsf}^{(0)} + \text{rhse}$
(8.2)	$\text{rhsp} + \text{rhsf}_3{}^{(1)}{:}\text{rhse} \rightarrow \text{rhsp:}\text{rhsf}^{(1)} + 2\,\text{rhsf}^{(0)} + \text{rhse}$
(8.3)	$\text{rhsp} + \text{rhsf}_3{}^{(1)}{:}\text{rhse} \rightarrow \text{rhsp:}\,\text{rhsf}^{(0)} + \text{rhsf}^{(1)} + \text{rhsf}^{(0)} + \text{rhse}$
(8.4)	$\text{rhsp} + \text{rhsf}_3{}^{(2)}{:}\text{rhse} \rightarrow \text{rhsp:}\text{rhsf}^{(1)} + \text{rhsf}^{(1)} + \text{rhsf}^{(0)} + \text{rhse}$
(8.5)	$\text{rhsp} + \text{rhsf}_3{}^{(2)}{:}\text{rhse} \rightarrow \text{rhsp:}\,\text{rhsf}^{(0)} + 2\,\text{rhsf}^{(1)} + \text{rhse}$
(8.6)	$\text{rhsp} + \text{rhsf}_3{}^{(3)}{:}\text{rhse} \rightarrow \text{rhsp:}\text{rhsf}^{(1)} + 2\,\text{rhsf}^{(1)} + \text{rhse}$
(9.1)	$\text{rhsp} \rightarrow \varnothing$
(10.1)	$\text{rprot} \rightarrow \text{rmfp}$
(11.1)	$\text{rhsp} + \text{rmfp} \rightleftarrows \text{rhsp:rmfp}$

Table 14.2: The list of reactions for the refined model - Continued

(12.1)   rhsp:rmfp $\rightarrow$ rhsp + rprot

An ODE-based model for the basic HSR model was introduced and analyzed in [263]. A similar model for the refined HSR model was introduced in [176].

14.4   Quantitative Refinement for ODE Models

The main problem quantitative model refinement of ODE-based models tackles is to identify the kinetic rate constants of the refined model that lead to a solution of the refinement condition (14.3). An attempt to obtain all solutions of (14.3) would require solving the system of ODEs corresponding to the mass-action model for the basic and the refined models; in general, this cannot be done analytically because the ODEs can be non-linear. An alternative was proposed in the form of a sufficient condition in [143]. We recall that result in the following. For two vectors of nonnegative integers $\alpha = (\alpha_1, \ldots, \alpha_m)$, $\alpha' = (\alpha'_1, \ldots, \alpha'_{m'})$, we denote

$$\binom{\alpha}{\alpha'} = \frac{\prod_{i=1}^{m} \alpha_i!}{\prod_{j=1}^{m'} \alpha'_j!}.$$

Theorem 14.1 ([143]). Let Σ and Σ' be two sets of species and $\rho \subseteq \Sigma \times \Sigma'$ a species refinement relation. Let $M = (\Sigma, R)$ be a reaction-based model and $M' = (\Sigma', R')$ be the full structural ρ-refinement of M; we use letters k (indexed with the reaction name) to indicate the kinetic rate constants of M and letters k' those of M'. If for every $\alpha \rightarrow \beta \in R$ and for any $\alpha' \in \rho(\alpha)$ we have that

$$\sum_{\beta' \in \rho(\beta)} k'_{\alpha' \rightarrow \beta'} = \binom{\alpha}{\alpha'} k_{\alpha \rightarrow \beta}, \tag{14.4}$$

then M' is a fit-preserving data refinement of M.

The solution thus obtained is evidently not unique. For example, in the heat shock response refined model the kinetic rate constants of all reactions involving at least one form of acetylated hsf could be set to zero; such a choice would cancel the refinement since the influence of all acetylated variables would be ignored in the model. Theorem 14.1 allows one to choose a multitude of different solutions. One can, for example, include in the solution some parameter values that were obtained from experiments or literature, while using condition (14.4) to choose suitable values for the remaining parameters. In fact, condition (14.4) can also be used to check if a set of given parameter values (for all parameters) leads to a fit-preserving refined model.

The main disadvantage of ODE-based models is that each species gets its own variable and then its own ODE. The framework does not allow for the implicit specification of some of its variables, even when the semantic difference between them is minor (as it could be between sibling subspecies). This leads to an explosion in the size of the refined ODE-based model with respect to the size of the basic ODE-based model. We discuss in the following sections the model refinement approach in three other widely used modeling frameworks. In each case, we focus on whether a more compact specification of the refined model is possible.

14.5 Quantitative Refinement for Rule-Based Models

Rule-based modeling is an approach for tackling the combinatorial explosion induced by expanding reaction-based models. The key feature is that rules only specify those aspects of the input species that are critical for that interaction, while omitting all their other attributes. The rules can be translated either into a set of ODEs, following a continuous, deterministic interpretation, or into a stochastic process, following a stochastic, discrete interpretation of the biological phenomena. Two description languages for the implementation of such models are *BioNetGen*, see [57], and *Kappa*, see [93]. We use BioNetGen in the following. We refer to [127, 128] for details on how models are represented in BioNetGen. A BioNetGen input file, for instance, is essentially a description of the molecular species and their components, reaction rules, kinetic rate constants, initial concentrations, and simulation commands. The reaction network generated by BioNetGen can be used to emulate system's dynamics deterministically or stochastically, see [309]. RuleBender is an open source editor for rule-based models which allows for the construction of large models. The simulation, based on a BioNetGen simulator, see [349], generates the reaction network, in *SBML* and *NET* format, corresponding to the given rule-based model. Simulations can run either deterministically, using ODEs, or stochastically, using SSA algorithms, see [309, 349].

A BioNetGen Implementation of the HSR Model

We discussed in [178] the implementation of the basic heat shock response model of [263] with BioNetGen and RuleBender. The model can be found in [177]. All reactions in our implementation follow the principle of mass action. The BioNetGen model consists of 12 rules, which produce 17 irreversible reactions; kinetic rate constants and initial values are set according to [263]. For example, the RuleBender implementation of the dimerization of hsf is illustrated in Figure 14.1. A deterministic simulation for the BioNetGen model revealed identical simulation results for DNA binding for a temperature of 42°C as the ODE-based model in [263].

To implement in BioNetGen the refinement described in Section 14.3 required only one change: the addition of a site to hsf, having two possible states: acetylated and non-acetylated. The initial concentrations were set conforming to [176]. For more details regarding the implementation, we refer the reader to [178].

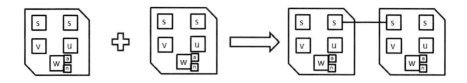

Figure 14.1: A graphical representation of the species hsf (containing sites 's','u','v','w') and of the rule showing the dimerization of hsf, illustrated by binding one of the 's' sites of the hsf species with one of the 's' site of the other hsf species. Note the two possible states of the site 'w', namely 'a' and 'n', which depict two possible states of the species, acetylated or non-acetylated respectively.

14.6   Quantitative Refinement for Petri Net Models

In this section we model the heat shock response and its refinement using the framework of Petri nets. We implemented our models using Snoopy, a visualization, modeling, and simulation tool with support for many types of Petri nets, see [162].

The Basic HSR Model as a Petri Net

For the general method of building a (standard) Petri net model from a given set of biochemical reactions we refer to [205]. We built our implementation of the heat shock response following the standard procedure: each species is represented as a place, and each irreversible reaction is represented as a transition having as pre-places the places corresponding to species on the left hand side of the reaction, and as post-places the places representing species on the right hand side of the reaction; arc multiplicities denote the stoichiometric coefficients of the species involved in the reaction.

We checked several properties of the model to ensure that our implementation is correct. For example, the P-invariants of the Petri net encode the three mass conservation relations of the biological model, as described in [263]. The net is covered by T-invariants, and all places except for the place representing species hsp are covered by P-invariants, which means they are bounded. Our PRISM implementation uses as bounds for the species (except hsp) the constants from the three mass conservation relations, namely the total amount of hsf, hse, and prot. We also simulated the model with the numerical setup of [263] and obtained the same DNA binding curve for a heat shock of 42°C as that shown in [263] for the corresponding ODE-based model. We refer to [178] for details.

The Refined HSR Model as a Colored Petri Net

Implementing a model as a (standard) Petri net means that each reaction is represented as a transition. In doing so, the refined model is inevitably larger than the initial model, similarly as in the case of ODE-based models. However, the framework of *colored Petri nets* allows for stacking more than one species in a colored place, and for identifying each species via a color, see Figure 14.2 for an example.

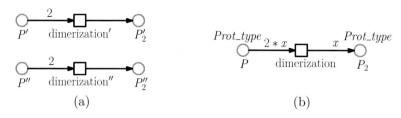

(a) (b)

Figure 14.2: Representing the dimerization of two different proteins, P' and P'' with (a) a transition for each of them, and (b) a single colored transition for both. In (b) we use a color set with two colors, $Prot\_type = \{1, 2\}$. The choice between colors 1 and 2 is done by the variable x; when $x = 1$ the reaction will consume two proteins with color 1 and produce one dimer with color 1, and when $x = 2$ the reaction will consume two proteins with color 2 and produce one dimer with color 2. In the figure, all places and transitions have identifiers, and in (b) we also list the color set for each place (italic text).

We implemented the refinement of the heat shock response model as a colored continuous Petri net, in order to maintain a compact representation. Multiple coloring strategies are possible; we considered two. One of them aimed at using as few colors as possible. In this approach, trimers are represented with four colors $0, 1, 2, 3$, denoting the number of acetylated sites. A possible refinement of reaction (7) with a single-acetylated trimer is

$$\mathsf{rhsp} + \mathsf{rhsf_3}^{(1)} \to \mathsf{rhsp\colon rhsf}^{(0)} + \mathsf{rhsf}^{(0)} + \mathsf{rhsf}^{(1)},$$

and another one is

$$\mathsf{rhsp} + \mathsf{rhsf_3}^{(1)} \to \mathsf{rhsp\colon rhsf}^{(1)} + 2\,\mathsf{rhsf}^{(0)}.$$

In order to differentiate between the two refined reactions, we had to use two transitions, thus increasing the number of transitions compared to that of the basic model. The second strategy was to preserve the structure of the network in terms of number of places, transitions, and the connections between them. We were able to do so by considering the color sets of places denoting some of the species (e.g., $\mathsf{hsf_2}$, $\mathsf{hsf_3}$, $\mathsf{hsf_3\colon hse}$, $\mathsf{hsp\colon hsf}$) as Cartesian products of the color sets of the places corresponding to the species they consist of. This corresponds to a refinement where the order of the acetylated monomers in a trimer is explicitly described. An example of the reversible dimerization reaction using the second coloring approach is presented in Figure 14.3. We refer to [144] for more details about the two modeling strategies based on colored Petri nets.

14.7 Quantitative Refinement for PRISM Models

PRISM is a free and open source guarded command language and probabilistic model checker. It can be used to model and analyze a wide range of probabilistic systems. PRISM supports various types of probabilistic models: probabilistic automata (PAs), probabilistic timed automata (PTAs), discrete-time Markov chains (DTMCs), continuous-time Markov chains (CTMCs), Markov decision processes (MDPs). A PRISM model consists of a keyword

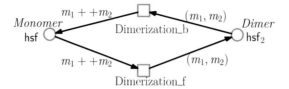

Figure 14.3: Modeling the hsf dimers using a compound color set $Dimer = Monomer \times Monomer$. The regular text next to places and transitions denotes their respective identifier, while the color sets are written in italic font. The hsf monomers are represented using the color set $Monomer = \{0, 1\}$. The preplaces of the forward reaction are two monomers, with colors m_1 and m_2. The result will be the production of one dimer with color (m_1, m_2). In the reverse reaction, one dimer with color (m_1, m_2) is split into the two monomers m_1 and m_2.

Table 14.3: PRISM code for the dimerization in (a) the basic and (b) the refined HSR models (N_{hsf} is the upper bound for hsf and N_{hsf_2} is the upper bound for hsf_2).

(a) Dimerization in the basic HSR model

$[]2 \leq \mathsf{hsf} \leq N_{\mathsf{hsf}} \wedge 0 \leq \mathsf{hsf}_2 \leq N_{\mathsf{hsf}_2} - 1 \to \mathsf{hsf} * \mathsf{hsf} *0.5 * k_1:$
$(\mathsf{hsf}' = \mathsf{hsf} -2) \wedge (\mathsf{hsf}_2{}' = \mathsf{hsf}_2 +1);$

(b) Dimerization in the refined HSR model

$[]2 \leq \mathsf{rhsf}^{(0)} \leq N_{\mathsf{hsf}} \wedge 0 \leq \mathsf{rhsf}_2{}^{(0)} \leq N_{\mathsf{hsf}_2} - 1 \to \mathsf{rhsf}^{(0)} * \mathsf{rhsf}^{(0)} *0.5 * k_1:$
$(\mathsf{rhsf}^{(0)}{}' = \mathsf{rhsf}^{(0)} -2) \wedge (\mathsf{rhsf}_2{}^{(0)}{}' = \mathsf{rhsf}_2{}^{(0)} +1);$
$[]1 \leq \mathsf{rhsf}^{(0)} \leq N_{\mathsf{hsf}} \wedge 1 \leq \mathsf{rhsf}^{(1)} \leq N_{\mathsf{hsf}} \wedge 0 \leq \mathsf{rhsf}_2{}^{(1)} \leq N_{\mathsf{hsf}_2} - 1 \to$
$\mathsf{rhsf}^{(0)} * \mathsf{rhsf}^{(1)} *k_1:(\mathsf{rhsf}^{(0)}{}' = \mathsf{rhsf}^{(0)} -1) \wedge (\mathsf{rhsf}^{(1)}{}' = \mathsf{rhsf}^{(1)} -1)$
$\wedge(\mathsf{rhsf}_2{}^{(1)}{}' = \mathsf{rhsf}_2{}^{(1)} +1);$
$[]2 \leq \mathsf{rhsf}^{(1)} \leq N_{\mathsf{hsf}} \wedge 0 \leq \mathsf{rhsf}_2{}^{(2)} \leq N_{\mathsf{hsf}_2} - 1 \to \mathsf{rhsf}^{(1)} * \mathsf{rhsf}^{(1)} *0.5 * k_1:$
$(\mathsf{rhsf}^{(1)}{}' = \mathsf{rhsf}^{(1)} -2) \wedge (\mathsf{rhsf}_2{}^{(2)}{}' = \mathsf{rhsf}_2{}^{(2)} +1);$

---

which describes the model type (e.g., CTMC) and a set of modules whose states are defined by the state of their finite range variables (e.g., hsf). The state of the variables in each module is specified by some commands including a guard and one or more updates, see [212].

The HSR Models as PRISM Implementations

We implemented the basic heat shock response as a CTMC model within a single module. The PRISM model consists of 10 variables, each of them corresponding to one of the reactants in the model, and 17 guards representing the 17 irreversible reactions of the system. For example, the guard corresponding to *dimerization*, reaction (1) in Table 14.1, is expressed in Table 14.3(a).

For the refined heat shock response model, the corresponding PRISM model was built in a similar way, following its reactions in Table 14.2. For example, the guards corresponding to *dimerization* are presented in Table 14.3(b). The complete models can be found at [177] and more details on how we built them can be found in [178].

Stochastic Model Checking of the PRISM HSR Models

The maximum number of states that PRISM can handle for CTMCs does not exceed 10^{10}, see [210], which leads to difficulties in handling the *state space explosion* problem, see [160]. To avoid this problem, we used *approximate verification*, see [221, 167, 160], to verify our two PRISM models.

We are interested in verifying two properties discussed in [263]: (i) the existence of three mass-conservation relations, and (ii) the level of DNA binding eventually returns to the basal values, both at $37°C$ and at $42°C$.

The following three properties are used to check whether the mass-conservation relations, corresponding to the level of hsf, hse, and prot, are valid in all states along the path:

- $p =?$ [G hsf $+2$ hsf$_2$ $+3$ hsf$_3$ $+3$ hsf$_3$:hse $+$ hsp:hsf $=$ hsf$_{\text{const}}$],

- $p =?$ [G hse $+$ hsf$_3$:hse $=$ hse$_{\text{const}}$],

- $p =?$ [G prot $+$ mfp $+$ hsp:mfp $=$ prot$_{\text{const}}$].

As expected, the value of p was confirmed to be 1 in all cases, with confidence level 95%, i.e., the three mass conservation relations are respected in the model.

We verified in PRISM that for time points larger than 14400, the value of hsf$_3$:hse complex returns to the initial value by formulating the property, $p =?$ $[F \geq 14400 \quad$ hsf$_3$:hse $= 3]$. We chose 14400 as a time point reference to correspond to the upper limit of the simulation time for the model in [263]. The probability value calculated by PRISM was 1 for this property as well, with confidence level 95%.

Finally, we also checked if the model confirms the experimental data of [203] on DNA binding. Due to the memory issues of PRISM, it was not possible to run the simulation many times and use the average run plot to verify the experimental data. Therefore, as an alternative approach, we checked the probability of having a data point within the interval $[0.9 \cdot d, 1.1 \cdot d]$ in the time period $[0.9 \cdot t, 1.1 \cdot t]$, where d is the experimental data point at time t. The results in both cases show high value probabilities, which confirms that our two PRISM models are in accordance with the experimental data of [203]; we refer to [144] for the numerical results.

14.8 Discussion

We discussed in this paper quantitative model refinement, an approach to step-wise construction of biological models. Preserving the quantitative behavior of models throughout the model construction process is at the forefront of this approach. This allows the modeler to avoid repeating the computationally expensive process of parameter estimation at every step of the process, thus avoiding the need of collecting larger and larger sets of high-quality time data. Quantitative refinement also allows the modeler to deal with partial and incomplete information about some of the parameters of the model-to-build, including such information when available, checking its consistency with the other parameters and with the data, and compensating for lack of information about parameters with an algorithmic solution.

We investigated in this paper the versatility of the fit-preserving refinement method with respect to four broadly used frameworks: reaction models (with ODEs), rule-based models (with BioNetGen), Petri net models (with Snoopy), and guarded command language models (with PRISM). Dealing with the combinatorial explosion to account for post-translational modifications was considerably different from one framework to another. We conclude that the method is cross-platform: it is implementable in all chosen frameworks, despite their distinct underlying modeling paradigms.

In our case study based on the heat shock response, the data structure provided by BioNetGen proves to be suitable for modeling the refinement of biological systems. One could effectively employ species, sites, links, etc., to produce a compact representation of the refined model. On the other hand, using the colored version of Petri nets provides the modeler with appropriate tools to introduce data types into the places of the network. The modeling choices in the definition of the new data types and associating biological meanings to them directly affect the compactness of the representation as well as the corresponding model's complexity. In contrast, PRISM supports only elementary data types for the variables in the model, which leads to an explicit detailing of all elements of the refined model, similarly as in the case of ODE-based models.

We show that our approach toward quantitative model refinement is a potentially suitable one to build a large biomodel which can be implemented within a wide spectrum of modeling frameworks. In this method the modeler is able to easily modify the level of details in the model in an algorithmic fashion, while ensuring that the model fit is preserved from one level of detail to another. It should also be noted that one can switch from one modeling framework to another in order to use the advantages that each model formulation offers in terms of fast simulations, model checking, or compact model representation.

Acknowledgments

The authors thank Monika Heiner for help on Snoopy and Charlie, James Faeder and Leonard Harris for advice on the BioNetGen implementation of the heat shock response, and Adam Smith for technical support regarding RuleBender. We gratefully acknowledge support from Academy of Finland through project 267915.

Chapter 15

Developing and Verifying User Interface Requirements for Infusion Pumps: A Refinement Approach

Rimvydas Rukšėnas

Queen Mary University of London

Paolo Masci

Queen Mary University of London

Paul Curzon

Queen Mary University of London

Abstract. When describing criteria for the acceptable safety of systems, it is common practice for the regulator to provide safety requirements that should be satisfied by the system. These requirements are typically described precisely but in natural language and it is often unclear how the regulator can be assured that the given requirements are satisfied. This chapter applies a rigorous refinement process to demonstrate that a precise requirement is satisfied by the specification of a given medical device. It focuses on a particular class of requirements that relate to the user interface of the device. For user interface requirements, refinement is made more complex by the fact that systems can use different interaction technologies that have very different characteristics. The described refinement process recognises the variety of interaction technologies and models them as an interface hierarchy.

15.1   Introduction

Demonstrating that interactive devices are acceptably safe is a significant and important element in their development in various domains. For example, interaction design errors in medical devices have an impact on patient safety and contribute to health-care costs. Because of this, medical device regulators require manufacturers to provide sufficient evidence that the risks associated with the device are "as low as reasonably practicable" as well as being fit for purpose before entering the market. This process is known as the premarket review process. For the majority of medical devices, the premarket approval relies on manufacturers demonstrating that the new device is as safe and effective as an already legally marketed device [336], or that it has been developed in accordance with recognised international standards [86].

In its current form, the premarket approval process involves the analysis of tens of thousands of printed pages [235] rather than a direct evaluation of the product. To reduce the amount of paperwork and enable the submission of more succinct and rigorous evidence, the use of formal methods is being promoted by the US Food and Drug Administration (FDA), the regulator for medical devices in the US. Their approach relies on *usage models* for the verification of software [185]. A usage model is a formal representation that describes the common characteristics and behaviour of software for broad classes of devices. The approach is based on the idea of developing usage models that satisfy core sets of safety requirements that are designed to mitigate typical hazards. This way, usage models can be used as a reference by manufacturers – if they are able to show that their product is compliant with the behaviours of the usage models, then regulators have evidence that the manufacturer's device meets minimum safety conditions. These models are developed manually starting from safety requirements, verifying the models against these requirements using model checking techniques.

Our approach shows how stepwise refinement and the Event-B/Rodin platform can be conveniently used to develop correct-by-construction usage models that are related to the interactive aspects of medical devices. It addresses two key points of the FDA's approach. The first is how to design safety requirements so that they are sufficiently precise to be effectively operationalised. The second is to provide, by operationalising requirements, the means for encompassing the range of input/output technologies that are likely to be encountered in interacting with the systems. Event-B is initially used here to express the high level requirements such as those proposed by the FDA. Stepwise refinement is then used both to make those high level requirements more precise and to demonstrate that the requirement can be cascaded into a hierarchy that encompasses potential input/output technologies.

To illustrate the approach, we focus specifically on infusion pumps. We take as a starting point a particular sample set of *user related* requirements specified by the FDA. The original FDA specifications are in natural language. We give abstract formal specifications of these requirements. We then show how they can be refined to more concrete versions. These versions can be verified against the formal specification of specific pump designs. Here we concentrate on a particular infusion pump design based on a commercially available pump.

This paper is based on and extends our earlier work [286] on user interface requirements for infusion pumps. In particular, it formally develops and verifies new requirements related to the safeguards against accidental tampering with infusion settings. Also, the specific pump design is modelled in more detail which reflects the actual device more truthfully.

15.1.1 Outline of the Approach

The proposed approach is based on three layers: requirements hierarchies, interface hierarchies, and concrete interfaces, each described below.

The requirements hierarchies layer, which is directly relevant to regulators, concerns the development of user interface requirements. The regulator will be interested in the satisfaction of these requirements to assure them of the device's safety. A minimal set of such requirements, relevant to some usability aspect of device interfaces, is developed. The aim is that these requirements should be sufficiently abstract to encapsulate the behaviour of the largest class of possible devices. Refinements are then used to detail these requirements in a sequence of steps. It is also possible that refinement can lead to alternative interface requirements that also provide assurance of the safety of the device. These modified requirements would be developed as a contract between regulator and manufacturer. The development of the requirements hierarchy layer is discussed in Section 15.4.

The concrete interface layer focuses on the user interfaces of specific devices. This layer is most relevant to manufacturers as they demonstrate that the user interfaces of their devices satisfy the requirements developed in the requirements hierarchy layer. There are several possible approaches when trying to produce such a demonstration.

The first is to verify a specific interface against the safety requirements directly. How complicated this is will depend on the extent the requirements were operationalised. An example of this approach is discussed in Section 15.5.2.

The second approach aims to simplify the process of demonstrating that a specific user interface adheres to the relevant set of user requirements. It facilitates the dialogue between regulators and manufacturers by means of an intermediate layer, the interface hierarchies. This layer essentially develops a refinement based hierarchy (classification) of user interfaces. The idea is that user requirements are verified once for most abstract classes of interfaces. More concrete classes of interfaces at the lower levels of this hierarchy are then guaranteed to satisfy the requirements by construction. Now, instead of directly verifying a specific interface against the requirements it suffices to demonstrate that it is an instance of some concrete class of user interfaces. This approach, briefly discussed in Section 15.5.3, correlates with the current FDA pre-market review process which involves providing evidence that a new device is 'substantially equivalent' to already approved and legally marketed medical devices.

15.2 Sample User Interface Requirements from FDA

The regulator's aim is to be assured that risks associated with the use of a device are as low as reasonably practicable. As previously discussed part of this assurance is achieved through a credible demonstration that safety requirements are true of the device. Before showing how this demonstration can be achieved in the proposed framework, we describe a subset of user related safety requirements developed by the FDA.

These requirements relate to two aspects of infusion pump designs: the usability of their data entry systems and the safeguards against inadvertent changes of or tampering with infusion settings. The subset considered is relevant to the volumetric infusion pump used by clinicians that forms the basis of the example contained in this paper. The safety requirements are taken from a larger set produced by the FDA [335]. This larger set is intended specifically for PCA (Patient Controlled Analgesic) pumps. As a result it has more emphasis on patient tampering than clinician errors, and therefore the overall focus is slightly different than is relevant to the volumetric infusion pump. The aim is to show how these independently determined properties can be framed in our framework.

The requirements from the FDA document, considered in the subsequent sections, are listed below:

R1 *The flow rate and vtbi (volume to be infused) for the pump shall be programmable.* This safety requirement aims to mitigate hazards due to incorrectly specified infusion parameters (e.g., flow rate is too high or low).

R2 *The vtbi settings shall cover the range from v_{min} to v_{max} ml.*

R3 *The user (clinician) shall be able to set the vtbi in j ml increments for volumes below x ml.*

R4 *The user (clinician) shall be able to set the vtbi in k ml increments for volumes above x ml.*

R5 *Clearing of the pump settings and resetting of the pump shall require confirmation.* This requirement aims to safeguard against clinicians changing infusion settings inadvertently.

R6 *To avoid accidental tampering of the infusion pump's settings such as flow rate/vtbi, at least two steps should be required to change the setting.*

15.3　Background

This section briefly discusses approaches to interface refinement and introduces the Event-B formalism used in our approach.

15.3.1　Interface Refinement Approaches

Several previous projects on formal refinement for user interface design had different foci to our work. For example, the main focus of Bowen and Reeves [64, 65] is on a description of the actions that the user can engage with and how these actions can be refined. The refinement process involves actions being replaced by more concrete actions in terms of more concrete structures. The refinement described by them is more akin to trace refinement.

Although they argue that their interest is in ensuring that requirements are true of the more refined system, there is less concern with how the requirements are transformed through the levels of refinement. Duke and Harrison [123] are concerned with data refinement. They note that abstract representations of objects can be refined in two directions, into what is perceivable and into the architecture of the device. Darimont and van Lamsweerde [96] are concerned with requirements described in terms of the refinement of goals using the KAOS language. The interesting innovation in their proposal is that the formal refinement process may be achieved through a set of patterns.

The approach we take here has most in common with the work of Yeganefard and Butler [350] who demonstrate a similar refinement process, in this case for control systems, using Event-B. They describe an approach to requirements structuring to facilitate refinement-based formalisation. Their work considers control systems consisting of plants, controllers and operators. In these terms, our focus is narrower, encompassing phenomena shared between controller and the operator. Also, we emphasise the formalisation of high level requirements and their clarification through refinement, whereas Yeganefard and Butler focus attention on requirements structuring. The structure developed is then mapped to a formal model in the stepwise refinement process. Moreover, their work is yet to address non-functional requirements, considered here.

In our previous work [235, 155], we explored a different approach to formalisation and refinement of user-related requirements. First, an abstract logic model is created that encapsulates key notions and relationships presented in the textual description of the requirements. Second, a mapping relation is established between the abstract logic model and a concrete model of a device being verified. This mapping relation is used as a basis to instantiate requirements for the concrete device model. The concrete model is then mechanically verified against the instantiated requirements.

15.3.2 Event-B/Rodin Framework

Event-B specifications are discrete models that consist of a state space and state transitions. A state includes constants and variables that describe the system. State transitions are specified as *events*. A specification of an event consists of two parts as seen in the example below. The first is a list of *guards*. Each guard is a predicate over the state variables and constants. All the guards are combined using logical conjunction which is implicit in the Event-B syntax. These guards together define the necessary conditions for the event to occur. The second part is a list of *actions* which describe how the state variables are modified as a result of event execution.

Specifications are structured in terms of *machines* and *contexts*. Machines specify the dynamic aspects of a system, whereas contexts specify its static aspects. A machine includes state variables and events. Invariant properties are expressed as machine invariants, i.e., predicates that must hold in all machine states. A context includes constants defined by a set of axioms. A machine may reference constants from the contexts it 'sees'.

Intuitively, machine execution means that one of the events, with all guards being true, is chosen. The machine variables are modified as specified by the actions of that event. The basic syntactic form of an event is given below, other features of Event-B are introduced when needed.

Event $E \mathrel{\widehat{=}}$ **when** $G(v)$ **then** $T(v)$ **end**

Here v is a list of variables. $G(v)$ denotes the guards of E and $T(v)$ denotes the actions associated with *E*. The formal semantics of events is given using before-after predicates that encode the relation between the machine variables before and after an event occurrence. A detailed description of Event-B can be found in [6].

15.4   The Requirement Hierarchies

In this section, the informal requirements from Section 15.2 are first formalised in Event-B then made more precise through gradual refinement. The requirements **R1–R4** related to data entry interfaces are considered first.

15.4.1   Requirements for Data Entry

The informal requirements **R1** and **R2** provide a basis for the abstract specification of user requirements relevant to data entry. **R3** and **R4** are introduced in a later refinement.

15.4.1.1   Requirements R1 and R2

The requirement **R1** (*The vtbi/flow rate for the pump shall be programmable*) is expressed as the following machine in Event-B. This abstract description simply requires that a variable called *data* has the attribute that it is programmable. The requirement asserts that *data* commences with a value named *source* and describes the event *programmable* as changing the value to *target*. The possible values of *data* are given as the set *Numbers*. All three constants, *Numbers*, *source*, and *target*, are defined in the context ReqParams1 below.

```
MACHINE  Reqs1  SEES  ReqParams1
VARIABLES   data   INVARIANTS   data ∈ Numbers
EVENTS
Initialisation begin data: = source end
Event   programmable ≙ begin data: = target end
END
```

The invariant of Reqs1 simply gives typing of *data*. The initialisation event assigns the *source* value to it. Since the *programmable* event expresses an abstract requirement, its guard is assumed to be always true, and the **when** clause is omitted in the above specification.

The requirement **R2** (*The vtbi settings shall cover the range from v_{min} to v_{max} ml*) is specified in the context ReqParams1 which defines the corresponding constants *Min*, *Max*. It is assumed that *Max* exceeds *Min* and that *Min* is non-negative. The set constant (type) *Numbers* is assumed to be the interval $0 .. Max$. The context defines a number of other constants: *RefValues*, *source*, and *target*. It is assumed that the *source* value belongs to the interval *Numbers* and it is assumed that *target* is a member of the set of reference values (*RefValues*) that covers the required range of settings. At this stage, no other assumptions are made as to what these values are.

```
CONTEXT   ReqParams1
CONSTANTS   Min   Max   Numbers   RefValues   source   target
AXIOMS
        Min ≥ 0      Max>Min      Numbers = 0 .. Max
        RefValues ⊆ Numbers ∩ {x | x ≥ Min}
        source ∈ Numbers      target ∈ RefValues
END
```

Because the **R1** requirement is specified in a non-operational form it is necessary to refine the machine. Informally, machine refinement means verifying three constraints. The first concerns event refinement: a concrete event must refine the corresponding abstract one (new events must refine an implicit event that does nothing). The second constrains new events: they must 'converge' (i.e., not run forever on their own). The third states that the concrete machine must not deadlock before the machine it refines.

The following refinement of Reqs1 provides guidance about how **R1** can be implemented. The operational version of **R1** has a number of new characteristics. Two new variables are introduced: *entry* and *disp*. Whether a number is being entered is indicated by *entry*, whereas *disp* gives the displayed value of the number entered. The initial state requires that *data* and *disp* are both initialised to the *source* value and *entry* is false, indicating that entry of the target number has not commenced. The new requirement decomposes the event representing **R1** into three events. The first one (*choose*) is used to elect to enter the target value, while the second one models the modification of the display value (this is not necessarily the data value). The final event is triggered when the display and target values are equal. At this step the data value is set to be equal to the display value and entry becomes false. This operational requirement indicates more about the programming process but says little about how the value is entered.

```
MACHINE   Reqs11   REFINES   Reqs1   SEES   ReqParams1
VARIABLES   data   disp   entry      INVARIANTS      disp ∈ Numbers   entry ∈ BOOL
EVENTS
Initialisation      begin   data: = source   disp: = source   entry: = FALSE   end
Event   choose ≙   Status anticipated
        when   entry = FALSE   then   disp: = data   entry: = TRUE   end
Event   modify ≙   Status anticipated   when   entry = TRUE   then   disp: ∈ Numbers   end
Event   set ≙   refines programmable
        when   disp = target   entry = TRUE   then   data: = disp   entry: = FALSE   end
END
```

The machine Reqs11 specifies that *set* refines the abstract event *programmable* (intuitively, both events assign *target* to *data*). The other two events, *choose* and *modify*, are new. For the machine Reqs11 to refine Reqs1 these newly introduced events must 'converge' (i.e., they must not execute forever). The specification does not attempt to prove that. Rather than requiring their convergence, the specification assumes, as indicated by the keyword 'anticipated', that *choose* and *modify* will not run forever. If necessary, this assumption can be proven later.

15.4.1.2   Requirements R3 and R4

In the case of **R3** (*The user shall be able to set the VTBI in j ml increments for volumes below x ml*) and, similarly, **R4**, the requirements are expressed in a sufficiently concrete form to proceed directly to their operationalised versions. They are captured in the following context ReqParams11 which extends ReqParams1 by adding three relevant constants— *Threshold* (x in **R3** and **R4**), j, and k—with three associated axioms:

CONTEXT   ReqParams11   **EXTENDS**   ReqParams1   **CONSTANTS**   *Threshold* j k
AXIOMS

   $Threshold \in Min + 1 .. Max - 1$ $j < Threshold$ $k \le Threshold$
   $RefValues \subseteq \{x \cdot x > 0 \land j * x \le Threshold \mid j * x\} \cup \{x \cdot x > 0 \mid Threshold + k * x\}$

END

The fourth axiom restricts the reference set (*RefValues*) to the values obtained using the increments j and k. This context is used by Reqs111 which is the same machine as Reqs11 otherwise:

MACHINE   Reqs111   **REFINES**   Reqs11   **SEES**   ReqParams11   ....

The last step in the refinement of requirements has a more technical nature. It decomposes Reqs111 so that the assumptions about the user behaviour are removed from the requirements for the pump interfaces. In particular, one guard (*disp = target*) in the event *set* encompasses the notion of a target. Though the latter is relevant to the user behaviour, it would be meaningless to apply it to the pump interface. The decomposition introduces a machine that replaces the constant *target* by a variable that represents the display value 'passed' to the user. The details are omitted here, since this does not affect the actual data entry.

15.4.2   Safeguards against Inadvertent Changes or Tampering

In this section, the requirements **R5** and **R6** are formally developed.

15.4.2.1   Requirement R5

The informal requirement **R5** (*Clearing of the pump settings and resetting of the pump shall require confirmation*) simply states that an attempt to clear the pump settings (such as vtbi and flow rate) cannot take its effect immediately but should result in a request for confirmation. The requirement is captured by the event *clear* in the following Event-B machine Reqs5. The variable *require* set to true represents a request for the confirmation of the clearing action. The event *clear* is enabled when the clearing action has not already been attempted (*require = FALSE*). In that case, it initiates a request to confirm the clearing action (*require := TRUE*). Note that this specification, faithful to the informal requirement, says nothing about the effect of the clearing action on the pump settings:

MACHINE   Reqs5

VARIABLES   *require*   **INVARIANTS**   *require* $\in Bool$

EVENTS
Initialisation begin *require*: $= FALSE$ **end**
Event   *clear* $\hat{=}$ **when**   *require = FALSE*   **then**   *require*: $= TRUE$ **end**
END

Though it is not stipulated explicitly, the requirement **R5** also implies that there should be some kind of acknowledgement for the clearing request. By making this assumption explicit in the following refinement of the machine Reqs5, the requirement becomes more precise and rigorous. In the refinement Reqs51, the acknowledgement is modelled as a new event, *acknowledge*. Reqs51 also introduces a new variable, *ack*. When set to true, this variable represents an acknowledgement of the clearing action:

MACHINE Reqs51 **REFINES** Reqs5

VARIABLES *require ack* **INVARIANTS** $ack \in BOOL$

EVENTS
Event Initialisation **extends** Initialisation **then** $ack := FALSE$ **end**
Event clear **extends** clear **end**
Event *acknowledge* $\widehat{=}$ **Status** anticipated
    when $ack = FALSE$ $require = TRUE$ **then** $ack := TRUE$ **end**
END

The event *acknowledge* is enabled when there is an unacknowledged request for confirmation and sets *ack* to true. For the remaining events, the keyword **extends** indicates that the event in question incorporates (and refines) the corresponding event in Reqs5. In addition, the refined initialisation event sets *ack* to false.

The requirement as modelled by Reqs51 captures mode transitions in the pump behaviour. However, it puts no constraints on the changes to the pump settings associated with those transitions. The need to be more rigorous about that is addressed in the next refinement of Reqs51. Not to be bound by any specific interpretation of the concept of pump settings, we assume that they are expressed as an abstract type (set), *Settings*. Constant *blank* from *Settings* represents the pump settings being cleared. These assumptions are modelled as the following context:

CONTEXT ReqParams5
SETS *Settings* **CONSTANTS** *blank*
AXIOMS
    $blank \in Settings$
END

The refinement Reqs511 introduces new variable *data*. It is an abstract representation of the pump settings. The requirement **R5** is once again made more precise by distinguishing two possibilities associated with the acknowledgement event. The first one is that the clearing request is actually confirmed by the pump user. This is modelled by the event *confirm* which extends *acknowledge* by updating *data* to *blank* (the pump settings are cleared). The second possibility is that the user changes their mind and cancels the clearing request. This is modelled by the event *quit* which leaves *data* unchanged:

MACHINE Reqs511 **REFINES** Reqs51 **SEES** ReqParams5
VARIABLES *require   ack   data*   **INVARIANTS**   *data* ∈ *Settings*
EVENTS
Event Initialisation **extends** Initialisation **then** *data*: ∈ *Settings* **end**
Event clear **extends** clear **end**
Event confirm **extends** acknowledge **then** *data*: = *blank* **end**
Event quit **extends** acknowledge **end**
Event *other* ≙ **Status** anticipated
         when *require* = *FALSE* **then** *data*: ∈ *Settings* **end**
END

A new event (*other*) is added to Reqs511 to avoid constraining changes to the pump settings when there is no request to clear them. In such states (*require* = *FALSE*), event *other* allows the pump settings to be updated in an arbitrary way (*data*: ∈ *Settings*).

15.4.2.2   Requirement R6

Expressed in natural language, the requirement **R6** (*To avoid accidental tampering of the infusion pump's settings such as flow rate / vtbi, at least two steps should be required to change the setting*) is open to several interpretations. For example, the 'two step' condition can be interpreted as one change of the setting followed by another. However, in the context of safeguarding against accidental tampering, a more plausible interpretation of **R6** seems to be that any attempt to change a setting like vtbi should require a separate confirmation step. Such an interpretation relates **R6** to **R5**. Therefore, it would be plausible to view **R6** as a part of the requirements hierarchy formally developed for the requirement **R5**, since a two step (request - acknowledgement) structure has already been specified in the machine Reqs51. In particular, **R6** can be formally introduced as a refinement of Reqs51.

At the same time, there are several aspects in which **R6** is different from **R5**. Firstly, it applies to *any* change to a particular pump's setting, not just clearing of that setting. Secondly, **R6** refers to the changes to a particular setting as opposed to the simultaneous clearing of all settings as stated in **R5**. Though, due to the abstract nature, the type *Settings* in our specification can be interpreted as both, a particular pump's setting and all the settings combined, it is not obvious what advantages a formal linkage of **R6** and **R5** would provide.

Taking these considerations into account, the requirement **R6** is formalised in a separate development. Its starting point captures the 'two step' condition, modelled as the following Event-B machine:

MACHINE Reqs6

VARIABLES *change*   **INVARIANTS**   *change* ∈ *Bool*

EVENTS
Initialisation begin *change*: ∈ *Bool* **end**
Event *update* ≙ **when** *change* = *TRUE* **then** *change*: = *change* **end**
Event *acknowledge* ≙ **when** *change* = *TRUE* **then** *change*: = *FALSE* **end**
END

The variable *change* represents the mode of pump operation where changes to a particular pump's setting can be made. When the pump is in such a mode, the event *update* (first step) stands for all the possible updates to that setting. These updates leave the mode of pump operation unchanged (*change*: = *change*). The event *acknowledge* (second step) stands for the confirmation of changes, which results in the pump exiting the change mode (*change*: = *FALSE*).

Next, we look in more detail at how the two steps specified in Reqs6 relate to the actual changes to the pump's setting considered. The informal requirement **R6** suggests that any changes to the relevant setting are 'provisional'. In that respect, however, **R6** can be interpreted in two ways at least. The first interpretation is that the changes, before they are confirmed, do not affect the actual pump's setting. Instead, they are recorded in a temporary pump's parameter. Only the confirmation step updates the relevant setting with the new value from the temporary parameter. The second interpretation is that the changes are applied to the relevant setting immediately. However, they still have to be confirmed in the confirmation step. Otherwise, the setting is restored to its old value. Both interpretations are below formalised as refinements of Reqs6. Both refinements introduce a new variable, *param*, that represents the relevant pump's setting. We start with the first interpretation.

In addition to *param*, the refinement Reqs61 introduces another variable, *new*, that models provisional changes to *param*. This variable is initialised to the value of *param*:

```
MACHINE  Reqs61  REFINES  Reqs6  SEES  ReqParams5
VARIABLES    change   param   new
INVARIANTS      param ∈ Settings   new ∈ Settings
EVENTS
Event  Initialisation   extends Initialisation  then   param: ∈ Settings    new: = param   end
Event  update   extends update   then    new: ∈ Settings   end
Event  confirm   extends acknowledge   then    param: = new   end
Event  cancel   extends acknowledge   end
END
```

The event *update* extends the same event from Reqs6. It guarantees that any changes to the setting are provisional and temporarily recorded in the variable *new*. Both events *confirm* and *cancel* refine the old event *acknowledge*. The confirmation of the changes to the setting is modelled by *confirm*. It extends *acknowledge* by updating *param* with the new value for this setting (*new*). Any changes are cancelled by the event *cancel*.

The second, perhaps less natural, interpretation of **R6** is formalised as the following refinement Reqs62. In addition to *param*, it introduces another variable, *old*. This variable stores the old setting and is initialised to *param*:

```
MACHINE  Reqs62  REFINES  Reqs6  SEES  ReqParams5
VARIABLES    change   param   old
INVARIANTS      param ∈ Settings   old ∈ Settings
EVENTS
Event  Initialisation   extends Initialisation  then   param: ∈ Settings   old: = param   end
Event  update   extends update   then    param: ∈ Settings   end
Event  confirm   extends acknowledge   end
Event  cancel   extends acknowledge   then   param: = old   end
END
```

According to this specification, the setting *param* is changed with each *update* event. However, the changes are disregarded by the event *cancel*, if this option is selected instead of *confirm*.

The difference between these formal interpretations of the requirement **R6** is quite subtle. If the changes to the setting are finished in a normal way by confirming or cancelling them, the two formal requirements make no difference. Only when the changes are abruptly interrupted (e.g., the pump is switched off before the confirmation step), Reqs61 and Reqs62 will result in different requirements for the pump design. It is up to the regulators of medical devices to decide which version of the requirement **R6** is preferable, or whether they both are equally acceptable. Formalising several alternative interpretations highlights the issue but also raises the possibility of exploring the consequences for safety of each choice formally.

15.5   Verification of Concrete Interfaces

Having produced an operational but abstract definition of the requirements, the next stage is to make sense of the requirement in terms of the particular device that the developer wishes to certify. The aim of this section is to show how an interface specification of a specific device can be shown to satisfy user related requirements. Ideally, such a specification would be provided by the device manufacturer. Alternatively, it can be reverse engineered by interactively exploring the actual device [328, 155].

To illustrate our approach, we consider the number entry module of the Alaris GP Volumetric Pump [77]. A specification of this module has been reverse engineered in PVS and SAL [236]. The specification given below is its direct translation to Event-B. The purpose of using this translation is to demonstrate two ways of verifying the relevant user requirements for the independently developed specifications of concrete interfaces.

15.5.1   Specification of the vtbi Entry in Alaris

The Alaris pump uses a chevron based number entry interface. In this type of interface, the current data value is updated by pressing the 'up' (increase) and 'down' (decrease) chevron keys. The fast versions of these keys are used to speed up data entry. For example, a fast 'up' chevron increases the current value by a larger amount compared to a slow 'up' one.

In the PVS and SAL versions, the behaviour of the Alaris chevrons (slow and fast up-/down keys) is captured using functions that specify how the current value is modified by pressing each chevron. In Event-B, the corresponding functions, *alaris_up*, *alaris_dn*, *alaris_UP*, and *alaris_DN*, are defined in the following context. It extends RealDefinitions which provides an Event-B model for the real numbers supported by the Alaris pump. The definitions of *alaris_dn*, *alaris_UP*, and *alaris_DN* (omitted here) are similar to that of *alaris_up*:

CONTEXT AlarisDefinitions **EXTENDS** RealDefinitions
CONSTANTS *trim* *alaris_up* *alaris_dn* *alaris_UP* *alaris_DN* *init* ...
AXIOMS

$\quad trim \in \mathbb{Z} \to real \quad alaris\_up \in real \to real \quad init \in real$
$\quad \forall x \cdot (x < minAlaris \Rightarrow trim(x) = minAlaris) \wedge$
$\qquad (x > maxAlaris \Rightarrow trim(x) = maxAlaris) \wedge$
$\qquad (x \geq minAlaris \wedge x \leq maxAlaris \Rightarrow trim(x) = x)$
$\quad \forall x \cdot x \in real \Rightarrow (x < r100 \Rightarrow alaris\_up(x) = trim((floor(x * 10) + r1)/10)) \wedge$
$\qquad\qquad (x \geq r100 \wedge x < r1000 \Rightarrow alaris\_up(x) = trim(x + r1)) \wedge$
$\qquad\qquad (x \geq r1000 \Rightarrow alaris\_up(x) = trim((floor(x/10) + r1) * 10)) \quad ...$

END

The behaviour of the four chevrons when entering vtbi values is described by the events *up*, *dn*, *UP*, and *DN*, e.g., *up* is specified below:

Event *up* $\widehat{=}$ **Status** anticipated

$\quad$ **when** $topline = dispvtbi \quad entrymode = vtmode$ **then** $display := alaris\_up(display)$ **end**

Here the condition *topline = dispvtbi* indicates that the pump is in the mode where the vtbi value can be changed, whereas *entrymode = vtmode* says that the changes are performed by updating the vtbi value with the chevron keys. The *display* variable represents the displayed value of vtbi. This event does not change the actual vtbi setting represented by the variable *vtbi*. The specifications of the remaining chevrons are similar.

The machine Alaris_vtbi1 includes these four chevron events and two events that model the acknowledgement (confirmation and cancellation) of the changes made to the vtbi value using the chevrons. The confirmation case is specified as follows:

Event *confirm* $\widehat{=}$

$\quad$ **when** $topline = dispvtbi \quad entrymode = vtmode$ **then** $vtbi := display \quad topline :=$
$\quad ptop(infstate) \quad entrymode := pentry(infstate)$ **end**

This event updates the vtbi setting with the entered value recorded by *display*. The mode of pump operation (*topline*) and the data entry mode (*entrymode*) go back to their previous values. These are given as the values of functions *ptop* and *pentry*, respectively. They depend on whether the pump is in the infusing state or not, which is modelled as the boolean variable *infstate*. The functions *ptop* and *pentry* are defined in the context AlarisDefinitions. The cancellation case is modelled similarly.

Now we illustrate two ways of verifying that the Alaris vtbi entry module satisfies the requirements formalised in Section 15.4. First, the requirement **R6** is verified directly for the machine Alaris_vtbi1.

15.5.2 Requirement R6

In this illustration, our first interpretation (machine Reqs61) is used for the informal requirement **R6**.

Since the structure of Alaris_vtbi1 is not that different from Reqs61, it is feasible to demonstrate the refinement relation between them directly. However, as a preparatory step, the abstract set *Settings* and constant *blank* used in Reqs61 must be instantiated to the concrete set *Numbers* and value *0*, respectively, used in Alaris_vtbi1. In Event-B, this is automatically done by applying the generic instantiation plugin to Reqs61 (we will use the same name for the instantiated machine).

To establish refinement between the instantiated machine Reqs61 and Alaris_vtbi1, one has to provide a 'glueing' invariant that relates the concrete variables in Alaris_vtbi1 and the abstract variables they replace in Reqs61. The abstract variables in question are *change*, *param*, and *new*. As discussed earlier, *change = TRUE* models the mode where a relevant pump's setting can be changed. In Alaris_vtbi1, the corresponding mode for the vtbi entry is defined by the condition *topline = dispvtbi ∧ entrymode = vtmode*. Assuming this condition is true, the vtbi setting *vtbi* and its provisional value *display* are identified with their counterparts in Reqs61. The resulting glueing invariant allows one to prove the following refinement:

MACHINE Alaris_vtbi1 **REFINES** Reqs61 **SEES** AlarisDefinitions   ...
INVARIANTS
  $(change = TRUE) \Leftrightarrow (topline = dispvtbi \wedge entrymode = vtmode)$

  $(topline = dispvtbi \wedge entrymode = vtmode) \Rightarrow (vtbi = param \wedge display = new)$ ...
EVENTS
Event $up \mathrel{\widehat{=}}$ **Status** anticipated **refines** *update*
    when $topline = dispvtbi$ $entrymode = vtmode$ **with** $new':new' = display'$ **then**
        $display := alaris\_up(display)$ **end**   ...
END

Similarly to *up*, the remaining chevron events (*dn*, *UP*, and *DN*) in Alaris_vtbi1 refine the requirement event *update*. Finally, the acknowledgement events *confirm* and *cancel* refine their counterparts in Reqs61.

15.5.3   Requirements R1-R4

This section illustrates an alternative approach for demonstrating that the data entry systems in infusion pumps satisfy relevant safety requirements. A number of such systems are already used in infusion pumps [254] and there is future scope for many more. The approach presumes that a refinement-based hierarchy of user interfaces has been previously developed that is relevant for various modes of data entry in infusion pumps [286]. It is also assumed that the relevant requirements have been verified for the classes at the top of the hierarchy. If so, then the interface classes at the lower levels are guaranteed to preserve them by construction. To verify a specific interface against those requirements, it then suffices to show that the interface is an instance of some class in the hierarchy. This principle is demonstrated for the Alaris vtbi entry system.

We will show that the Alaris vtbi entry is an instance of the class of interfaces with four chevron keys. This class, represented by the machine Chevron_Entry11, has already been shown to satisfy the formalisation of the requirements **R1-R4** [286]. Thus, the demonstration that the Alaris vtbi entry interface is an instance of that class boils down to proving refinement between Chevron_Entry11 and Alaris_vtbi1. For such a proof, the generic parameters (such as *j*, *k*, and *Threshold*) used by Chevron_Entry11 must be instantiated with the concrete values from the Alaris specification (context AlarisDefinitions) as, for example, shown below:

CONTEXT ChevronAlarisParams **EXTENDS** ChevronDefinitions11 AlarisDefinitions

AXIOMS

$$Min = minAlaris \qquad Max = maxAlaris$$
$$j = r01 \qquad k = r1 \qquad Threshold = r100 \quad ...$$

END

To specify the behaviour of four chevron keys, Chevron_Entry11 already includes the events *up*, revtdn, *UP*, and *DN*. These must be refined by the corresponding events in Alaris_vtbi1. Finally, the invariants of Alaris_vtbi1 must include a glueing invariant that specifies the connection between the state spaces of both machines:

MACHINE Alaris_vtbi1 **REFINES** Chevron_Entry11 **SEES** ChevronAlarisParams ...

INVARIANTS

$$(entry = TRUE) \Leftrightarrow (topline = dispvtbi \wedge entrymode = vtmode)$$

$$(vtbi = data \wedge display = disp) \quad ...$$

EVENTS

Event $up \mathrel{\widehat{=}}$ **Status** anticipated **refines** up

    when $topline = dispvtbi \quad entrymode = vtmode$ **with** $disp':disp' = display'$ **then**

        $display := alaris\_up(display)$ **end** ...

END

15.6 Conclusions

We have demonstrated how Event-B can be used to support manufacturers as they aim to demonstrate that the regulator's requirements are satisfied by their products. All the refinements described have been proven using the Rodin platform. The refinement hierarchies thus developed for requirements and user interfaces enable developers to trace the regulator requirements down to the specialised classes that match the physical characterisation of their device. Such an approach fits well with the FDA pre-market review process which involves providing evidence that a new device is 'substantially equivalent' to already approved and legally marketed medical devices.

We envisage showing that a device satisfies a full set of requirements by developing specification fragments. Each fragment would address one or more requirements and would provably demonstrate that the requirements in question are satisfied. It then remains an open question as to how one proves that these components are consistent with each other and how they might fit into a larger specification. This is future work. It would explore work on composition [304] and product lines [141] in Event-B being carried out at Southampton. The advantage of using Event-B is that the approach is tool supported. It is feasible that standard refinement processes such as these can be made easier for developers to use.

Acknowledgments

This work is supported by EPSRC as part of CHI+MED (Computer-Human Interaction for Medical Devices, EPSRC research grant [EP/G059063/1]). We are grateful to Michael Harrison who collaborated in various aspects of the work described.

Chapter 16

Self-Assembling Interactive Modules: A Research Programme

Gheorghe Stefanescu

University of Bucharest

Abstract. In this paper we propose a research programme for getting structural characterisations for 2-dimensional languages generated by self-assembling tiles. This is part of a larger programme on getting a formal foundation of parallel, interactive, distributed systems.

16.1   Tiling: A Brief Introduction

Tiling is an old and popular subject [148]. While our focus is on 2-dimensional tiles/tiling and 2-dimensional regular expressions, the formalism itself can be easily extended to cope with 3, 4, or more dimensions.

An interesting tiling problem was proposed by Wang in 1961. *Wang tiles* are: (1) finite sets of unit squares, with a color on each side; (2) which cannot be rotated or reflected; and (3) such that there is an infinite number of copies from each tile. A *tiling* is a side-by-side arrangement of tiles, such that neighbouring tiles have the same color on the common borders. The main problem of interest here is the following: *"Given a set of tiles, is there a tiling of the whole 2-dimensional plane?"* Wang has conjectured that only regular periodic tilings are possible, but this conjecture was refuted later and aperiodic tilings of the plane were found; the smallest known set for which an aperiodic tiling does exist consists in 13 tiles [87, 199]. See also [124] for some recent results, seen from a mathematical perspective.

Another interesting model using tiles is the Winfree's abstract Tile Assembly Model; see [261] for a recent survey. This is a theoretic model aiming to capture the basic features of self-assembling systems from physics, chemistry, or biology. The main problem of interest here is: *"Find a set of tiles such that any finite tiling yields the same specified final configuration."* Reformulated in rewriting language terms, the problem of interest is to find confluent and terminating sets of tiling rules. The interest in this problem comes from practical reasons: to use self-assembling for producing complex substances with no errors and in a fast way.

The approach presented in this paper comes from a different perspective: how to use tiling to describe the syntax and the semantics of open, interactive, distributed programs, and computing systems. The problem of interest here is: *"Find all finite tiling configurations with a given color on each west/north/east/south border"*. This is an abstract formulation of the basic fact that we are interested in running scenarios of distributed systems, corresponding to tiling configurations, which start from initial states (on north), initial interaction classes (on west) and, in a finite number of steps, reach final states (on south) and final interaction classes (on east). To conclude, while formally our tiles are somehow similar with Wang or Winfree tiles, the problem of interest is different.

Tiling has also been used for the study of 2-dimensional languages[1], a topic closer to our approach here. The field of 2-dimensional languages started in 1960s, mostly related to "picture languages". The field has got a renewed interest in 1990s, when a robust class of *regular* 2-dimensional languages has been identified; good surveys from that period are [140, 228]. Comparisons between our approach and some known results in the area of 2-dimensional languages are included in the text below, as well as in our previous papers on a new type of 2-dimensional regular expressions to be introduced in the next section, see, e.g., [44, 46, 45].

We end this brief introduction with a comment on the use of the "self-assembling" term here. According to some conventions, a (chemical) self-assembling system is an assembling system with the following distinctive features related to the *order*, the *interactions*, and the used *building blocks*: (1) usually, the resulting configurations have higher order; (2) they use "weak interactions" for coordination; and (3) larger or heterogeneous building blocks may be used. A property as (1) is present in the statement of the proposed research programme presented in Section 1.3. Property (3) is a basic ingredient in our new type of 2-dimensional regular expressions, based on scenario composition. Property (2) is more related to physical systems - a slightly similar one may be considered in our setting, making a difference between computing (on a machine) and coordinating activities. In short, there are strong enough reasons to use the "self-assembling" term for constructing scenarios in the current distributed computing formalism. Both, the term and the self-assembling way of thinking may be useful when considering distributed software services, as well.

---

[1]This approach considers an abstract notion of dimension. Nevertheless, our work on tiling originates from a study of interactive systems (i.e., the register-voice interactive systems model [318, 319]), leading to a model with 2 dimensions and a particular interpretation of them: the vertical dimension represents *time*, while the horizontal dimension is used for *space*. More on this distinction between abstract tiling and developing scenarios out of interactive modules is included in the last section.

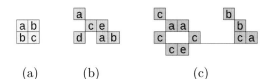

(a)   (b)   (c)

Figure 16.1: 2-dimensional words

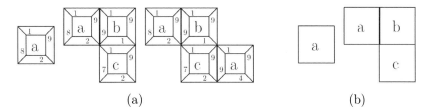

(a)   (b)

Figure 16.2: Scenarios and accepted words

16.2 Two-Dimensional Languages: Local vs. Global Glueing Constraints

16.2.1 Words and Languages in Two Dimensions

A *2-dimensional word* is a finite area of unit cells, in the lattice $\mathbb{Z} \times \mathbb{Z}$, labelled with letters from a finite alphabet. A *2-dimensional letter* is a 2-dimensional word consisting of a unique cell. A *2-dimensional language* is a set of 2-dimensional words.

These 2-dimensional words are invariant with respect to translation by integer offsets, but not with respect to mirror or rotation. A word may have several disconnected components. Examples of words are presented in Figure 16.1: in (a) it is a rectangular word; in (b) it is a word of arbitrary shape having no holes and 1 component; in (c) there is a word with 1 hole and 2 components.

16.2.2 Local Constraints: Tiles

A *tile* is a letter enriched with additional information on each border. This information is represented abstractly as an element from a finite set[2] and is called a *border label*. The role of border labels is to impose local glueing constraints on self-assembling tiles: two neighbouring cells, sharing a horizontal or a vertical border, should agree on the label on that border. A *scenario* is similar to a 2-dimensional word, but: (1) each letter is replaced by a tile; and (2) horizontal or vertical neighbouring cells have the same label on the common border. Examples of tiles and scenarios are presented in Figure 16.2(a).

A *self-assembling tile system*[3] (shortly, *SATS*) is defined by a finite set of tiles, together

---

[2]Often, we use sets of numbers or sets of colors.

[3]Tiling is a popular research subject, see e.g. [148]. The model we are using here (with a finite label set for borders) was introduced in the context of interactive programming [317, 319] using a different terminology. In that context, the concept is called *finite interactive system (FIS)*. The vertical dimension represents

with a specification of what border labels are to be used on the west/north/east/south external borders. An *accepting scenario* of a SATS F is a scenario, obtained by self-assembling tiles from F, having the specified labels on the external borders. Finally, *the 2-dimensional language recognized by a SATS F*, denoted $\mathcal{L}(\mathcal{F})$, is the set of 2-dimensional words obtained from the accepting scenarios of F, dropping the border labels.

Example 16.1. An example of a SATS is F defined by:

(1) tiles: 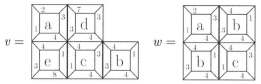 and

(2) labels for external w/n/e/s borders: $\{7,8\}/\{1\}/\{9\}/\{2\}$.

All scenarios in Figure 16.2(a) are correct scenarios of F. The first two are accepting scenarios, while the last one is not (there is a label 4 on the south border). Dropping the border labels in the accepting scenarios in Figure 16.2(a) we get the recognized words in Figure 16.2(b).

Tiles constrain the letters of horizontal and vertical neighbouring cells. They control the letters of those cells in a word laying in a *horizontally-vertically connected component* (shortly *hv-component*)[4] of a word and do not affect separated components. Consequently, our focus below will be on the structure of hv-components of the words.

Projected on one dimension, this model produces classical *finite automata*. For instance, this can be done by considering different labels for the west and the east borders of the tiles, inhibiting horizontal growth of the scenarios. Then each accepted hv-connected scenario is a 1-column scenario which may be seen as an acceptance witness of a usual 1-dimensional word by the corresponding finite automaton.

It is not at all obvious how to define 2-dimensional word composition, extending usual 1-dimensional word concatenation. We start with the simpler definition of scenario composition.

Scenario composition. For two scenarios v and w, the *scenario composite $v.w$* consists of all valid scenarios resulting from putting v and w together, without overlapping. This actually means that, if v and w share some borders in a particular placement, then the labels on the shared borders should be the same.

Example 16.2. We consider two scenarios:

$$v = \qquad\qquad\qquad\qquad w = \qquad\qquad\qquad$$

The composite $v.w$ has three results, sharing at least one cell border; they are presented in Figure 16.3.

Word composition, a preliminary solution. The above scenario composition guides us towards what word composition should be able to achieve.

---

time, while the horizontal dimension represents *space*. The labels on the north and south borders represent (abstract) *memory states*, while the ones on the west and east borders represents (abstract) *interaction classes*. The selected labels on the external borders are called *initial* for west and north borders, and *final* for east and south borders.

[4]The words in Figure 16.1(a),(b),(c) have 1,3,7 hv-components, respectively.

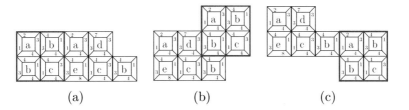

Figure 16.3: Scenario composition

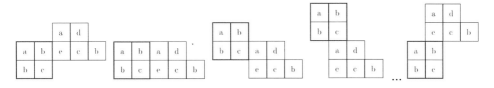

Figure 16.4: Border agnostic word composition

One possibility to define word composition is to mimic scenario composition using enriched alphabets and renaming. Notice that tiles may be codified using letters in appropriate enriched alphabets - one possibility is presented is the next paragraph. There are three strong criticisms to this solution. First, composition would depend on the underlying tile system and we want a word composition definition depending only on the words themselves. Second, the use of tiles (border labels) does not scale to large systems[5]. Finally, and perhaps the most important reason, renaming does not behave properly in combination with intersection, a key operation used in classical regular expressions [140, 228]; for instance, by renaming letters in an expression representing square words one gets an expression producing arbitrary rectangular words - see [44].

The drawbacks of this "word composition via scenario composition" is inherited by "regular expressions" including renaming operators [140, 228][6]. Indeed, with renaming we can mimic the border labels of the tiles as additional information in the cells letters; e.g., we can enrich letter a to become $\overline{a} =_{def} (a, 1, 2, 3, 4)$, including into the cell letter the labels 1,2,3,4 of the side borders. Then, we can use this extra information on neighbouring cells to control the shape of the words and, finally, using a renaming, we can remove the additional information. Actually, with renaming the border labels are reintroduced in the model.

To conclude:

> *Scenario composition is based on a set of tiles, while word composition (and the associated regular expressions) should not have any underlying tile system behind.*

Word composition. The solution to the problems identified in the previous paragraphs is to define a restricted composition using relevant information of the contours of the words [44, 46, 45]. A few elements of interest on the contours are:

---

[5]In one particular case, this is a well-known problem. When the tiling is on the vertical direction only, we get finite automata with labels for controlling sequential program executions. Large programs are difficult to cope with using "go-to programs" corresponding to finite automata. The well-known solution to avoid using labels in sequential programming is to use "structured programming" based on a particular class of regular expressions.

[6]Renaming is also used in getting regular expressions for Petri nets [137] and timed automata [18].

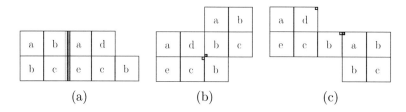

Figure 16.5: Restricted word composition

- *side borders* (w = west, i.e., vertical side border such that the cell at the right is inside the word; e = east; etc.),

- *land corners* (nw = north-west corner seen from inside the word, i.e., the cell at the bottom-right of the point is inside the word, while the other three cells around are not inside the word; ne = north-east; etc.) and

- *golf corners* (nw' = north-west corner seen from outside the word, i.e., the cell at the bottom-right of the point is outside the word, while those at the top-right and bottom-left are inside the word; ne' = north-east golf corner; etc.).

Example 16.3. To get the words corresponding to the scenario composition in Figure 16.3, we can use compositions based on the following restrictions:

1. v (w=e) w - the west border of v is equal to the east border of w; this yields a result similar to the one presented in Figure 16.3(a);

2. v (sw'=sw) w - the south-west golf corners of v (there is only one) are identified with the south-west land corners of w (there is only one); this restriction is good for Figure 16.3(b);

3. v (ne>nw) w - the north-east corners of v (there are two) include the north-west corners of w (there is only one); this is good for Figure 16.3(c).

These restricted compositions are illustrated in Figure 16.5.

Remark: Corner composition should be used with care. For instance, we can constrain the order of elements on a diagonal as in $(a$ (se=nw) $b)$ (se=nw) c, which is not possible using tiles alone, as they only constrain horizontally-vertically connected cells. However, it can be done if we have cells near diagonal to ensure hv-connections between the cells on the diagonal.

General restricted word composition: A general *restricted composition* is defined as follows. Suppose we are given:

1. Two words v, w; a subset Y of elements of the contour of v (emphasized elements on the contour of v as in Figure 16.6); a subset G of elements of the contour of w (emphasized elements on the contour of w); the subset B of actual contact elements after composing v with w via the points indicated by the little arrows (the emphasized elements in the composed $R(Y, G, B)$ picture in Figure 16.6).

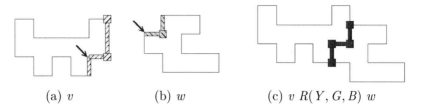

(a) v (b) w (c) $v \ R(Y, G, B) \ w$

Figure 16.6: General restricted composition

2. a relation $R(Y, G, B)$ between the above three subsets[7].

The resulted restricted composition is denoted by $v \ R(Y, G, B) \ w$. The iterated version of $\_R(Y, G, B)\_$ is denoted by $\_ * R(Y, G, B)$.

16.2.3 Global Constraints: Regular Expressions

Regular expressions. A new approach for defining classes of two-dimensional regular expressions has been introduced in [44, 46, 45]. It is based on words with arbitrary shapes and classes of restricted composition operators.

The basic class *n2RE* uses compositions and iterated compositions corresponding to the following restrictions (see [45] for more details and examples):

1. the selected elements of the word contours are: *side borders*, *land corners*, and *golf corners*;

2. the atomic comparison operators are: *equal-to* '=', *included-in* '<', *non-empty intersection* '#';

3. the general comparison formulas are boolean formulas built up from the atomic formulas defined in 2.

An enriched class *x2RE* [45] is obtained adding "extreme cells" glueing control. A cell is *extreme* in a word if it has at most one neighbouring cell in that word, considering all vertical, horizontal, and diagonal directions. The restricted composition may use elements of interest of the contours belonging to extreme cells. They are denoted by prefixing the normal *n2RE* restrictions with an 'x'; e.g., xw, xse, xnw′, etc. For instance, v (e>xw) w is true if the west borders of the extreme cells in w are included in the east borders of v.

16.2.4 Systems of Recursive Equations

A system of recursive equations is defined using variables representing sets of (two-dimensional) words and regular expressions. Formally, a *system of recursive equations* is represented by:

$$\begin{cases} X_1 &= \sum_{i_1=1,k_1} E_{1i_1}(X_1, \ldots, X_n) \\ &\cdots \\ X_n &= \sum_{i_n=1,k_n} E_{ni_n}(X_1, \ldots, X_n) \end{cases}$$

---

[7]For the example in Figure 16.6, a relation R making the restricted composition valid may be: $G \subseteq Y \wedge G \subseteq B$ (after composition, all the emphasized elements on the contour of w are on the common border and included in the set of emphasized elements of v).

$$X = \boxed{x} + \begin{array}{c} \texttt{a a a a a} \\ \texttt{a} \quad\quad \texttt{a} \\ \texttt{a} \quad X \quad \texttt{a} \\ \texttt{a} \quad\quad \texttt{a} \\ \texttt{a a a a a} \end{array} \; ; \qquad U = X + \begin{array}{c} \texttt{c c c c c} \\ \texttt{c c} \;\; \texttt{c c} \\ \texttt{c} \;\; V \;\; \texttt{c} \\ \texttt{c c} \;\; \texttt{c c} \\ \texttt{c c c c c} \end{array}$$

$$Y = \boxed{y} + \begin{array}{c} \texttt{b} \\ \texttt{b} \;\; \texttt{b} \\ \texttt{b} \;\; Y \;\; \texttt{b} \\ \texttt{b} \;\; \texttt{b} \\ \texttt{b} \end{array} \; ; \qquad V = Y + \begin{array}{c} \texttt{d} \\ \texttt{d d d} \\ \texttt{d} \quad\quad \texttt{d} \\ \texttt{d d} \;\; U \;\; \texttt{d d} \\ \texttt{d} \quad\quad \texttt{d} \\ \texttt{d d d} \\ \texttt{d} \end{array}$$

Figure 16.7: Recursive equations

where X_i are variables (denoting sets of words) and E_{ij} are regular expressions over a given alphabet, extended with occurrences of variables X_i.

Comments: A basic operation here is the substitution of 2-dimensional word languages for the variables X_i used in these equations. A similar operation, but restricted to rectangular words, has been used in [81, 80] to define tile grammars. Our formalism, using arbitrary shape words, may be seen as a generalization of this mechanism.

Example 16.4. The language consisting of square words filled with a, except for the center which contains x, may be represented by the following system/equation, illustrated in Figure 16.7:

(*) $X = x + E(X)$

where:

$$E_r = \big({}_3({}_1 a\star (\texttt{e=w}))\big)_1 (\texttt{se=ne}) \big({}_2 a\star (\texttt{s=n}) \big)_2\big)_3$$
$$\big({}_5 (\texttt{sw=ne}) \big)_5 \big({}_4({}_1 a\star (\texttt{e=w}))\big)_1 (\texttt{nw=sw}) \big({}_2 a\star (\texttt{s=n}) \big)_2\big)_4$$
$$E_{rect} = \big(E_r({}_6 (\texttt{nw>ne}) \,\&\, (\texttt{nw>sw}))_6 a\big) ({}_7 (\texttt{se>ne}) \,\&\, (\texttt{se>sw}))_7 \, a$$
$$E(X) = X({}_8 (\texttt{n<s}) \,\&\, (\texttt{e<w}) \,\&\, (\texttt{s<n}) \,\&\, (\texttt{w<e}))_8 E_{rect}$$

The expression within the parentheses with index 1 in E_r generates horizontal bars of a's. The one in 2 produces vertical bars. The restrictions in 3 and 4 orthogonally glue the bars in corners, yielding '⌐' and '∟' shapes, respectively. Finally, 5 constrains these two corners to produce a rectangle, without the nw and se corners. Then, the constraints 6 and 7 in E_{rect} fill in a's in these corners. For $E(X)$, the restriction 8 requires X to have all borders included in those of E_{rect}, hence X has to be a rectangle itself and to fill the whole interior part of the rectangle specified by E_{rect}. Finally, the recursive procedure (*) starts with a square x, so we get precisely the required square words.

Example 16.5. The example above can be adapted to produce square diamonds of b's, having y in the center (Y in Figure 16.7). Indeed, instead of the restrictions (e=w) and (s=n), used in 1 and 2, one can use (se=nw) and (sw=ne), respectively, to produce diagonal bars. To produce orthogonal diagonals, as in 3 and 4, is slightly more complicated: use extreme cells to locate the corners in the heads of the diagonal bars to be connected. The remaining part is similar. As a last remark, notice that the resulting expression is in *x2RE*, not in *n2RE*.

Example 16.6. Finally, U and V in Figure 16.7 describe a mutually recursive construction built on top of the languages in the previous examples.

16.3 Structural Characterisation for Self-Assembling Tiles

In this section we present the proposed research plan.

16.3.1 From Local to Global Constraints

A main question in understanding the tiling procedure is the connection between: (1) the local representation using tiles, scenarios, and labels on borders; and (2) the global representation making use of regular expressions and systems of recursive equations.

Here, we are particularly interested in one direction:

Q1: *Is it possible to get representations by systems of recursive equations for all tile systems?*

Q2: *What is a minimal set of regular expressions expressive enough to represent tile systems?*

Q3: *Can we find a kind of "normal form representation" for these systems of recursive equations?*

And finally,

Q4: *Is it possible to develop a (correct and complete) algebraic approach for modelling tile systems?*

16.3.2 Languages Generated by Two-Colors Border Tiles

A minimal, non-trivial SATS should have at least two labels for each vertical and horizontal dimension. Up to a bijective representation, the tiles of a SATS using two distinct labels for each vertical and horizontal dimension can be seen as elements in the following set

In this representation, the labels for the vertical and the horizontal borders are denoted by 0/1. The letter associated to a tile is the hexadecimal number obtained from the binary representation of the sequence of its west-north-east-south 0/1 digits, in this order; for instance, the label of the tile b is the number represented by 1011, hence b.

Notation: *(1) Tiles:* If $A = \{t_1, \ldots, t_k\}$ is a subset of $\{0, 1, \ldots, f\}$, then a SATS defined by the tiles in A is denoted by $Ft_1 t_2 \ldots t_k$. To have an example, $F02ac$ consists of tiles 0 2 a c.

(2) External labels: A SATS also uses labels for the external borders of its accepting scenarios. The recursive specifications below will use variables X_{wnes} denoting the set of words recognised by scenarios with $w/n/e/s$ on their west/north/east/south borders, respectively. Here, for simplicity, *we add the restriction* to have *only one* label for each west/north/east/south external border. Then, $Ft_1 t_2 \ldots t_k$ completely defines a SATS by specifying the labels used for the external borders. There are at most 16 possibilities, each one denoted by a hexadecimal number representing the sequence of the west-north-east-south 0/1 labels used for the external borders.

(3) Final notation: To conclude, the SATSs to be investigated are represented as $Ft_1 t_2 \ldots t_k.z$, where $t_1, t_2, \ldots, t_k$ are the tiles and the binary digits of z specify the labels used for the external borders.

As an example, $F02ac.c$ consists of: (1) tiles ▢0 ▢2 ▢a ▢c and (2) labels for external borders: 1 on west, 1 on north, 0 on east, and 0 on south.

The research programme: The goal of the proposed research programme is to:

P1 *Find representations by systems of recursive equations for the languages generated by all SATSs $Ft_1 \ldots t_k.z$ (there are 1048576 cases to consider - 2^{16} subsets of tiles and 2^4 combinations of border labels for each subset);*

P2 *Extend these representations to SATSs with any number of labels for borders (and any number of labels for external borders).*

16.3.3   A Case Study: $F02ac.c$

To start the analysis, note that we can construct any shape of ▢0 and ▢a, and vertical bars of ▢2. The last tile ▢c can be composed with itself, in a connected word, only along the diagonal. There are four possible direction changes for ▢c, denoted by letters X, Y, Z, T. A direction change in an hv-connected component is possible if: (1) one can insert one tile in interior, as in $X3$; or (2) one can insert two tiles on exterior areas as in $X1$ and $X2$. Three interior direction changes via $Y3, Z3, T3$ are not possible; the last one, $X3$, may be filled with ▢2. On the other hand, via exterior connections only the combination $Z1\&Z2$ is possible using, say, ▢a and ▢0.

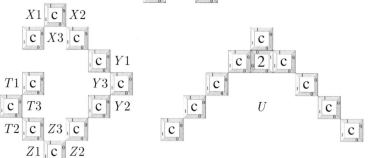

To fill in the area U we need a horizontal passing from border 0 to 1, and this may be done by horizontal words from the expressions 0^*2a^*.

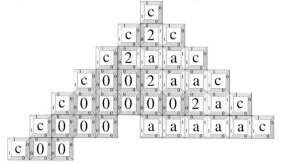

The north selected label is 1, hence the top of each column should start with an c. Hence, in a first approximation, an hv-connected component of an accepted word should have the *hat-form* above.

The general format of an hv-component is slightly more complicated: two hat-forms, as before, can be connected via the cells of their extreme bottom legs as in

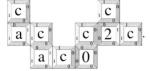

To get a recursive specification for the hat-form we proceed as follows:

1. construct a shape $C1$ for c's as two diagonals connected on the top cells;

2. construct horizontal bars with 0's on the left, a's on the right, and separated by a 2;

3. iteratively fill with these bars the shape $C1$, requiring to completely connect their west, north, and east borders; let $C2$ be the resulting word;

4. finally, iteratively connect $C2$ with horizontal bars of 0's on west-north and horizontal bars of a's on north-east, completely connecting their west-north and north-east borders, respectively.

It is not difficult to see that this procedure may be formalized by a system of recursive equations using expressions in $n2RE$. One possibility is $X11$ defined by the following system of recursive equations:

$$X1 = c + c \text{ (sw=ne) } X1$$
$$X2 = X1 \text{ (ne=nw) } 2$$
$$X3 = c + c \text{ (se=nw) } X3$$
$$X4 = X2 \text{ (ne=nw) } X3$$
$$X5 = c \text{ (s<n) \& ! (sw\#nw) \& ! (se\#ne) } X4$$
$$X6 = ((0 * \text{(e=w)) (e=w) } 2) \text{(e=w) } (a * \text{(e=w)})$$
$$X7 = (0 * \text{(e=w)})$$
$$X8 = (a * \text{(e=w)})$$
$$X9 = X5 + X6 \text{ (n<s) \& (w<e) \& (e<w) } X9$$
$$X10 = X9 + X7 \text{ (n<s) \& (w<e) } X10$$
$$X11 = X10 + X8 \text{ (n<s) \& (e<w) } X11$$

The patterns defined by $X1, \ldots, X11$ are illustrated in Figure 16.8.

For the general format, consisting in connected hat-forms, one can use an $x2RE$ variation of the previous system. Use an additional $X5'$ defined by (!x means non-extreme)

$$X5' = X5 + X5 \, (\, (\text{xne<xsw}) \, \& \, ! \, (\, (!\text{x}) \, \text{nw\#sw}) \,) \, X5'$$

and change $X9$ to

$$X9 = X5' + \ldots$$

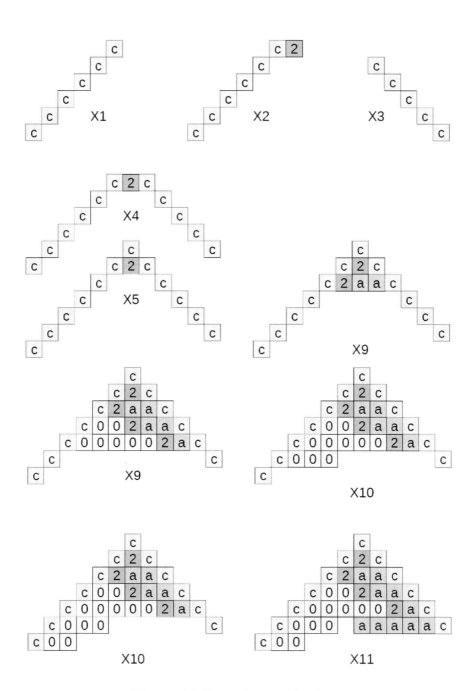

Figure 16.8: Recursive specifications

16.4 Interactive Programs

A classical slogan states that "program = control + data". The control part for simple sequential programs is provided by finite automata. We can extend this saying that "distributed program = (control & interaction) + data". The control & interaction part may be specified by SATSs. In this section we briefly show how data can be added to SATSs to get completely specified distributed programs. Actually, one can use either SATSs or regular expressions for specifying the control & interaction part of a distributed program.

16.4.1 Words and Traces in Two Dimensions

Recently, 2-dimensional languages have been used to study *parallel, interactive, distributed systems* [319, 122, 121, 45, 110]; we simply call them *interactive systems*. In these studies, the approach is less syntactical considering words up to a *graph-isomorphism equivalence* [109, 110]. This means, one uses two types of letters: *connectors* in a set C and (uninterpreted) *statements* in a set X. Then, two words over $C \cup X$ are considered equivalent if they have the same occurrences of letters in X connected in the same way via the elements in C. In other words, the placement of the letters in X in the cells of a word does not matter, as long as elements in C are used to ensure the connecting structure between the elements in X is the same.

The model of SATSs is not universal. To compare this class with 1-dimensional languages, one can consider a 1-dimensional projection, for instance considering only the top row of the accepted rectangular 2-dimensional words. This way one gets a class of usual languages; by a theorem in [222], this class coincides with the class of context-sensitive languages. A universal class can be easily obtained putting more information around the letters of the words. Scenarios for real computations, as the ones used in rv-systems [319] or in Agapia programming [121, 268], have complex spatial and temporal data attached to the cells. They are universal.

In most studies on 2-dimensional languages there is no distinction between the vertical and the horizontal dimension. The interpretation used in interactive systems is somehow different, considering the vertical dimension to represent the *temporal evolution* of the system, while the horizontal dimension takes into account the *spatial aspects* of the system. This way, a natural notion of *2-dimensional trace* occurs: a column represents a process run (in an interacting environment), while a row describes a "transaction", i.e., a chain of process interactions via message passing (in a state dependent processing environment).

A notion of *trace-based refinement* for (structured) interactive systems has been recently presented in [110]. *Traces* represent running scenarios modulo graph-isomorphism, projected on classes and states. In this interpretation one focuses on data stored in or flowing through the cells of the interactive system.

16.4.2 Interactive Modules and Programs

In a SATS, a tile , linearly represented as $X:\langle w \mid n \rangle \to \langle e \mid s \rangle$, uses elements from an abstract finite set to label the borders. These labels are used for defining the control and the interaction used in interactive systems. In concrete, executable interactive systems the control/interaction labels are enriched with data. The data on the north and south borders are represented as spatial data implemented on memory, while the data on the west and east borders are temporal data implemented on streams. The former spatial data represent the states of the interactive processes, while the latter temporal data represents the messages flowing between processes.

An *interactive module* is a cell with: (1) control/interaction labels and temporal/spatial data on its borders, and (2) a specification of a relation between border data. The relation describing the module functionality may be specified with a program, or in another way.

Example 16.7. We present a distributed program for a communication protocol between two parties using a channel which, due to time constrains, may decide to drop some data. During the first attempt of sending data, the sender keeps all the messages, while the receiver keeps two sets: one with received messages and one with the indices of missing ones. Then, the receiver sends indices of missing elements one by one to the sender waiting to receive those missing elements.

We use the convention that $\frown$ separate data on west or east borders coming from different modules, while $\smile$ is similarly used, but for north or south borders. The overall behaviour of the protocol, corresponding to the scenario in Figure 16.9, is:

$$Protocol{:}\langle a^\frown b^\smile c \mid 0, \varnothing \rangle \to \langle a^\frown b^\frown c \mid \varnothing \rangle$$

First we give the specification of the modules, then we present the scenario.

Modules:
    Send-and-Keep:
$SK{:}\langle x \mid i, Y \rangle \to \langle (i,x) \mid i+1, Y \cup \{(i,x)\} \rangle$;
    Communicate-Yes/No:
$CY{:}\langle (i,x) \mid \ \rangle \to \langle (i,x) \mid \ \rangle$; $CN{:}\langle (i,x) \mid \ \rangle \to \langle (i,?) \mid \ \rangle$;
    Receive-and-Keep:
$RK{:}\langle (i,x) \mid U, V \rangle \to \langle \ \mid U, V \cup \{(i,x)\} \rangle$; $RK{:}\langle (i,?) \mid U, V \rangle \to \langle \ \mid U \cup \{i\}, V \rangle$
    Send-first-bunch-End:
$SEnd{:}\langle \ \mid i, U \rangle \to \langle (i, end) \mid U \rangle$;
    Receive-first-bunch-End:
$REnd{:}\langle (n, end) \mid U, V \rangle \to \langle i \mid U \backslash \{i\}, V \rangle$, for $i \in U$
$REnd{:}\langle (n, end) \mid \varnothing, V \rangle \to \langle OK \mid V \rangle$
    Receive-and-Keep-at-Request:
$RKR{:}\langle (i,x) \mid U, V \rangle \to \langle j \mid U \backslash \{i\}, V \cup \{(i,x)\} \rangle$, for $i \in U$ and $j \in U \backslash \{i\}$;
$RKR{:}\langle (i,x) \mid U, V \rangle \to \langle OK \mid V \cup \{(i,x)\} \rangle$, if $U = \{i\}$;
$RKR{:}\langle (i,?) \mid U, V \rangle \to \langle i \mid U, V \rangle$, for $i \in U$;
    Output-Stream:
$OS{:}\langle \ \mid V \rangle \to \langle x \mid V \backslash \{(i,x)\} \rangle$, if $(i,x) \in V$ and $i = min\{j : (j,y) \in V\}$;

Scenario: An example of a running scenario[8] is presented in Figure 16.9.

Interactive programs may be introduced on top of interactive modules in two ways: (1) either in an unstructured way using tiles with labels, or (2) in a structured way using (particular) regular 2-dimensional expressions. The rv-programs in [318, 319] use the first option.

---

[8]The drawn "back-arrows" represent a short notation for diagonal composition [121, 122]. The '0' cells may be omitted if we do not stick to rectangular words/scenarios.

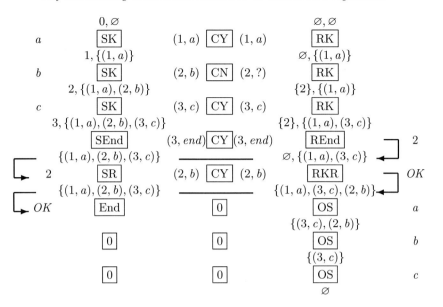

Figure 16.9: A scenario for the communication protocol example

On the other hand, Agapia structured interactive programming [121, 268] is based on the second possibility, using rectangular 2-dimensional words/scenarios and three particular composition operators, together with their iterated versions: *horizontal*, *vertical*, and *diagonal composition*.

16.4.3 Refinement of Structured Interactive Programs

The presented formalism, based on self-assembling interactive modules, can be used as a basis for a refinement-based design. The approach consists of the following steps:

Basic step: A starting simple specification $S0$ may be introduced using independent specifications for process behaviours (columns) and for transactions (chains of interactions used in rows). In this step finite automata are used.

Matching control and interaction: This step refines $S0$ to a specification $S1$ using scenarios with a finite number of border labels (i.e., memory states and interaction classes). For $S1$ use a SATS[9] for control and interaction. Refinement correctness here requires the rows and the columns of $S1$ satisfy specification $S0$.

Adding data: In this step, a new specification $S2$ is obtained by adding to the border labels concrete data for memory states and interaction messages around each scenario cell. This way one gets running scenarios describing concrete computations. The correctness criterion in this step is easy to formulate: by dropping the additional data and keeping the labels only, one must get simple patterns of control and interaction satisfying $S1$.

Iterated data and computation refinement: Iterated refinement of data and computation, preserving trace semantics.

---

[9]Recall that what we call here a "*SATS*" was introduced and used under the name "*finite interactive system*" in this context; see [318, 319] and the subsequent papers.

This trace-based refinement design is presented in an abstract way here, no programming language being involved. Actually, it was introduced in combination with the programming language Agapia for describing running scenarios in structured interactive programs [110]. An open problem is to lift this trace-based refinement definition to a refinement definition on Agapia programs themselves.

16.5   Conclusions

Sequential computation is already a mature research subject. A witness is the rich algebraic theories based on regular expressions and the associated regular algebra [202, 289, 85, 209, 315, 206]. Recent extensions of regular algebra to network algebra [314, 316] show a broader area of applications and deep connections with classical mathematics, especially via the particular instance provided by trace monoidal categories [197, 297]. The present approach is an attempt to extend these formalisms to open, distributed, interactive programs.

There are a few models for parallel/distributed computation based on regular expressions. We mention two of them: regular expressions for Petri nets [137] and for timed automata [18]. Both are based on renaming and intersection, two questionable operations.

Our work on this subject started with the exploration of *space-time duality* and its role in organizing the space of interactive computation: (1) A space-time invariant model extending *flowcharts* was shown in [318, 319]. (2) A space-time invariant extension of (structured) *while programs* was presented in [121]. (3) An enriched version supporting *recursion* was presented in [268]. (4) Space-time invariant (2-dimensional) *regular expressions* are presented in [44, 46, 45] and in the current paper. (5) Verification methods lifting *Floyd and Hoare logics* to 2-dimensions have been described, too. (6) A notion of *refinement*, based on this space-time invariant model, was introduced in [109, 110].

To conclude: the proposed research programme on clarifying the structure of self-assembling (2-dimensional) tiles is not only of general interest, but it is also very important in understanding open, distributed, interactive computation.

Bibliography

[1] M. Abadi and L. Lamport. An old-fashioned recipe for real time. In J. W. de Bakker, C. Huizing, W. P. de Roever, and G. Rozenberg, editors, *Real-Time: Theory in Practice, REX Workshop, Mook, The Netherlands, June 3-7, 1991, Proceedings*, volume 600 of *Lecture Notes in Computer Science*, pages 1–27. Springer, 1991.

[2] J. Abrial, M. J. Butler, S. Hallerstede, T. S. Hoang, F. Mehta, and L. Voisin. Rodin: an open toolset for modelling and reasoning in event-b. *STTT*, 12(6):447–466, 2010.

[3] J. Abrial, D. Cansell, and D. Méry. A mechanically proved and incremental development of IEEE 1394 tree identify protocol. *Formal Aspects of Computing*, 14(3):215–227, 2003.

[4] J. Abrial and L. Mussat. Introducing dynamic constraints in B. In D. Bert, editor, *B'98: Recent Advances in the Development and Use of the B Method, Second International B Conference, Montpellier, France, April 22-24, 1998, Proceedings*, volume 1393 of *Lecture Notes in Computer Science*, pages 83–128. Springer, 1998.

[5] J.-R. Abrial. *The B-Book: Assigning Programs to Meanings*. Cambridge University Press, 1996.

[6] J.-R. Abrial. *Modeling in Event-B: System and Software Engineering*. Cambridge University Press, 2010.

[7] ABS tools. `http://tools.hats-project.eu`.

[8] G. A. Agha. *ACTORS - A Model of Concurrent Computation in Distributed Systems*. MIT Press series in artificial intelligence. MIT Press, 1990.

[9] W. Ahrendt and M. Dylla. A system for compositional verification of asynchronous objects. *Science of Computer Programming*, 77(12):1289–1309, 2012.

[10] M. Åkerfelt, R. I. Morimoto, and L. Sistonen. Heat shock factors: integrators of cell stress, development, and lifespan. *Nature Reviews Molecular Cell Biology*, 11(8):545–555, 2010.

[11] E. Albert, P. Arenas, A. Flores-Montoya, S. Genaim, M. Gómez-Zamalloa, E. Martin-Martin, G. Puebla, and G. Román-Díez. SACO: static analyzer for concurrent objects. In E. Ábrahám and K. Havelund, editors, *Proc. 20th International Conference on Tools and Algorithms for the Construction and Analysis of Systems (TACAS'14)*, volume 8413 of *Lecture Notes in Computer Science*, pages 562–567. Springer, 2014.

[12] E. Albert, F. S. de Boer, R. Hähnle, E. B. Johnsen, and C. Laneve. Engineering virtualized services. In M. A. Babar and M. Dumas, editors, *2nd Nordic Symp. Cloud Computing & Internet Technologies*, pages 59–63. ACM, 2013.

[13] E. Albert, F. S. de Boer, R. Hähnle, E. B. Johnsen, R. Schlatte, S. L. T. Tarifa, and P. Y. H. Wong. Formal modeling and analysis of resource management for cloud architectures: an industrial case study using Real-Time ABS. *Service Oriented Computing and Applications*, 8(4):323–339, 2014.

[14] B. Alpern and F. B. Schneider. Defining liveness. *Information Processing Letters*, 21(4):181 – 185, 1985.

[15] R. Alur and D. L. Dill. A theory of timed automata. *Theoretical Computer Science*, 126(2):183 – 235, 1994.

[16] K. R. Apt, F. S. de Boer, and E.-R. Olderog. *Verification of Sequential and Concurrent Systems*. Texts and Monographs in Computer Science. Springer, 3rd edition, 2009.

[17] F. Arnold, A. Belinfante, F. Van der Berg, D. Guck, and M. Stoelinga. DFTCalc: A tool for efficient fault tree analysis. In F. Bitsch, J. Guiochet, and M. Kaaniche, editors, *Computer Safety, Reliability, and Security*, volume 8153 of *Lecture Notes in Computer Science*, pages 293–301. Springer Berlin Heidelberg, 2013.

[18] E. Asarin, P. Caspi, and O. Maler. Timed regular expressions. *Journal of the ACM*, 49:172–206, 2002.

[19] P. Asirelli, M. H. ter Beek, A. Fantechi, and S. Gnesi. A logical framework to deal with variability. In *Integrated Formal Methods - 8th International Conference, IFM 2010, Nancy, France, October 11-14, 2010. Proceedings*, volume 6396 of *Lecture Notes in Computer Science*, pages 43–58. Springer-Verlag, 2010.

[20] P. Asirelli, M. H. ter Beek, S. Gnesi, and A. Fantechi. Formal description of variability in product families. In *SPLC '11: Proceedings of the 15th International Software Product Line Conference, Volume 2*, pages 130–139. ACM, 2011.

[21] A. Aswani, N. Master, J. Taneja, D. E. Culler, and C. Tomlin. Reducing transient and steady state electricity consumption in HVAC using learning-based model-predictive control. *Proceedings of the IEEE*, 100(1):240–253, 2012.

[22] A. Avizienis, J.-C. Laprie, B. Randell, and C. Landwehr. Basic concepts and taxonomy of dependable and secure computing. *IEEE Transactions on Dependable and Secure Computing*, 1(1):11–33, 2004.

[23] R. J. Back. *On the Correctness of Refinement in Program Development*. PhD thesis, Department of Computer Science, University of Helsinki, 1978.

[24] R. J. Back. Correctness preserving program refinements: Proof theory and applications. *Mathematical Center Tracts*, 131, 1980.

[25] R. J. Back. Refinement calculus ii: parallel and reactive programs. In J. de Bakker, W.-P. de Roever, and G. Rozenberg, editors, *Stepwise Refinement of Distributed Systems: Models, Formalisms, Correctness*, volume 430 of *Lecture Notes in Computer Science*, pages 67–93. Springer-Verlag, 1990.

[26] R. J. Back. Atomicity refinement in a refinement calculus framework, reports on computer science and mathematics 141. Technical report, Åbo Akademi, 1993.

[27] R. J. Back and R. Kurki-Suonio. Decentralization of process nets with centralized. In *Proceedings of the 2nd ACM SIGACT-SIGOPS Symp. on Principles of Distributed Computing*, pages 131–142, 1983.

[28] R. J. Back and R. Kurki-Suonio. Decentralization of process nets with centralized control. *Distributed Computing*, 3(2):73–87, 1983.

[29] R. J. Back and R. Kurki-Suonio. Distributed cooperation with action systems. *ACM Transactions on Programming Languages and Systems*, 10(4):513–554, 1988.

[30] R. J. Back, L. Petre, and I. Porres. Continuous action systems as a model for hybrid systems. *Nordic Journal of Computing*, 8(1):2–21, 2001.

[31] R. J. Back and K. Sere. Stepwise refinement of action systems. In J. L. A. van de Snepscheut, editor, *Mathematics of Program Construction, 375th Anniversary of the Groningen University, International Conference, Groningen, The Netherlands, June 26-30, 1989, Proceedings*, volume 375 of *Lecture Notes in Computer Science*, pages 115–138. Springer, 1989.

[32] R. J. Back and K. Sere. Stepwise refinement of action systems. *Structured Programming*, 12(1):17–30, 1991.

[33] R. J. Back and K. Sere. From action systems to modular systems. In M. Naftalin, T. Denvir, and M. Bertran, editors, *FME '94: Industrial Benefit of Formal Methods*, volume 873 of *Lecture Notes in Computer Science*, pages 1–25. Springer Berlin Heidelberg, 1994.

[34] R. J. Back and K. Sere. From action systems to modular systems. *Software - Concepts and Tools*, 17:26–39, 1996.

[35] R. J. Back and K. Sere. Superposition refinement of reactive systems. *Formal Aspects of Computing*, 8(3):324–346, 1996.

[36] R. J. Back and J. von Wright. Refinement calculus i: Sequential nondeterministic programs. In J. de Bakker, W.-P. de Roever, and G. Rozenberg, editors, *Stepwise Refinement of Distributed Systems: Models, Formalisms, Correctness*, volume 430 of *Lecture Notes in Computer Science*, pages 42–66. Springer-Verlag, 1990.

[37] R. J. Back and J. von Wright. Trace refinement of action systems. In B. Jonsson and J. Parrow, editors, *CONCUR-94:Concurrency Theory*, volume 836 of *Lecture Notes in Computer Science*, pages 367–384, Uppsala, Sweden, Aug 1994. Springer-Verlag.

[38] R. J. Back and J. von Wright. *Refinement Calculus: A Systematic Introduction*. Springer-Verlag, 1998.

[39] C. Baier and J. Katoen. *Principles of Model Checking*. MIT Press, 2008.

[40] R. Banach and M. Bozzano. The mechanical generation of fault trees for reactive systems via retrenchment I: combinational circuits. *Formal Aspects of Computing*, 25(4):573–607, 2013.

[41] R. Banach and M. Bozzano. The mechanical generation of fault trees for reactive systems via retrenchment II: clocked and feedback circuits. *Formal Aspects of Computing*, 25(4):609–657, 2013.

[42] R. Banach and M. Butler. Cruise control in hybrid event-b. In Z. Liu, J. Woodcock, and H. Zhu, editors, *Theoretical Aspects of Computing - ICTAC 2013*, volume 8049 of *Lecture Notes in Computer Science*, pages 76–93. Springer-Verlag Berlin Heidelberg, 2013.

[43] R. Banach, H. Zhu, W. Su, and R. Huang. Continuous kaos, asm, and formal control system design across the continuous/discrete modeling interface: a simple train stopping application. *Formal Aspects of Computing*, 26(2):319–366, 2014.

[44] I. Banu-Demergian, C. Paduraru, and G. Stefanescu. A new representation of two-dimensional patterns and applications to interactive programming. In *FSEN 2013*, volume 8161 of *Lecture Notes in Computer Science*, pages 172–206. Springer, 2013.

[45] I. Banu-Demergian and G. Stefanescu. Towards a formal representation of interactive systems. *Fundamenta Informaticae*, 131:313–336, 2014.

[46] I. Banu-Demergian and G. Stefanescu. On the contour representation of two-dimensional patterns. *Carpathian Journal Mathematics*, 2016. To appear.

[47] G. Behrmann, A. David, and K. G. Larsen. A tutorial on UPPAAL. In M. Bernardo and F. Corradini, editors, *Formal Methods for the Design of Real-Time Systems*, volume 3185 of *Lecture Notes in Computer Science*, pages 200–237. Springer Verlag, 2004.

[48] N. Benes and J. Kretínský. Process algebra for modal transition systems. In *Sixth Doctoral Workshop on Mathematical and Engineering Methods in Computer Science, MEMICS 2010, Selected Papers, October 22-24, 2010, Mikulov, Czech Republic*, pages 9–18, 2010.

[49] N. Benes, J. Kretínský, K. G. Larsen, M. H. Møller, and J. Srba. Parametric modal transition systems. In *Automated Technology for Verification and Analysis, 9th International Symposium, ATVA 2011, Taipei, Taiwan, October 11-14, 2011. Proceedings*, pages 275–289, 2011.

[50] N. Benes, J. Kretínský, K. G. Larsen, and J. Srba. Exptime-completeness of thorough refinement on modal transition systems. *Information and Computation*, 218:54–68, 2012.

[51] N. Beneš. *Disjunctive Modal Transition Systems*. PhD thesis, Faculty of Informatics, Masaryk University, Brno, 2012.

[52] J. Bengtsson, K. G. Larsen, F. Larsson, P. Petterson, and W. Yi. UPPAAL — a tool-suite for the automatic verification of real-time systems. In R. Alur, T. Henzinger, and E. D. Sontag, editors, *Hybrid Systems III*, volume 1066 of *Lecture Notes in Computer Science*, pages 232–243. Springer, 1996.

[53] J. Bergstra, A. Ponse, and S. Smolka, editors. *Handbook of Process Algebra*. Elsevier Science Inc., New York, NY, USA, 2001.

[54] J. Berthing, P. Boström, K. Sere, L. Tsiopoulos, and J. Vain. Refinement-based development of timed systems. In J. Derrick, S. Gnesi, D. Latella, and H. Treharne, editors, *Integrated Formal Methods – 9th International Conference, IFM 2012, Pisa, Italy, June 18-21, 2012. Proceedings*, volume 7321 of *Lecture Notes in Computer Science*, pages 69–83. Springer, 2012.

[55] J. Bicarregui, A. Arenas, B. Aziz, P. Massonet, and C. Ponsard. Towards modelling obligations in Event-B. In *Abstract State Machines, B and Z*, volume 5238 of *Lecture Notes in Computer Science*, pages 181–194. Springer, 2008.

[56] J. Bjørk, F. S. de Boer, E. B. Johnsen, R. Schlatte, and S. L. Tapia Tarifa. User-defined schedulers for real-time concurrent objects. *Innovations in Systems and Software Engineering*, 9(1):29–43, 2013.

[57] M. L. Blinov, J. R. Faeder, B. Goldstein, and W. S. Hlavacek. BioNetGen: Software for rule-based modeling of signal transduction based on the interactions of molecular domains. *Bioinformatics*, 20(17):3289–3291, 2004.

[58] E. Boiten and J. Derrick. Modelling divergence in relational concurrent refinement. In M. Leuschel and H. Wehrheim, editors, *IFM 2009: Integrated Formal Methods*, volume 5423 of *Lecture Notes in Computer Science*, pages 183–199. Springer Verlag, February 2009.

[59] E. Boiten and J. Derrick. Incompleteness of relational simulations in the blocking paradigm. *Science of Computer Programming*, 75(12):1262–1269, 2010.

[60] E. Boiten, J. Derrick, and G. Schellhorn. Relational concurrent refinement II: internal operations and outputs. *Formal Aspects of Computing*, 21(1-2):65–102, 2009.

[61] M. M. Bonsangue and J. N. Kok. The weakest precondition calculus: recursion and duality. *Formal Aspects of Computing*, 6A:788–800, 1994.

[62] M. M. Bonsangue, J. N. Kok, and K. Sere. An approach to object-orientation in action systems. In J. Jeuring, editor, *Mathematics of Program Construction (MPC'98)*, volume 1422 of *Lecture Notes in Computer Science*, pages 68–95. Springer, 1998.

[63] P. Bouyer, F. Laroussinie, and P.-A. Reynier. Diagonal constraints in timed automata: forward analysis of timed systems. In P. Pettersson and W. Yi, editors, *FORMATS'05*, volume 3829 of *Lecture Notes in Computer Science*, pages 112–126. Springer- Heidelberg, 2005.

[64] J. Bowen and S. Reeves. Refinement for user interface designs. *Formal Aspects of Computing*, 21:589–612, 2009.

[65] J. Bowen and S. Reeves. Modelling safety properties of interactive medical systems. In *Proceedings of the 5th ACM SIGCHI Symposium on Engineering Interactive Computing Systems*, pages 91–100. ACM, 2013.

[66] M. Bozzano, A. Cimatti, and F. Tapparo. Symbolic fault tree analysis for reactive systems. In *Automated Technology for Verification and Analysis*, volume 4762 of *Lecture Notes in Computer Science*, pages 162–176. Springer Berlin Heidelberg, 2007.

[67] M. Bozzano and A. Villafiorita. The FSAP/NuSMV-SA safety analysis platform. *International Journal on Software Tools for Technology Transfer*, 9(1):5–24, 2007.

[68] M. Bravetti and G. Zavattaro. Towards a Unifying Theory for Choreography Conformance and Contract Compliance. In *Proceedings of 6th Symposium on Software Composition, SC 2007*, pages 34–50. Springer, 2007.

[69] P. Bulychev, A. David, K. G. Larsen, M. Mikučionis, D. B. Poulsen, A. Legay, and Z. Wang. UPPAAL-SMC: Statistical model checking for priced timed automata. In H. Wiklicky and M. Massink, editors, *Quantitative Aspects of Programming Languages and Systems*, Proceedings of the 10th Workshop on, volume 85 of *Electronic Proceedings in Theoretical Computer Science*, pages 1–16. Open Publishing Association, 2012.

[70] L. Burdy, Y. Cheon, D. R. Cok, M. D. Ernst, J. R. Kiniry, G. T. Leavens, K. R. M. Leino, and E. Poll. An overview of JML tools and applications. *International Journal on Software Tools for Technology Transfer*, 7(3):212–232, 2005.

[71] M. Butler. External and internal choice with event groups in Event-B. *Formal Aspects of Computing*, 24(4-6):555–567, 2012.

[72] M. Butler and I. Maamria. Practical theory extension in event-b. In *Theories of Programming and Formal Methods - Essays Dedicated to Jifeng He on the Occasion of His 70th Birthday*, volume 8051 of *Lecture Notes in Computer Science*, pages 67–81. Springer, 2013.

[73] M. Butler, E. Sekerinski, and K. Sere. An Action System Approach to the Steam Boiler Problem. In *Formal Methods for Industrial Applications: Specifying and Programming the Steam Boiler Control*, volume 1165 of *Lecture Notes in Computer Science*, pages 129–148. Springer-Verlag, 1996.

[74] R. Buyya, C. S. Yeo, S. Venugopal, J. Broberg, and I. Brandic. Cloud computing and emerging IT platforms: Vision, hype, and reality for delivering computing as the 5th utility. *Future Generation Computer Systems*, 25(6):599–616, 2009.

[75] D. Cansell, D. Méry, and J. Rehm. Time constraint patterns for Event B development. In *B 2007: Formal specification and development in B*, volume 4355 of *Lecture Notes in Computer Science*, pages 140–154. Springer Heidelberg, 2007.

[76] L. Cardelli. On process rate semantics. *Theoretical Computer Science*, 391(3):190–215, 2008.

[77] Cardinal Health Inc. Alaris GP volumetric pump: directions for use. Technical report, Cardinal Health, 1180 Rolle, Switzerland, 2006.

[78] G. Castagna, N. Gesbert, and L. Padovani. A Theory of Contracts for Web Services. *ACM Transactions on Programming Languages and Systems*, 31(5):19:1–19:61, 2009.

[79] Z. Chaochen and M. Hansen. *Duration Calculus: A Formal Approach to Real-Time Systems*. Springer, Heidelberg, 2004.

[80] A. Cherubini and M. Pradella. Picture Languages: From Wang Tiles to 2D Grammars. In *Algebraic Informatics*, volume 5725 of *Lecture Notes in Computer Science*, pages 13–46. Springer-Verlag, 2009.

[81] A. Cherubini, S. C. Reghizzi, M. Pradella, and P. S. Pietro. Picture languages: tiling systems versus tile rewriting grammars. *Theoretical Computer Science*, 356:90–103, 2006.

[82] W.-N. Chin, C. David, H.-H. Nguyen, and S. Qin. Enhancing modular OO verification with separation logic. In *Proceedings of the 35th annual ACM SIGPLAN-SIGACT symposium on Principles of programming languages*, pages 87–99. ACM, 2008.

[83] F. Ciocchetta and J. Hillston. Bio-pepa: A framework for the modelling and analysis of biological systems. *Theoretical Computer Science*, 410(33–34):3065–3084, 2009.

[84] P. Clements and L. Northrop. *Software Product Lines: Practices and Patterns*. Addison-Wesley Professional, 2001.

[85] J. Conway. *Regular Algebra and Finite Machines*. Chapman and Hall, 1971.

[86] Council of the European Communities. Council directive 93/42/EEC of 14 June 1993 concerning medical devices. http://eur-lex.europa.eu/LexUriServ/LexUriServ.do?uri=CONSLEG:1993L0042:20071011:EN:PDF, 2007.

[87] K. Culik II. An Aperiodic Set of 13 Wang Tiles. *Discrete Mathematics*, 160:245–251, 1996.

[88] O.-J. Dahl. Can program proving be made practical? In M. Amirchahy and D. Néel, editors, *Les Fondements de la Programmation*, pages 57–114. Institut de Recherche d'Informatique et d'Automatique, Toulouse, France, Dec. 1977.

[89] O.-J. Dahl. *Verifiable Programming*. International Series in Computer Science. Prentice Hall, New York, N.Y., 1992.

[90] O.-J. Dahl. The birth of object orientation: the simula languages. In O. Owe, S. Krogdahl, and T. Lyche, editors, *From Object-Orientation to Formal Methods, Essays in Memory of Ole-Johan Dahl*, volume 2635 of *Lecture Notes in Computer Science*, pages 15–25. Springer, 2004.

[91] O.-J. Dahl and O. Owe. Formal development with ABEL. In S. Prehn and H. Toetenel, editors, *Proc. Formal Software Development Methods (VDM'91)*, volume 552 of *Lecture Notes in Computer Science*, pages 320–362. Springer, 1991.

[92] P. H. Dalsgaard, T. L. Guilly, D. Middelhede, P. Olsen, T. Pedersen, A. P. Ravn, and A. Skou. A toolchain for home automation controller development. In *39th Euromicro Conference on Software Engineering and Advanced Applications, SEAA 2013*, pages 122–129. IEEE, 2013.

[93] V. Danos, J. Feret, W. Fontana, R. Harmer, and J. Krivine. Rule-based modelling, symmetries, refinements. In *Formal Methods in Systems Biology*, volume 5054 of *Lecture Notes in Computer Science*, pages 103–122. Springer, 2008.

[94] V. Danos, J. Feret, W. Fontana, R. Harmer, and J. Krivine. Rule-based modelling and model perturbation. In *Transactions on Computational Systems Biology XI*, volume 5750 of *Lecture Notes in Computer Science*, pages 116–137. Springer, 2009.

[95] V. Danos and C. Laneve. Formal molecular biology. *Theoretical Computer Science*, 325(1):69–110, 2004.

[96] R. Darimont and A. van Lamsweerde. Formal refinement patterns for goal-driven requirements elaboration. In *Proceedings 4th ACM Symposium on the Foundations of Software Engineering (FSE'03)*, pages 179–190. ACM Press, 1996.

[97] A. David, K. G. Larsen, A. Legay, U. Nyman, and A. Wasowski. Timed I/O Automata: A complete specification theory for real-time systems. In *HSCC'10, Proceedings of the 13th ACM International Conference on Hybrid Systems: Computation and Control*, pages 91–100. ACM, 2011.

[98] F. S. de Boer. Reasoning about histories in object-based distributed systems. In P. Ciancarini, A. Fantechi, and R. Gorrieri, editors, *Formal Methods for Open Object-Based Distributed Systems*, volume 10 of *IFIP - The International Federation for Information Processing*, pages 35–49. Springer US, 1999.

[99] F. S. de Boer, R. Hähnle, E. B. Johnsen, R. Schlatte, and P. Y. H. Wong. Formal modeling of resource management for cloud architectures: An industrial case study. In *Proc. European Conference on Service-Oriented and Cloud Computing (ESOCC)*, volume 7592 of *Lecture Notes in Computer Science*, pages 91–106. Springer, 2012.

[100] B. de Bono, P. Grenon, M. Helvenstijn, J. Kok, and N. Kokash. ApiNATOMY: Towards multiscale views of human anatomy. In *Processings of the 13th International Symposium on Intelligent Data Analysis (IDA)*, volume 8819 of *Lecture Notes in Computer Science*, pages 72–83. Springer-Verlag, 2014.

[101] B. de Bono and P. Hunter. Integrating knowledge representation and quantitative modelling in physiology. *Biotechnology Journal*, 7(8):958–972, 2012.

[102] W.-P. de Roever and K. Engelhardt. *Data Refinement: Model-Oriented Proof Methods and their Comparison*. Cambridge University Press, 2008.

[103] E. Deelman, G. Singh, M. Livny, G. B. Berriman, and J. Good. The cost of doing science on the cloud: The Montage example. In *Proceedings of the Conference on High Performance Computing (SC'08)*, pages 1–12. IEEE/ACM, 2008.

[104] J. Derrick and E. Boiten. Relational concurrent refinement. *Formal Aspects of Computing*, 15(1):182–214, 2003.

[105] J. Derrick and E. Boiten. *Refinement in Z and Object-Z*. Springer-Verlag, 2nd edition, 2014.

[106] J. Derrick and E. Boiten. Relational concurrent refinement III: traces, partial relations, and automata. *Formal Aspects of Computing*, 26(2):407–422, 2014.

[107] J. Derrick and E. A. Boiten. Relational concurrent refinement with internal operations. In B. Aichernig, E. A. Boiten, J. Derrick, and L. Groves, editors, *BCS-FACS Refinement Workshop*, volume 187 of *Electronic Notes in Theoretical Computer Science*, pages 35–53, 2007.

[108] J. Derrick and G. Smith. Temporal-logic property preservation under Z refinement. *Formal Aspects of Computing*, 24(3):393–416, 2012.

[109] D. Diaconescu, I. Leustean, L. Petre, K. Sere, and G. Stefanescu. Refinement-preserving translation from Event-B to register-voice interactive systems. In *Proceedings IFM 2012, Integrated Formal Methods*, volume 7321 of *Lecture Notes in Computer Science*, pages 221–236. Springer-Verlag, 2012.

[110] D. Diaconescu, L. Petre, K. Sere, and G. Stefanescu. Refinement of structured inter-active systems. In *Theoretical Aspects of Computing? ICTAC 2014*, volume 8687 of *Lecture Notes in Computer Science*, pages 133–150. Springer, 2014.

[111] H. Dierks, S. Kupfersmid, and K. G. Larsen. Automatic abstraction refinement for timed automata. In J.-F. Raskin and P. S. Thiagarajan, editors, *FORMATS'07 Proceedings of the 5th International Conference on Formal Modeling and Analysis of Timed Systems*, volume 4763 of *Lecture Notes in Computer Science*, pages 114–129, 2007.

[112] E. W. Dijkstra. Guarded commands, nondeterminacy and formal derivation of programs. *Communications of the ACM*, 18(8):453–457, 1975.

[113] E. W. Dijkstra. *A Discipline of Programming*. Prentice–Hall International, 1976.

[114] C. C. Din. *Verification of Asynchronously Communicating Objects*. PhD thesis, Department of Informatics, University of Oslo, Norway, 2014.

[115] C. C. Din, J. Dovland, E. B. Johnsen, and O. Owe. Observable behavior of distributed systems: component reasoning for concurrent objects. *Journal of Logic and Algebraic Programming*, 81(3):227–256, 2012.

[116] C. C. Din and O. Owe. Compositional reasoning about active objects with shared futures. *Formal Aspects of Computing*, 27:1–22, 2014.

[117] J. Dovland. *Incremental Reasoning about Distributed Object-Oriented Systems*. PhD thesis, University of Oslo, Dept. of Computer Science, 2009.

[118] J. Dovland, E. B. Johnsen, O. Owe, and M. Steffen. Lazy behavioral subtyping. Research Report 368, Dept. of Informatics, University of Oslo, Nov. 2007. Available from `heim.ifi.uio.no/creol`.

[119] J. Dovland, E. B. Johnsen, O. Owe, and M. Steffen. Lazy behavioral subtyping. *Journal of Logic and Algebraic Programming*, 79(7):578–607, 2010.

[120] J. Dovland, E. B. Johnsen, O. Owe, and M. Steffen. Incremental reasoning with lazy behavioral subtyping for multiple inheritance. *Science of Computer Programming*, 76(10):915 – 941, 2011.

[121] C. Dragoi and G. Stefanescu. Agapia v0. 1: a programming language for interactive systems and its typing system. *Electronic Notes in Theoretical Computer Science*, 203(3):69–94, 2008.

[122] C. Dragoi and G. Stefanescu. On compiling structured interactive programs with registers and voices. In *Proceedings SOFSEM 2008*, volume 4910 of *Lecture Notes in Computer Science*, pages 259–270. Springer, 2008.

[123] D. J. Duke and M. D. Harrison. Mapping user requirements to implementations. *Software Engineering Journal*, 10(1):13–20, 1995.

[124] S. Eigen, J. Navarro, and V. S. Prasad. An aperiodic tiling using a dynamical system and beatty sequences. In B. Hasselblatt, editor, *Dynamics, Ergodic Theory, and Geometry*, volume 54 of *Mathematical Sciences Research Institute Publications*, pages 223–241. Cambridge University Press, 2007.

[125] E. A. Emerson and E. M. Clarke. Using branching time temporal logic to synthesize synchronization skeletons. *Science of Computer Programming*, 2(3):241 – 266, 1982.

[126] K. Etessami, M. Z. Kwiatkowska, M. Y. Vardi, and M. Yannakakis. Multi-objective model checking of Markov decision processes. *Logical Methods in Computer Science*, 4(4), 2008.

[127] J. R. Faeder, M. L. Blinov, B. Goldstein, and W. S. Hlavacek. Rule-based modeling of biochemical networks. *Complexity*, 10(4):22–41, 2005.

[128] J. R. Faeder, M. L. Blinov, and W. S. Hlavacek. Graphical rule-based representation of signal-transduction networks. In *Proceedings of the 2005 ACM Symposium on Applied Computing*, pages 133–140. ACM, 2005.

[129] A. Fantechi and S. Gnesi. A behavioural model for product families. In *Proceedings of the 6th Joint Meeting of the European Software Engineering Conference and the ACM SIGSOFT International Symposium on Foundations of Software Engineering*, pages 521–524, 2007.

[130] A. Fantechi and S. Gnesi. Formal modeling for product families engineering. In *Software Product Lines, 12th International Conference, SPLC 2008, Proceedings*, pages 193–202. IEEE, 2008.

[131] H. Fecher and H. Schmidt. Comparing disjunctive modal transition systems with an one-selecting variant. *Journal of Logic and Algebraic Programming*, 77(1-2):20–39, 2008.

[132] D. Fischbein, S. Uchitel, and V. A. Braberman. A foundation for behavioural conformance in software product line architectures. In *Proceedings of the 2006 Workshop on Role of Software Architecture for Testing and Analysis, held in conjunction with the ACM SIGSOFT International Symposium on Software Testing and Analysis (ISSTA 2006), ROSATEA 2006, Portland, Maine, USA, July 17-20, 2006*, pages 39–48, 2006.

[133] A. Flores-Montoya, E. Albert, and S. Genaim. May-happen-in-parallel based deadlock analysis for concurrent objects. In D. Beyer and M. Boreale, editors, *Proc. International Conference on Formal Techniques for Distributed Systems (FMOODS/FORTE 2013)*, volume 7892 of *Lecture Notes in Computer Science*, pages 273–288. Springer, 2013.

[134] S. Fowler and A. Wellings. Formal analysis of a real-time kernel specification. In B. Jonsson and J. Parrow, editors, *FTRTFT'96, Proceedings of 4th International Symposium on Formal Techniques in Real Time and Fault Tolerant Systems, Uppsala*, volume 1135 of *Lecture Notes in Computer Science*, pages 440–458. Springer, 1996.

[135] C. A. Furia, M. Rossi, D. Mandrioli, and A. Morzenti. Automated compositional proofs for real-time systems. *Theoretical Computer Science*, 376(3):164–184, 2007.

[136] A. Fürst. Design patterns in Event-B and their tool support. Master's thesis, ETH, Eidgenössische Technische Hochschule Zürich, Department of Computer Science, 2009.

[137] V. Garg and M. Ragunath. Concurrent regular expressions and their relationship to Petri nets. *Theoretical Computer Science*, 96:285–304, 1992.

[138] E. Gasparis. LePUS: A formal language for modeling design patterns. In *Design Pattern Formalization Techniques*, pages 357–372. IGI Global, 2007.

[139] C. Ghezzi, D. Mandrioli, and A. Morzenti. TRIO, a logic language for executable specifications of realtime systems. *Journal of Systems and Software*, 12(2):255–307, 1990.

[140] D. Giammarresi and A. Restivo. Two-dimensional languages. In *Handbook of Formal Languages*, volume 3, pages 215–267. Springer, 1997.

[141] A. Gondal, M. Poppleton, and C. Snook. Feature composition - towards product lines of Event-B models. In *1st International Workshop on Model-Driven Product Line Engineering (MDPLE'09)*. CTIT Workshop Proceedings, 2009.

[142] M. Gordon. A mechanized Hoare logic of state transitions. In A. W. Roscoe, editor, *A Classical Mind: Essays in Honour of C.A.R. Hoare*, pages 143–159. Prentice Hall, 1994.

[143] C. Gratie and I. Petre. Fit-preserving data refinement of mass-action reaction networks. In A. Beckmann, E. Csuhaj-Varjú, and K. Meer, editors, *Language, Life, Limits*, volume 8493 of *Lecture Notes in Computer Science*, pages 204–213. Springer, 2014.

[144] D.-E. Gratie, B. Iancu, S. Azimi, and I. Petre. Quantitative model refinement in four different frameworks, with applications to the heat shock response. Technical Report 1067, TUCS, 2013.

[145] C. Grioli. Improvement and analysis of behavioural models with variability. Master's thesis, University of Pisa, Italy, 2013.

[146] J. Groslambert. Verification of LTL on B Event Systems. In *B 2007: Formal Specification and Development in B*, volume 4355 of *Lecture Notes in Computer Science*, pages 109–124. Springer, 2006.

[147] R. Grossman, A. Nerode, A. Ravn, and H. Rischel. *Hybrid Systems*. Springer-Verlag, 1993.

[148] B. Grunbaum and G. Shephard. *Tilings and Patterns*. W.H. Freeman and Co., 1990.

[149] V. H. Haase. Real-time behavior of programs. *IEEE Transactions on Software Engineering*, 7(5):594–501, Sept. 1981.

[150] G. Haddad, F. Hussain, and G. T. Leavens. The design of SafeJML, a specification language for SCJ with support for WCET specifications. In *Proceedings of the 8th International Workshop on Java Technologies for Real-Time and Embedded Systems, JTRES'10*, pages 155–163. ACM, 2010.

[151] R. Hähnle and E. B. Johnsen. Designing resource-aware cloud applications. *IEEE Computer*, 48:72–75, 2015.

[152] S. Hallerstede. On the purpose of Event-B proof obligations. *Formal Aspects of Computing*, 23(1):133–150, 2011.

[153] S. Hallerstede, M. Leuschel, and D. Plagge. Validation of formal models by refinement animation. *Science of Computer Programming*, 78(3):272 – 292, 2013.

[154] K. M. Hansen, A. P. Ravn, and V. Stavridou. From safety analysis to software requirements. *Software Engineering, IEEE Transactions on*, 24(7):573–584, 1998.

[155] M. D. Harrison, P. Masci, J. C. Campos, and P. Curzon. Demonstrating that medical devices satisfy user related safety requirements. In *4th International Symposium on Foundations of Healthcare Information Engineering and Systems (FHIES2014)*, 2014.

[156] Ø. Haugen, K. E. Husa, R. K. Runde, and K. Stølen. STAIRS towards formal design with sequence diagrams. *Journal of Software and Systems Modeling*, 4:355–367, 2005.

[157] I. Hayes, M. Jackson, and C. Jones. Determining the specification of a control system from that of its environment. In K. Araki, S. Gnesi, and D. Mandrioli, editors, *FME 2003: Formal Methods*, volume 2805 of *Lecture Notes in Computer Science*. Springer-Verlag, Berlin, 2003.

[158] He Jifeng and C.A.R. Hoare. Prespecification and data refinement. In *Data Refinement in a Categorical Setting*, Technical Monograph, number PRG-90. Oxford University Computing Laboratory, Nov. 1990.

[159] He Jifeng, C.A.R. Hoare, and J. Sanders. Data refinement refined. In B. Robinet and R. Wilhelm, editors, *ESOP 86, European Symposium on Programming*, volume 213 of *Lecture Notes in Computer Science*, pages 187–196. Springer-Verlag, 1986.

[160] J. Heath, M. Kwiatkowska, G. Norman, D. Parker, and O. Tymchyshyn. Probabilistic model checking of complex biological pathways. In *Computational Methods in Systems Biology*, volume 4210 of *Lecture Notes in Computer Science*, pages 32–47. Springer, 2006.

[161] G. Heineman and W. Councill (editors). *Component-Based Software Engineering: Putting the Pieces Together*. Addison-Wesley, 2001.

[162] M. Heiner, M. Herajy, F. Liu, C. Rohr, and M. Schwarick. Snoopy – a unifying Petri net tool. In S. Haddad and L. Pomello, editors, *Application and Theory of Petri Nets*, volume 7347 of *Lecture Notes in Computer Science*, pages 398–407. Springer Berlin Heidelberg, 2012.

[163] C. Heinzemann, C. Brenner, S. Dziwok, and W. Schäfer. Automata-based refinement checking for real-time systems. *Computer Science - Research and Development*, 30:255–283, 2015.

[164] T. A. Henzinger. The theory of hybrid automata. In *Proceedings, 11th Annual IEEE Symposium on Logic in Computer Science*, pages 278–292. IEEE Computer Society, 1996.

[165] T. A. Henzinger, R. Majumbar, and J. F. Raskin. A classification of symbolic transition systems. *ACM Transactions on Computational Logic*, 6(1):1–32, 2005.

[166] J. Hillston. *A Compositional Approach to Performance Modelling*. Cambridge University Press, 1996.

[167] A. Hinton, M. Kwiatkowska, G. Norman, and D. Parker. PRISM: A tool for automatic verification of probabilistic systems. In *Proceedings of the 12th International Conference on Tools and Algorithms for the Construction and Analysis of Systems*, TACAS'06, pages 441–444. Springer, 2006.

[168] T. S. Hoang and J.-R. Abrial. Reasoning about liveness properties in Event-B. In *International Conference on Formal Engineering Methods*, volume 6991 of *Lecture Notes in Computer Science*, pages 456–471. Springer, 2011.

[169] T. S. Hoang, A. Fürst, and J. Abrial. Event-B patterns and their tool support. *Software and Systems Modeling*, 12(2):229–244, 2013.

[170] C. A. R. Hoare. *An axiomatic basis for computer programming.* Communications of the ACM, 12(10):576-580, 1969.

[171] C. A. R. Hoare. *Communicating Sequential Processes.* Prentice Hall, 1985.

[172] J. Hooman. Extending Hoare logic to real-time. *Formal Aspects of Computing*, 6(6A):801–826, 1994.

[173] J. Hooman. Compositional verification of real-time applications. In W.-P. de Roever, H. Langmaack, and A. Pnueli, editors, *Compositionality: The Significant Difference (Compos '97)*, volume 1536 of *Lecture Notes in Computer Science*, pages 276–300. Springer, 1998.

[174] J. E. Hopcroft. An n log n algorithm for minimizing states in a finite automaton. In Z. Kohavi and A. Paz, editors, *Proceedings of an International Symposium on the Theory of Machines and Computations*, pages 189–196. Academic Press, 1971.

[175] S. Hudon and T. S. Hoang. Systems design guided by progress concerns. In *Integrated Formal Methods, 10th International Conference, IFM 2013. Proceedings*, volume 7940 of *Lecture Notes in Computer Science*, pages 16–30, 2013.

[176] B. Iancu, E. Czeizler, E. Czeizler, and I. Petre. Quantitative refinement of reaction models. *International Journal of Unconventional Computing*, 8(5-6):529–550, 2012.

[177] B. Iancu, D.-E. Gratie, S. Azimi, and I. Petre. Computational modeling of the eukaryotic heat shock response: the BioNetGen implementation, the Petri net implementation and the PRISM implementation, 2013. Available at: http://combio.abo.fi/research/computational-modeling-of-the-eukaryotic-heat-shock-response/.

[178] B. Iancu, D.-E. Gratie, S. Azimi, and I. Petre. On the implementation of quantitative model refinement. In A.-H. Dediu, C. Martín-Vide, and B. Truthe, editors, *Algorithms for Computational Biology*, volume 8542 of *Lecture Notes in Computer Science*, pages 95–106. Springer International Publishing, 2014.

[179] A. Iliasov. Use case scenarios as verification conditions: Event-B/flow approach. In *SERENE 2011, Software Engineering for Resilient Systems*, volume 6968 of *Lecture Notes in Computer Science*, pages 9–23. Springer-Verlag, 2011.

[180] A. Iliasov, L. Laibinis, E. Troubitsyna, A. Romanovsky, and T. Latvala. Augmenting Event-B modelling with real-time verification. Technical Report 1006, TUCS, 2011.

[181] A. Iliasov, E. Troubitsyna, L. Laibinis, A. Romanovsky, K. Varpaaniemi, D. Ilic, and T. Latvala. Developing mode-rich satellite software by refinement in Event B. In S. Kowalewski and M. Roveri, editors, *Formal Methods for Industrial Critical Systems - 15th International Workshop, FMICS 2010. Proceedings*, volume 6371 of *Lecture Notes in Computer Science*, pages 50–66. Springer, 2010.

[182] A. Iliasov, E. Troubitsyna, L. Laibinis, A. Romanovsky, K. Varpaaniemi, D. Ilic, and T. Latvala. Supporting reuse in event b development: modularisation approach. In M. Frappier, U. Glässer, S. Khurshid, R. Laleau, and S. Reeves, editors, *Abstract State Machines, Alloy, B and Z, Second International Conference, ABZ 2010. Proceedings*, volume 5977 of *Lecture Notes in Computer Science*, pages 174–188. Springer, 2010.

[183] A. Iliasov, E. Troubitsyna, L. Laibinis, A. Romanovsky, K. Varpaaniemi, D. Ilic, and T. Latvala. Developing mode-rich satellite software by refinement in Event-B. *Science of Computer Programming*, 78(7):884–905, 2013.

[184] A. Iliasov, E. Troubitsyna, L. Laibinis, A. Romanovsky, K. Varpaaniemi, P. Väisänen, D. Ilic, and T. Latvala. Verifying mode consistency for on-board satellite software. In E. Schoitsch, editor, *Computer Safety, Reliability, and Security, 29th International Conference, SAFECOMP 2010. Proceedings*, volume 6351 of *Lecture Notes in Computer Science*, pages 126–141. Springer, 2010.

[185] R. Jetley, S. Purushothaman Iyer, and P. L. Jones. A formal methods approach to medical device review. *Computer*, 39(4):61–67, 2006.

[186] E. B. Johnsen, J. C. Blanchette, M. Kyas, and O. Owe. Intra-object versus inter-object: concurrency and reasoning in Creol. *Electronic Notes in Theoretical Computer Science*, 243:89–103, July 2009.

[187] E. B. Johnsen, R. Hähnle, J. Schäfer, R. Schlatte, and M. Steffen. ABS: a core language for abstract behavioral specification. In B. Aichernig, F. S. de Boer, and M. M. Bonsangue, editors, *Proc. 9th International Symposium on Formal Methods for Components and Objects (FMCO 2010)*, volume 6957 of *Lecture Notes in Computer Science*, pages 142–164. Springer, 2011.

[188] E. B. Johnsen and O. Owe. An asynchronous communication model for distributed concurrent objects. *Software and System Modeling*, 6(1):35–58, 2007.

[189] E. B. Johnsen, O. Owe, and E. W. Axelsen. A run-time environment for concurrent objects with asynchronous method calls. In *Proc. 5th International Workshop on Rewriting Logic and its Applications (WRLA'04)*, volume 117 of *Electronic Notes in Computer Science*. Elsevier, 2004.

[190] E. B. Johnsen, O. Owe, D. Clarke, and J. Bjørk. A formal model of service-oriented dynamic object groups. *Science of Computer Programming*, 115C:3–22, Jan. 2016.

[191] E. B. Johnsen, O. Owe, and I. Simplot-Ryl. A dynamic class construct for asynchronous concurrent objects. In M. Steffen and G. Zavattaro, editors, *Proc. 7th International Conference on Formal Methods for Open Object Based Distributed Systems (FMOODS'05)*, volume 3535 of *Lecture Notes in Computer Science*, pages 15–30. Springer, June 2005.

[192] E. B. Johnsen, O. Owe, and I. C. Yu. Creol: a type-safe object-oriented model for distributed concurrent systems. *Theoretical Computer Science*, 365(1–2):23–66, Nov. 2006.

[193] E. B. Johnsen, R. Schlatte, and S. L. Tapia Tarifa. A formal model of object mobility in resource-restricted deployment scenarios. In F. Arbab and P. Ölveczky, editors,

Proc. 8th International Symposium on Formal Aspects of Component Software (FACS 2011), volume 7253 of *Lecture Notes in Computer Science*, pages 185–202. Springer, 2012.

[194] E. B. Johnsen, R. Schlatte, and S. L. Tapia Tarifa. Modeling resource-aware virtualized applications for the cloud in real-time ABS. In *Proc. Formal Engineering Methods (ICFEM'12)*, volume 7635 of *Lecture Notes in Computer Science*, pages 71–86. Springer, Nov. 2012.

[195] E. B. Johnsen, R. Schlatte, and S. L. Tapia Tarifa. Integrating deployment architectures and resource consumption in timed object-oriented models. *Journal of Logical and Algebraic Methods in Programming*, 2014. Available online.

[196] B. Jonsson and Y.-K. Tsay. Assumption/guarantee specifications in linear-time temporal logic. *Theoretical Computer Science*, 167(2):47–72, 1996.

[197] A. Joyal, R. Street, and D. Verity. Traced monoidal categories. In *Proceedings of the Cambridge Philosophical Society*, volume 119, pages 447–468, 1996.

[198] J. Jürjens. Secrecy-preserving refinement. In *Proceedings of Formal Methods Europe (FME'01)*, volume 2021 of *Lecture Notes in Computer Science*, pages 135–152. Springer, 2001.

[199] J. Kari. A small aperiodic set of Wang tiles. *Discrete Mathematics*, 160:259–264, 1996.

[200] R. M. Keller. Formal verification of parallel programs. *Commun. ACM*, 19(7):371–384, July 1976.

[201] H. Kitano. Systems biology: a brief overview. *Science*, 295(5560):1662–1664, 2002.

[202] S. Kleene. Representation of events in nerve nets and finite automata. *Automata Studies*, 1956.

[203] M. P. Kline and R. I. Morimoto. Repression of the heat shock factor 1 transcriptional activation domain is modulated by constitutive phosphorylation. *Molecular and Cellular Biology*, 17(4):2107–2115, 1997.

[204] E. Klipp, R. Herwig, A. Kowald, C. Wierling, and H. Lehrach. *Systems Biology in Practice: Concepts, Implementation, and Application*. Wiley-VCH, 2005.

[205] I. Koch, W. Reisig, and F. Schreiber. *Modeling in Systems Biology: The Petri Net Approach*. Springer, 2010.

[206] D. Kozen. A completeness theorem for Kleene algebras and the algebra of regular events. In *LICS'91*, pages 214–225. IEEE, 1991.

[207] J. Krone, W. F. Ogden, and M. Sitaraman. Modular verification of performance constraints. In *ACM OOPSLA Workshop on Specification and Verification of Component-Based Systems (SAVCBS)*, pages 60–67, 2001.

[208] J. Krone, W. F. Ogden, and M. Sitaraman. Profiles: a compositional mechanism for performance specification. Technical Report RSRG-04-03, Department of Computer Science, Clemson University, Clemson, SC 29634-0974, June 2004.

[209] W. Kuich and A. Salomaa. *Semirings, automata, and languages.* Springer-Verlag, Berlin, 1985.

[210] M. Kwiatkowska, G. Norman, and D. Parker. Quantitative analysis with the probabilistic model checker PRISM. *Electronic Notes in Theoretical Computer Science*, 153(2):5–31, 2006.

[211] M. Kwiatkowska, G. Norman, and D. Parker. PRISM: probabilistic model checking for performance and reliability analysis. *SIGMETRICS Perform. Eval. Rev.*, 36(4):40–45, 2009.

[212] M. Kwiatkowska, G. Norman, and D. Parker. PRISM 4.0: Verification of probabilistic real-time systems. In G. Gopalakrishnan and S. Qadeer, editors, *Proceedings of the 23rd International Conference on Computer Aided Verification (CAV'11)*, volume 6806 of *Lecture Notes in Computer Science*, pages 585–591. Springer, 2011.

[213] L. Lamport. Hybrid systems in TLA$^+$. In R. L. Grossman, A. Nerode, A. P. Ravn, and H. Rischel, editors, *Workshop on Theory of Hybrid Systems*, volume 736 of *Lecture Notes in Computer Science*, pages 77–102. LNCS 736, Springer, 1993.

[214] L. Lamport. Introduction to TLA. Technical report, SRC Research Center, Dec. 1994. Technical Note.

[215] L. Lamport. Real-time model checking is really simple. In D. Borrione and W. Paul, editors, *Correct hardware design and verification methods, CHARME2005*, volume 3725 of *LNCS*, 2005.

[216] J. Laprie. Resilience for the scalability of dependability. In *Fourth IEEE International Symposium on Network Computing and Applications*, pages 5–6, 2005.

[217] K. G. Larsen and A. Legay. Statistical model checking past, present, and future - (track introduction). In *Leveraging Applications of Formal Methods, Verification, and Validation. Specialized Techniques and Applications - 6th International Symposium, ISoLA 2014, Imperial, Corfu, Greece, October 8-11, 2014, Proceedings, Part II*, pages 135–142, 2014.

[218] K. G. Larsen, U. Nyman, and A. Wasowski. Modal I/O automata for interface and product line theories. In *Programming Languages and Systems, 16th European Symposium on Programming, ESOP 2007, Held as Part of the Joint European Conferences on Theory and Practices of Software, ETAPS 2007, Braga, Portugal, March 24 - April 1, 2007, Proceedings*, pages 64–79, 2007.

[219] K. G. Larsen and B. Thomsen. A modal process logic. In *Proceedings of the Third Annual Symposium on Logic in Computer Science (LICS '88), Edinburgh, Scotland, UK, July 5-8, 1988*, pages 203–210, 1988.

[220] K. G. Larsen and L. Xinxin. Equation solving using modal transition systems. In *Proceedings of the Fifth Annual Symposium on Logic in Computer Science (LICS '90), Philadelphia, Pennsylvania, USA, June 4-7, 1990*, pages 108–117, 1990.

[221] R. Lassaigne and S. Peyronnet. Approximate verification of probabilistic systems. *Process Algebra and Probabilistic Methods: Performance Modeling and Verification*, 2399:277–295, 2002.

[222] M. Latteux and D. Simplot. Context-sensitive string languages and recognizable picture languages. *Information and Computation*, 138:160–169, 1997.

[223] M. Lavielle. *Mixed Effects Models for the Population Approach: Models, Tasks, Methods, and Tools*. Chapman and Hall/CRC, 2014.

[224] G. T. Leavens, K. R. M. Leino, and P. Müller. Specification and verification challenges for sequential object-oriented programs. *Formal Aspects of Computing*, 19(2):159–189, 2007.

[225] K. R. M. Leino and P. Müller. A verification methodology for model fields. In P. Sestoft, editor, *Programming Languages and Systems*, volume 3924 of *Lecture Notes in Computer Science*, pages 115–130. Springer Berlin Heidelberg, 2006.

[226] M. Leuschel and M. J. Butler. ProB: an automated analysis toolset for the B method. *STTT*, 10(2):185–203, 2008.

[227] M. Leuschel, J. Falampin, F. Fritz, and D. Plagge. Automated property verification for large scale B models. In *FM*, volume 5850 of *LNCS*, pages 708–723. Springer, 2009.

[228] K. Lindgren, C. Moore, and M. Nordahl. Complexity of two-dimensional patterns. *Journal of Statistical Physics*, 91(5-6):909–951, 1998.

[229] B. Liskov and L. Shrira. Promises: linguistic support for efficient asynchronous procedure calls in distributed systems. In D. S. Wise, editor, *Proc. SIGPLAN Conference on Programming Language Design and Implementation (PLDI'88)*, pages 260–267. ACM Press, June 1988.

[230] B. Liskov and J. M. Wing. A behavioral notion of subtyping. *ACMTransactions on Programming Languages and Systems*, 16(6):1811–1841, Nov. 1994.

[231] Z. Liu and M. Joseph. Specification and verification of fault-tolerance, timing, and scheduling. *ACM Trans. Program. Lang. Syst.*, 21(1):46–89, 1999.

[232] I. Lopatkin, A. Iliasov, A. Romanovsky, Y. Prokhorova, and E. Troubitsyna. Patterns for representing FMEA in formal specification of control systems. In *IEEE 13th International Symposium on High-Assurance Systems Engineering*, pages 146–151. IEEE, 2011.

[233] C. Luo and S. Qin. Separation logic for multiple inheritance. *Electronic Notes in Theoretical Computer Science*, 212:27–40, 2008.

[234] N. Lynch and F. Vaandrager. Forward and backward simulations - part ii: Timing-based systems. *Information and Computation*, 128, 1995.

[235] P. Masci, A. Ayoub, P. Curzon, M. D. Harrison, I. Lee, and H. Thimbleby. Verification of interactive software for medical devices: PCA infusion pumps and FDA regulation as an example. In *EICS 2013, Proceedings of the 5th ACM SIGCHI Symposium on Engineering Interactive Computing Systems*, pages 81–90. ACM New York, NY, USA, 2013.

[236] P. Masci, R. Rukšėnas, P. Oladimeji, A. Cauchi, A. Gimblett, Y. Li, P. Curzon, and H. Thimbleby. On formalising interactive number entry on infusion pumps. *Electronic Communications of the EASST*, 45, 2011.

[237] W. Materi and D. S. Wishart. Computational systems biology in drug discovery and development: methods and applications. *Drug Discovery Today*, 12(7):295–303, 2007.

[238] S. Matsuoka and A.Yonezawa. Analysis of inheritance anomaly in object-oriented concurrent programming languages. In G. Agha, P. Wegner, and A. Yonezawa, editors, *Research Directions in Concurrent Object-Oriented Programming*, pages 107–150. The MIT Press, 1993.

[239] C. McDowell and O. Owe. Towards a light-weight approach for concurrent active objects in java, 2016. Submitted for publication.

[240] A. McIver, C. Morgan, and E. Troubitsyna. The probabilistic steam boiler: a case study in probabilistic data refinement. In *IRW/FMP 98*, Proceedings of, 1998.

[241] J. McLean. A general theory of composition for trace sets closed under selective interleaving functions. In *Proceedings of the IEEE Symposium on Research in Security and Privacy*, pages 79–93. IEEE Computer Society, 1994.

[242] H. Messer, A. Zinevich, and P. Alpert. Environmental monitoring by wireless communication networks. *Science*, 312:713, 2006.

[243] B. Meyer. *Object-Oriented Software Construction*. Prentice Hall, 1988.

[244] B. Meyer. Design by contract: the Eiffel method. In *Proceedings of Tools (26)*, page 446. IEEE Computer Society, 1998.

[245] R. Milner. An algebraic definition of simulation between programs. In *Proceedings of the 2nd International Joint Conference on Artificial Intelligence. London, UK, September 1971*, pages 481–489, 1971.

[246] R. Milner. *Communication and Concurrency*. International Series in Computer Science. Prentice Hall, 1989.

[247] A. Mizera, E. Czeizler, and I. Petre. Self-assembly models of variable resolution. *LNBI Transactions on Computational Systems Biology*, 7625:181–203, 2011.

[248] C. Morgan, A. McIver, K. Seidel, and J. Sanders. Refinement-oriented probability for CSP. *Formal Aspects of Computing*, 8(6):617–647, 1996.

[249] C. C. Morgan. *Programming from Specifications*. Prentice Hall, 1990.

[250] J. M. Morris. A theoretical basis for stepwise refinement and the programming calculus. *Science of Computer Programming*, 9(3):287–306, Dec. 1987.

[251] E. Murphy, V. Danos, J. Feret, J. Krivine, and R. Harmer. *Elements of Computational Systems Biology*, chapter Rule Based Modelling and Model Refinement, pages 83–114. Wiley Book Series on Bioinformatics. John Wiley & Sons, Inc., 2010.

[252] H. R. Nielson. *Hoare-Logic for Run-Time Analysis of Programs*. PhD thesis, Edinburgh University, 1984.

[253] H. R. Nielson. A Hoare-like proof-system for run-time analysis of programs. *Science of Computer Programming*, 9, 1987.

[254] P. Oladimeji, H. Thimbleby, and A. Cox. Number entry and their effects on error detection. In P. Campos et al., editors, *Interact 2011*, number 6949 in *Lecture Notes in Computer Science*, pages 178–185. Springer Verlag, 2011.

[255] E.-R. Olderog and C. A. R. Hoare. Specification-oriented semantics for communicating processes. *Acta Informatica*, 23:9–66, 1986.

[256] P. Olsen, J. Foederer, and J. Tretmans. Model-based testing of industrial transformational systems. In *Testing Software and Systems - 23rd IFIP WG 6.1 International Conference, ICTSS 2011, Paris, France, November 7-10, 2011. Proceedings*, pages 131–145, 2011.

[257] O. Owe and I. Ryl. On combining object orientation, openness, and reliability. In *Proc. of the Norwegian Informatics Conference (NIK'99)*. Tapir, Nov. 1999.

[258] O. Owe and I. Ryl. Reasoning control in presence of dynamic classes. In *12th Nordic Workshop on Programming Theory*, Bergen, 2000. On-line proceedings http://www.ii.uib.no/~nwpt00.

[259] M. J. Parkinson and G. M. Bierman. Separation logic, abstraction, and inheritance. *SIGPLAN Not.*, 43(1):75–86, Jan. 2008.

[260] D. L. Parnas and J. Madey. Functional documents for computer systems. *Sci. Comput. Program.*, 25(1):41–61, 1995.

[261] M. Patitz. An introduction to tile-based self-assembly and a survey of recent results. *Natural Computing*, 2014.

[262] I. Pereverzeva, E. Troubitsyna, and L. Laibinis. Formal goal-oriented development of resilient MAS in event-b. In M. Brorsson and L. M. Pinho, editors, *Reliable Software Technologies - Ada-Europe 2012 - 17th Ada-Europe International Conference on Reliable Software Technologies, Stockholm, Sweden, June 11-15, 2012. Proceedings*, volume 7308 of *Lecture Notes in Computer Science*, pages 147–161. Springer, 2012.

[263] I. Petre, A. Mizera, C. L. Hyder, A. Meinander, A. Mikhailov, R. I. Morimoto, L. Sistonen, J. E. Eriksson, and R.-J. Back. A simple mass-action model for the eukaryotic heat shock response and its mathematical validation. *Natural Computing*, 10(1):595–612, 2011.

[264] C. Pierik and F. S. de Boer. A proof outline logic for object-oriented programming. *Theoretical Computer Science*, 343(3):413–442, 2005.

[265] D. Plagge and M. Leuschel. Seven at one stroke: LTL model checking for high-level specifications in B, Z, CSP, and more. *STTT*, 12(1):9–21, 2010.

[266] A. Platzer. *Logical Analysis of Hybrid Systems: Proving Theorems for Complex Dynamics*. Springer, Heidelberg, 2010.

[267] K. Pohl, G. Böckle, and F. J. v. d. Linden. *Software Product Line Engineering: Foundations, Principles, and Techniques*. Springer-Verlag New York, Inc., Secaucus, NJ, USA, 2005.

[268] A. Popa, A. Sofronia, and G. Stefanescu. High-level structured interactive programs with registers and voices. *J. UCS*, 13:1722–1754, 2007.

[269] C. Priami. Stochastic pi-calculus. *Computer Journal*, 38(7):578–589, 1995.

[270] C. Priami. Algorithmic systems biology. *Communications of the ACM*, 52(5):80–88, 2009.

[271] P. Puschner and A. Burns. A review of worst-case execution time analysis (editorial). *Real-Time Systems*, 18(2/3):115–128, 2000.

[272] S. Quinton and S. Graf. Contract-based verification of hierarchical systems of components. In *Proc. of SEFM 08*, pages 377–381. IEEE Computer Society, 2008.

[273] K. Raman and N. Chandra. Systems biology. *Resonance*, 15(2):131–153, 2010.

[274] F. Redmill, M. Chudleigh, and J. Catmur. *System Safety: HAZOP and Software HAZOP*. Wiley, 1999.

[275] A. Regev and E. Shapiro. Cellular abstractions: cells as computation. *Nature*, 419(6905), 2002.

[276] C. Rieger, D. Gertman, and M. McQueen. Resilient control systems: next generation design research. In *Proceedings of Human System Interactions*, pages 632–636, 2009.

[277] RODIN modularisation plug-in. http://wiki.event-b.org/index.php/Modularisation_Plug-in.

[278] M. Rönkkö, V. Kotovirta, and M. Kolehmainen. Classifying environmental monitoring systems. In J. Hrebícek, G. Schimak, M. Kubásek, and A. E. Rizzoli, editors, *Environmental Software Systems. Fostering Information Sharing - 10th IFIP WG 5.11 International Symposium, ISESS 2013, Neusiedl am See, Austria, October 9-11, 2013. Proceedings*, volume 413 of *IFIP Advances in Information and Communication Technology*, pages 533–542. Springer, 2013.

[279] M. Rönkkö and X. Li. Linear hybrid action systems. *Nordic Journal of Computing*, 8(1):159–177, 2001.

[280] M. Rönkkö and A. Ravn. Action systems with continuous behavior. In P. Antsakilis, W. Kohn, M. Lemmon, A. Nerode, and S. Sastry, editors, *Hybrid System V*, volume 1567 of *Lecture Notes in Computer Science*, pages 304–323. Springer, 1999.

[281] M. Rönkkö, A. Ravn, and K. Sere. Hybrid action systems. *Theoretical Computer Science*, 290:937–973, 2003.

[282] M. Rönkkö and K. Sere. Refinement and continuous behaviour. In F. W. Vaandrager and J. H. van Schuppen, editors, *Hybrid Systems: Computation and Control, Second International Workshop, HSCC'99, Berg en Dal, The Netherlands, March 29-31, 1999, Proceedings*, volume 1569 of *Lecture Notes in Computer Science*, pages 223–237. Springer, 1999.

[283] M. Rönkkö, M. Stocker, M. Neovius, L. Petre, and M. Kolehmainen. Designing resilience mediators for control systems. In *Proceedings of the IASTED International Conference on Modelling, Identification, and Control*, pages 147–154. ACTA Press, 2014.

[284] A. W. Roscoe. *The Theory and Practice of Concurrency*. Prentice Hall, 1998.

[285] A. W. Roscoe. *Understanding Concurrent Systems*. Springer, 2010.

[286] R. Rukšėnas, P. Masci, M. D. Harrison, and P. Curzon. Developing and verifying user interface requirements for infusion pumps: a refinement approach. *Electronic Communications of the EASST*, 69, 2013.

[287] R. K. Runde, Ø. Haugen, and K. Stølen. Refining UML interactions with explicit and implicit nondeterminism. *Nordic Journal of Computing*, 12:157–158, 2005.

[288] R. K. Runde, A. Refsdal, and K. Stølen. Relating computer systems to sequence diagrams: the impact of underspecification and inherent nondeterminism. *Formal Aspects of Computing*, 25(2):159–187, 2013.

[289] A. Salomaa. Two complete axiom systems for the algebra of regular events. *Journal of the ACM (JACM)*, 13(1):158–169, 1966.

[290] M. R. Sarshogh and M. Butler. Specification and refinement of discrete timing properties in Event-B. Technical report, Electronic and Computer Science, University of Southampton, 2011.

[291] A. Schäfer. Combining real-time model-checking and fault tree analysis. In *FME 2003: Formal Methods*, volume 2805 of *Lecture Notes in Computer Science*, pages 522–541. Springer Berlin Heidelberg, 2003.

[292] F. B. Schneider. Enforceable security policies. *ACM Transactions on Information and Systems Security*, 3(1):30–50, Feb. 2000.

[293] S. Schneider, H. Treharne, and H. Wehrheim. A CSP account of Event-B refinement. In *Proceedings 15th International Refinement Workshop, Refine 2011, Limerick, Ireland, 20th June 2011*, pages 139–154, 2011.

[294] S. Schneider, H. Treharne, and H. Wehrheim. The behavioural semantics of Event-B refinement. *Formal Asp. Comput.*, 26(2):251–280, 2014.

[295] S. Schneider, H. Treharne, H. Wehrheim, and D. Williams. Managing LTL properties in Event-B refinement. arXiv:1406:6622, June 2014.

[296] M. Schoeberl and R. Pedersen. WCET analysis for a Java processor. In *Proceedings of the 4th International Workshop on Java Technologies for Real-Time and Embedded Systems (JTRES'06)*, pages 202–211, 2006.

[297] P. Selinger. A survey of graphical languages for monoidal categories, volume *Lecture Notes in Physics 813*, pages 289–355. Springer, 2011.

[298] K. Sere. *Stepwise Derivation of Parallel Algorithms*. PhD thesis, Åbo Akademi, Department of Computer Science, 1990.

[299] K. Sere and E. Troubitsyna. Safety analysis in formal specication. In J. M. Wing, J. Woodcock, and J. Davies, editors, *FM 99 – Formal Methods*, volume 1709 of *Lecture Notes in Computer Science*, pages 1564–1583. Springer Berlin Heidelberg, 1999.

[300] K. Sere and M. Waldén. Reverse engineering distributed algorithms. *Journal of Software Maintenance: Research and Practice*, 8:117–144, 1996.

[301] L. Shargel, A. Yu, and S. Wu-Pong. *Applied Biopharmaceutics & Pharmacokinetics.* McGraw Hill Professional, 6th edition, 2012.

[302] A. C. Shaw. Reasoning about time in higher-level language software. *IEEE Transactions on Software Engineering*, 15(7):875–889, 1989.

[303] M. Shaw. A formal system for specifying and verifying program performance. Technical Report CMU-CS-79-129, Carnegie Mellon University, June 1979.

[304] R. Silva and M. Butler. Supporting reuse mechanisms for developments in event-b: composition. Technical report, University of Southampton, 2009.

[305] R. Silva and M. Butler. Shared event composition/decomposition in Event-B. In B. K. Aichernig, F. S. de Boer, and M. M. Bonsangue, editors, *Formal Methods for Components and Objects*, volume 6957 of *Lecture Notes in Computer Science*, pages 122–141. Springer, 2012.

[306] M. Sitaraman. Compositional performance reasoning. In *Proceedings of the Fourth ICSE Workshop on Component-Based Software Engineering: Component-Certification and System Prediction*, may 2001.

[307] M. Sitaraman, G. Kulczycki, J. Krone, W. F. Ogden, and A. L. N. Reddy. Performance specification of software components. In *ACM Sigsoft Symposium on Software Reuse*, 2001.

[308] J. Skön, M. Johansson, O. Kauhanen, M. Raatikainen, K. Leiviskä, and M. Kolehmainen. Wireless building monitoring and control system. *World Academy of Science, Engineering, and Technology*, 65:706–711, 2012.

[309] A. M. Smith, W. Xu, Y. Sun, J. R. Faeder, and G. E. Marai. RuleBender: integrated modeling, simulation, and visualization for rule-based intracellular biochemistry. *BMC Bioinformatics*, 13(Suppl 8):S3, 2012.

[310] C. Snook and M. Butler. UML-B: A plug-in for the Event-B tool set. In *Proceedings of the 1st International Conference on Abstract State Machines, B and Z, ABZ'08*, page 344. Springer-Verlag, 2008.

[311] N. Song, S. Zhang, and C. Liu. Overview of factors affecting oral drug absorption. *Asian Journal of Drug Metabolism and Pharmacokinetics*, 4:167–176, 2004.

[312] N. Soundararajan. A proof technique for parallel programs. *Theoretical Computer Science*, 31(1-2):13–29, May 1984.

[313] FP6 EU project SPEEDS: SPEculative and Exploratory Design in Systems Engineering. http://www.speeds.eu.com/.

[314] G. Stefanescu. *Feedback Theories (A Calculus for Isomorphism Classes of Flowchart Schemes)*. Number 24 in Preprint Series in Mathematics. INCREST, 1986. Also in: *Revue Roumaine de Mathematiques Pures et Applique*, 35:73–79, 1990.

[315] G. Stefanescu. On flowchart theories: part II. The nondeterministic case. *Theoretical Computer Science*, 52(3):307–340, 1987.

[316] G. Stefanescu. *Network Algebra*. Springer Verlag, 2000.

[317] G. Stefanescu. Interactive systems: from folklore to mathematics. In *Relmics'01*, pages 197–211. LNCS 2561, Springer, 2002.

[318] G. Stefanescu. Interactive systems with registers and voices. *Draft*, 2004. National University of Singapore, 2004.

[319] G. Stefanescu. Interactive systems with registers and voices. *Fundamenta Informaticae*, 73(1):285–305, 2006.

[320] M. Stoelinga and F. Vaandrager. Root contention in IEEE 1394. In J.-P. Katoen, editor, *Formal Methods for Real-Time and Probabilistic Systems*, pages 53–74. Springer Berlin Heidelberg, 1999.

[321] L. Strigini. Resilience assessment and dependability benchmarking: challenges of prediction. In *Proceedings of DSN Workshop Resilience Assess. Depend. Benchmarking*, 2008.

[322] W. Su, J.-R. Abrial, and H. Zhu. Formalizing hybrid systems with event-b and the rodin platform. *Science of Computer Programming*, 94:164–202, 2014.

[323] C. A. Szyperski. *Component Software*. Addison Wesley, 1998.

[324] T. Taibi and D. Ngo. Information and software technology. *Formal Specification of Design Pattern Combination using BPSL*, 45(3):157–170, 2003.

[325] A. Tarasyuk, E. Troubitsyna, and L. Laibinis. From formal specification in Event-B to probabilistic reliability assessment. In *Dependability (DEPEND), 2010 Third International Conference on*, pages 24–31, July 2010.

[326] A. Tarasyuk, E. Troubitsyna, and L. Laibinis. Formal modelling and verification of service-oriented systems in probabilistic Event-B. In *IFM 2012, Integrated Formal Methods*, volume 7321 of *Lecture Notes in Computer Science*, pages 237–252. Springer, 2012.

[327] The RODIN platform. http://rodin-b-sharp.sourceforge.net/.

[328] H. Thimbleby. Interaction walkthrough: evaluation of safety critical interactive systems. In G. Doherty and A. Blandford, editors, *Interactive Systems: Design, Specification, and Verification*, number 4323 in *Lecture Notes in Computer Science*, pages 52–66. Springer Verlag, 2007.

[329] A. Thums and G. Schellhorn. Model checking FTA. In *FME 2003: Formal Methods*, volume 2805 of *Lecture Notes in Computer Science*, pages 739–757. Springer Berlin Heidelberg, 2003.

[330] K. Trivedi, K. Dong Seong, and R. Ghosh. Resilience in computer systems and networks. In *IEEE International Conference on Computer-Aided Design - Digest of Technical Papers*, pages 74–77. IEEE, 2009.

[331] E. Troubitsyna. Reliability assessment through probabilistic refinement. *Nordic Journal of Computing*, 6(3):320–342, 1999.

[332] E. Troubitsyna. *Stepwise Development of Dependable Systems*. PhD thesis, Ph.D. thesis No.29. TUCS - Turku Centre for Computer Science, 2000.

[333] E. Troubitsyna. Dependability-explicit engineering with Event-B: overview of recent achievements. *CoRR*, abs/1210.7032, 2012.

[334] Y.-K. Tsay. Compositional verification in linear-time temporal logic. In J. Tiuryn, editor, *Proceedings of FoSSaCS 2000*, volume 1784 of *Lecture Notes in Computer Science*, pages 344–358. Springer, 2000.

[335] Safety requirements for the generic PCA pump. `http://rtg.cis.upenn.edu/gip-docs/Safety_Requirements_GPCA.doc`. Accessed: 04.04.2013.

[336] US Food and Drug Administration. Guidance for the content of premarket submissions for software contained in medical devices, May 2005.

[337] R. van Glabbeek. The linear time – branching time spectrum II; the semantics of sequential systems with silent moves (extended abstract). In E. Best, editor, Proceedings *CONCUR'93*, 4th International Conference on *Concurrency Theory*, Hildesheim, Germany, August 1993, volume 715 of *Lecture Notes in Computer Science*, pages 66–81. Springer, 1993.

[338] R. van Glabbeek. The linear time - branching time spectrum I. The semantics of concrete sequential processes. In Bergstra et al. [53], pages 3–99.

[339] M. Verhoef, P. G. Larsen, and J. Hooman. Modeling and validating distributed embedded real-time systems with VDM++. In J. Misra, T. Nipkow, and E. Sekerinski, editors, *Proceedings of the 14th International Symposium on Formal Methods (FM'06)*, volume 4085 of *Lecture Notes in Computer Science*, pages 147–162. Springer, 2006.

[340] J. von Wright and K. Sere. Program transformations and refinements in HOL. In *International Workshop on the HOL Theorem Proving System and Its Applications*, pages 231–239. IEEE, 1991.

[341] M. Waldén and K. Sere. Reasoning about action systems using the B-Method. *Formal Methods in System Design*, 13:5–35, 1998.

[342] A. Wang. Generalized types in high-level programming languages. Research Report in Informatics 1, Institute of Mathematics, University of Oslo, 1974. Cand. Real thesis.

[343] F. Wang. Efficient verification of timed automata with bdd-like data structures. *International Journal on Software Tools for Technology Transfer*, 6(1):77–97, 2004.

[344] R. Ward, J. Loftis, and G. McBride. The "data-rich but information-poor" syndrome in water quality monitoring. *Environmental Management*, 10(3):291–297, 1986.

[345] G. Wiederhold. Mediators in the architecture of future information systems. *Computer*, 25(3):38–49, 1992.

[346] M. Williams, D. Cornford, L. Bastin, R. Jones, and S. Parker. Automatic processing, quality assurance and serving of real-time weather data. *Computers & Geosciences*, 37:351–362, 2011.

[347] A. Wills. Capsules and types in fresco: program verification in smalltalk. In P. America, editor, *European Conference on Object-Oriented Programming*, volume 512 of *Lecture Notes in Computer Science*, pages 59–76. Springer, 1991.

[348] J. Woodcock and J. Davies. *Using Z: Specification, Refinement, and Proof.* Prentice Hall, 1996.

[349] W. Xu, A. M. Smith, J. R. Faeder, and G. E. Marai. RuleBender: a visual interface for rule-based modeling. *Bioinformatics*, 27(12):1721–1722, 2011.

[350] S. Yeganefard and M. Butler. Structuring functional requirements of control systems to facilitate refinement-based formalisation. In *Proceedings of the 11th International Workshop on Automated Verification of Critical Systems (AVoCS 2011)*, volume 46. Electronic Communications of the EASST, 2011.

[351] A. Yonezawa, J.-P. Briot, and E. Shibayama. Object-oriented concurrent programming in ABCL/1. In *Conference on Object-Oriented Programming Systems, Languages, and Applications (OOPSLA'86). Sigplan Notices*, 21(11):258–268, Nov. 1986.

[352] M. Zhang, Z. Liu, C. Morisset, and A. P. Ravn. Design and verification of fault-tolerant components. In *Methods, Models, and Tools for Fault Tolerance*, volume 5454 of *Lecture Notes in Computer Science*, pages 57–84. Springer Berlin Heidelberg, 2009.

[353] Q. Zhu and T. Basar. A dynamic game-theoretic approach to resilient control system design for cascading failures. In *Proceedings of International Conference on High Confidence Networked Systems*, pages 41–46, 2012.

Index